Fluid Machinery for the Oil, Petrochemical and Related Industries

Conference Organizing Committee

L E Smith, *CEng, FIMechE* (Chairman)
Consultant, Sulzer Bros (UK) Limited
Farnborough
Hampshire

Ir R M J Bouwman
Stork Pumps
Brussels
Belgium

A A Grant, *CEng, MIMechE, MIMarE*
Weir Pumps Limited
Glasgow

D Redpath, *CEng, MIMechE*
BP Group Research and Engineering
Uxbridge
Middlesex

D Reeves, *CEng, FIMechE, ACGI*
Pump Engineering Consultant
Collingham
Nottinghamshire

Eur Ing F Rhodes, *CEng, FIMechE, FInstPet*
Engineering Consultant
Uxbridge
Middlesex

Proceedings of the Institution of Mechanical Engineers

I MECH E

Fluid Machinery for the Oil, Petrochemical and Related Industries

Fifth European Congress

17–19 May 1993
The Netherlands Congress Centre, The Hague

Sponsored by
Fluid Machinery Committee of the
Power Industries Division of the
Institution of Mechanical Engineers

Supported by
Associazione Italiana di Meccanica Teorica Ed Applicata
Verein Deutscher Ingenieure
Koninklijk Instituut van Ingenieurs
Conseil National des Ingénieurs et des Scientifiques de France
Norwegian Society of Chartered Engineers

IMechE 1993–3

MEP

Published for IMechE by
Mechanical Engineering Publications Limited

First Published 1993

ISBN 0 85298 855 9

A CIP catalogue record for this book is available from the British Library.

Printed by Waveney Print Services Ltd, Beccles, Suffolk

Contents

C449/041/93

Development of main oil line and seawater injection pumps for oil industry application

P A SMITH
Weir Pumps Limited, Manchester

SYNOPSIS

Offshore production continues to provide a significant portion of total oil production, and the ongoing development of offshore technology is certain to facilitate further exploration for both land and sub-sea reserves.

This market is providing an interesting challenge to pump manufacturers, reliability and economy being of paramount importance. This paper reviews recent developments in centrifugal pump design including vertical multistage arrangements, high-speed geared configurations and application of advanced materials for oil transfer and water injection duties. The range of applicability of these new developments and established concepts are put into context.

INTRODUCTION

The offshore market is providing a challenge to the pump manufacturer in the areas of oil transfer and water injection. Installation constraints and customer preference in the selection of driver, maximum speeds and form of construction, leads to a diverse selection of machines to meet similar applications.

This variation in user requirement implies that a pump manufacturer servicing this market must not be restricted to a single pump type. The size of the machinery involved and the variation in duties to be accommodated result in the use of Engineered Products, i.e. pumps specifically designed to meet the requirements of an individual contract, rather than the application of a standard range.

Difficulty in access to the unit modules for major maintenance work, a shortage of space for the installation of standby equipment and the high cost of a loss in production, place great emphasis on reliability. This emphasis is reflected in the purchaser's specification, which generally includes stringent inspection requirements at all stages of manufacture and testing.

Comprehensive testing is of paramount importance. The manufacturer must have facilities to run machinery at full speed and load, both to demonstrate the hydraulic performance and to prove the mechanical integrity.

If maintenance work is required then the operation must be simple with a minimum of disruption.

PUMP DUTIES

The development of an oilfield takes several years, with production rising sharply to a peak, after which continued exploitation leads to a gradual decline in output, until the field becomes uneconomic. Initial production is often achieved by utilising the natural water or gas pressure within the oil field. As oil extraction continues the well pressure decays. If the oil recovery rate is to be maintained or increased then the reservoir pressure must be artificially elevated by the injection of water or gas. The injection rate and pressure are functions of the geological conditions of the individual well being exploited and the productivity required. The duty of the injection pump is consequently only established through operational experience and can alter substantially with oil extraction. This results in the use of series and parallel pumping systems and necessitates a wide operational range from the injection pump.

Flexibility of operational range is achieved by careful hydraulic and mechanical design. The ability to adapt further is often necessary and can be achieved by destaging or using different hydraulic bundles or cartridges, one for the initial injection duties and a second when the field has been fully developed. Speed control is one means of accommodating the changing field characteristics, and horizontal machines of higher power rating are likely to have this facility by virtue of being driven by a gas turbine.

Typical pressures range from 100 bar in the Gulf of Mexico to 350 bar in the Ekofisk field of the North Sea.

To match oil transfer rates with achieved production levels, Main-Oil-Line pumps are often driven through combined speed increasing gearboxes and Turbo couplings.

SELECTION CRITERIA

Operators and Consulting Engineers demand that M.O.L and injection pumps must satisfy the following requirements:

(1) Proven Reliability - Designs to be of proven pedigree.

Parentage of mechanical and hydraulic designs to be traceable in similar operating conditions.

Material selection to be demonstrated as suitable for the site fluid and operational conditions. On injection applications this translates to being capable of withstanding cavitation, corrosion, high velocity erosion and perhaps also abrasive wear.

Rotor dynamics to be insensitive to pump internal wear and drive train dynamics.

(2) Economy

High efficiency over operating range.

Compact design, low weight and small deck area requirement. The current cost of deck area is £234 thousand/m^2, with every tonne of equipment carried "deckside" requiring approximately 1.4 tonne to support it.

Ease of maintenance with minimal loss of production.

Competitive initial cost.

Long life due to a simple design.

(3) Adaptability

Wide operational range coupled with ease of modification to accommodate inevitable changes in well characteristics.

As stated in the Introduction, installation constraints and customer preference has led to a diverse selection of machine constructions to meet similar applications. The fact that they exist along side each other, without one being dominant, is testimony that each type has real or perceived advantages for certain applications and hydraulic duties, whilst having short-comings in others. The following review of pump constructions, currently available commercially, attempts to define their range of applicability and hence put the new developments and established concepts into context.

PUMP TYPES

Horizontal Axial Split Casing Multi Stage Volute Pumps (Fig 1)

High heads may be developed in relatively low speed pumps by the use of multi-stage axial split casing designs. A natural progression from the single-stage pump is the use of multi-stages arranged in opposition. Such an arrangement has the advantage of allowing the designer to offset the axial thrust generated in each impeller without having to resort to the use of elaborate thrust compensation devices.

Volutes are usually employed for axial split casing pumps in preference to diffusers as they reduce the complexity of the machine.

This type of machine, mounted horizontally, became universally used during the development of both oil transfer and water injection techniques. First designed in the USA and under American influence applied as first line M.O.L and injection pumps throughout the world, at a time when operating pressures were relatively modest.

Axial split casing pumps may be designed to withstand the high test pressures, necessitated by high discharge pressure, although the barrel casing pump, with circumferential joints, is arguably better suited. However, by use of Finite Element Analysis coupled with higher strength steel alloys and improvements in foundry methods, axial split casings can now operate successfully at working pressures up to 330 Barg. This implies test pressures up to 495 Barg.

The ability to inspect the rotating element whilst still within the bottom half casing, aligned and coupled to the driver is a major advantage of this construction. However, top half removal is time consuming if mechanised stud tensioning is not employed.

Barrel Casing Pumps (Figs 2 & 3)

Traditionally the barrel case pump was considered more difficult to maintain with significant work being required to dismantle or rebuild a pump on site. Manufacturers have developed their latest range of barrel casing pumps to resolve this traditional difficulty. The introduction of the total withdrawal cartridge and the shear ring design has now made the maintenance of such machines simple in the field and enables the more complex work to be carried out in a workshop, where good maintenance facilities are available.

This type of pump utilises a replaceable cartridge construction with access for cartridge withdrawal from the non-drive end on horizontal units, and from the drive end on vertical units. This mode of construction employs radial diffusers in preference to a volute arrangement, although volute designs may be used for the double entry first stages.

Later developments in water injection applications have seen the emergence of this type of unit to meet the need for higher pressures and flows, and also for use in M.O.L applications where space limitations have necessitated vertical configurations. Its pedigree can be traced to the Power Generation Industry and in particular to Boiler Feed Pumping. Unlike the modern Horizontal Axial Split Pump which was an oil industry related unit applied to water duties, occasionally as standard design, but normally with significant modification, (split casing in a Barrel), the Diffuser Barrel was a purpose designed water pump.

Developments in Boiler Feed duties had already included investigation into impeller life, hydraulic effects at part capacities, shaft stiffness, etc. It is not surprising therefore that this type of unit was initially easier to apply to the harsh Water Injection duties.

Horizontal versus Vertical Construction (Fig 4)

In certain applications by adoption of vertical barrel casings, it has been possible to eliminate the need for a booster pump. This is due to the possibility of mounting the unit at a lower level, because of its reduced deck area requirement, (only circa 24% of a equivalent horizontal set), coupled with the inherent advantage that by virtue of the units construction, the suction impeller has a higher N.P.S.H available relative to the units suction branch elevation.

A weight advantage is also achieved as vertical sets are approximately 6 to 8% lighter than the comparable horizontal set. Its construction and maintenance are simplified as no lube oil system and only one mechanical seal is required. This allows further simplication by minimising the need for instrumentation. Motor to pump alignment is also unaffected by pipework forces or deck movement (1) however these advantages have to be considered against the following limitations :-

1) Pump Running Speed dictated by drive motor slip speed, as insertion of a gearbox, whilst not impossible, is not currently practical because of the units power and ease of maintenance requirements. It would also increase the units height significantly.

2) Size of pumpsets limited by present 2-pole induction motor technology. Maximum power currently available 7 MW at 3600 R.P.M.

3) To maintain the pump cartridge, the drive motor has to be disconnected from its power supply, mechanically removed from the set and safely stored on the site or platform until the pump is fully re-assembled.

4) Maintenance lift heights increased.

5) Increased complexity in the Analysis of Set Structural movement due to "mast" type construction, necessitating interactive design process between Motor, Pump and Foundation Designer. (Fig 5).

The choice between vertical or horizontal construction is therefore not simple. The reduced weight, and apparent simplicity plus a smaller foundation area of the vertical set has to be offset against increased overhaul difficulty and lay-down space requirements. The potential conflict is between third-party purchasers favouring reduced initial purchase costs whilst plant operators favour ease of overhaul.

High Speed Geared Configuration (Fig 6 & 7)

These units, with absorbed powers in excess of 1 MW, were first introduced in 1987, and consisted of two single stage, double volute pumps, direct mounted on the output shafts of a speed increasing gearbox, with a separately driven booster stage. They have now been refined to incorporate a booster stage driven by step down gears in the same gearbox.

In comparison to conventional horizontal multi-stage pumpsets they offer a reduced footprint with virtually nil maintenance or laydown area required for the pump related components. Maintenance weights for pump casings are only circa 250 kg in comparison to 2 to 3000 kg for a total withdrawal cartridge.

Maintenance times are short, with one impeller and its respective mechanical seal being able to be removed and replaced in 2 hrs. The high pressure casings are interchangeable but impellers are not due to differing effective rotational directions.

Reduced mechanical complexity is claimed, however this depends upon the specific application. The number of components within the pumping element is reduced and the high-speed drive coupling is eliminated when compared to a conventional long coupled gear driven unit. However, for many applications a direct drive motor can be employed, without the need for the complication of a gearbox and extra coupling. In comparison to a vertical unit, the need for a lube oil system and multiple mechanical seals also negates this claim.

One cause of concern has been possible oil contamination in the event of a catastrophic mechanical seal failure, leading to internal damage of the gearbox. Due to the application of very conservative PV values, (circa 30% of the recommended maximum), by use of pressure breakdown bushes and regulated seal chamber pressure, the risk of seal failure is minimised. Should it occur, double carbon labyrinths with atmospheric drains upstream and downstream of the first element, prevent ingress of water into the gearbox. In comparison to conventional units the proximity of the mechanical seal to the bearing housing seal is similar. However the bearing seal system is superior to many conventional units currently operating in the field.

If one compares a High Speed Geared Configuration with a vertical unit for the following water injection duty:-

Nett Generated Head	1760 m
Rated Flow	228 m^3/h
N.P.S.H.A (Deck Level)	10 m

then the High Speed Geared Configuration has a deck footprint of 4.6 x 2.1 m and an operational weight of 18 000 kg, whilst the vertical unit is 1.78 m (over motor terminal box) x 1.2 m and 20 310 kg respectively, i.e. 4.5 times the deck area but only 89% the weight of the vertical unit.

Of greater significance is the difference in power consumption between the conventional and the High Speed Geared units. The conventional unit would have a rated efficiency of 72% with a peak efficiency of 74%. The High Speed Pump would have efficiencies of 62.5% and 65.8% respectively. Due to additional gear losses within the High Speed Geared Configuration this unit would consume almost 21% additional power and also require a 1.9 MW rated motor in comparison to 1.69 MW rating for the conventional units. As the absorbed power increases the difference in efficiency reduces and the additional power consumed by the High Speed Geared Unit becomes approximately 5% at 4.3 MW absorbed power.

PRODUCED WATER APPLICATIONS (Fig 8 & 9)

On water injection applications where groundwater or produced water is reinjected into the formation rock, it is often physically or environmentally difficult to remove the suspended solids from the injection fluid. This necessitates the injection pump handling a very abrasive and corrosive fluid, that may become more abrasive and corrosive as the field is further worked.

The original technology supplied for such services in the early to mid 1980's was Duplex Stainless Steel Impellers and Diffusers with hard faced wear surfaces, in cobalt based alloys, on wear rings and balance drums .

In certain applications, in Alaska, the quality of the injection fluid deteriorated and the typical operating time between major overhauls reduced to between 6 and 8 months. The units suffered from severe erosion, tripping out on high vibration levels due to excessive wear ring and balance drum clearances. The rear shrouds on all impellers, with the exception of the delivery stage, were found to be very heavily eroded at the hub/shroud intercept, with perforations into the impeller's flow passage in several places (Fig 8). The impeller vanes at inlet also showed signs of erosion at the hub. Impeller wear rings were loose due to erosion of the securing pins, with the hub ring clearance at 9 times the Maximum Design Clearance and the balance drum clearance at 3 times the Maximum Design Clearance.

In late 1989, Weir Material Ltd's Super Duplex Stainless Steel "Zeron 100" and Boarts Hot Isostatic Pressed Sintered Tungsten Carbides "S6" and "S10" were brought together to solve the erosion, corrosion and wear problems associated with pumping produced water containing large quantities of formation and or "frac" sand.

Using installation techniques pioneered for the retention of Silicon Carbide sleeves in product lubricated bearings, coupled with Finite Element Analysis of assembly, transportation, and operational stress levels in both carbide and metallic carriers, stationary and rotating wear surfaces have been fitted with solid carbide components. The advantage of this being the available depth of wear surface. Carbide coatings have a more limited life due to their radial thickness being only circa 0.8mm after finish machining. With solid carbide however the minimim radial wear depth available is 3mm.

After 24 months operational experience on site, it is projected that by utilising this new development, service life between overhauls should be circa 4 years, i.e. 35 000 hours. This represents an increase in average operational life of 600%.

MATERIAL SELECTION

As wells become older and more marginal fields are exploited, M-O-L and water injection equipment have to resist more erosive and corrosive pumped fluids. M-O-L applications are often not the gentle service that they were, with produced or injection water carryover and silicates now within many hydrocarbons. Because of this and higher fluid temperatures, many M-O-L pumps supplied in the early 1990's have had identical materials of construction to that of water injection pumps, i.e. full Duplex or Super Duplex Stainless Steel construction for all metallic components in contact with the pumped fluid.

To guarantee the corrosion performance of stainless steels and hence that of the pumpset, the steels used must be of known Pitting Resistance Equivalent (PRE_N) value, as this parameter is used to control those elements which determine the corrosion resistance of the material. Currently, few suppliers control the minimum PRE_N value of their alloys. Many quote typical values which can be significantly higher than the minimum values which can occur in materials as supplied. Thus equipment can be purchased which has inferior corrosion, erosion performance and cavitation resistance. Control of the ferrite/austenite phase balance in Duplex materials is also necessary to ensure that their low temperature impact toughness is maximised.

NOISE CONTROL

The close proximity of the living quarters of rig personnel to high powered equipment poses a noise control problem.

Pumps, gearboxes and drivers generate high sound levels and in the case of pumps the pipework helps to transmit noise to locations in the rig far removed from the equipment module. It is possible however, to meet the set parameters. Two courses of action may be taken. The individual item of equipment may be supplied with acoustic enclosures, or the rig module can be acoustically treated. The latter approach is now often preferred by the operators as it simplifies maintenance. None of the pump constructions considered here are simpler or easier to attenuate than any other. Double case pumps should offer lower emissions, but in practice this does not appear to be the case, consequently there is no advantage in selecting one particular configuration in preference to another.

CONDITION MONITORING

One of the major developments in offshore technology has been in condition monitoring. In the 1970's bearing, pump casing, motor winding and lubricating oil temperatures were monitored along with shaft vibration level, lubricating oil and suction and discharge pressures. Alarm and trip functions were active on most of these parameters, thereby protecting plant from catastrophic failure in the event of a malfunction. It was in fact, a damage limitation system for "after the event".

With advances in electronics and the continual striving by operators to achieve the unmanned rig the Main-Oil-Line and Seawater Injection package in the latter half of the 1980's and early 90's is now computer monitored and controlled. In addition to the functions of the 1970's the computer systems are capable of continually updating history files so trends can be identified and possible malfunctions detected before the event. This enables preventive maintenance to be instigated and minimises costs associated with unscheduled loss of production. It also means that simplification and reduction in instrumentation levels on pumpsets is not as advantageous as perhaps it may have been prior to computerised monitoring.

DISCUSSION AND CONCLUSION

The possibilities open to the designer to meet a given duty are considerable and as such, the role of the Pump Applications Engineer at the tender stage of a project is important. The user may specify his preferred pump type but specifications are often open to modification if an improved system can be offered. This may result in more than one proposal from each manufacturer tendering, thus leading to a highly competitive market situation. Offshore applications place great emphasis on reliability, with efficiency, weight and space considerations being also of paramount importance. With the emphasis on reliability, considerable credence is placed by the user on application lists citing similar designs in service. If an innovative solution is necessitated, then previous experience cannot be used to validate designs. In this situation more extensive analysis is required, coupled with a more stringent test programme.

The configurations reviewed here all offer viable solutions to water injection and oil transfer pumping applications. Each has advantages, disadvantages and constraints which need to be evaluated on an application specific basis.

The relative cost of each configuration is however difficult to assess. The conventional horizontal units should be more expensive to purchase than both the High-Speed Geared and Vertical pumpsets, due to its higher weight and traditional design. The inclusion of a gearbox in the high-speed unit should give the vertical pumpset the lowest initial purchase price. Marketing strategies and fluctuations in exchange rates however, will in all probability, continue to alter their respective purchase prices and make such simple comparisons very difficult.

REFERENCE

(1) HOFMANN, H.; A vertical heavy-duty pump concept opens up new platform dimensions in offshore technology - topside cost savings. I.Mech.E 1990 C403/024.

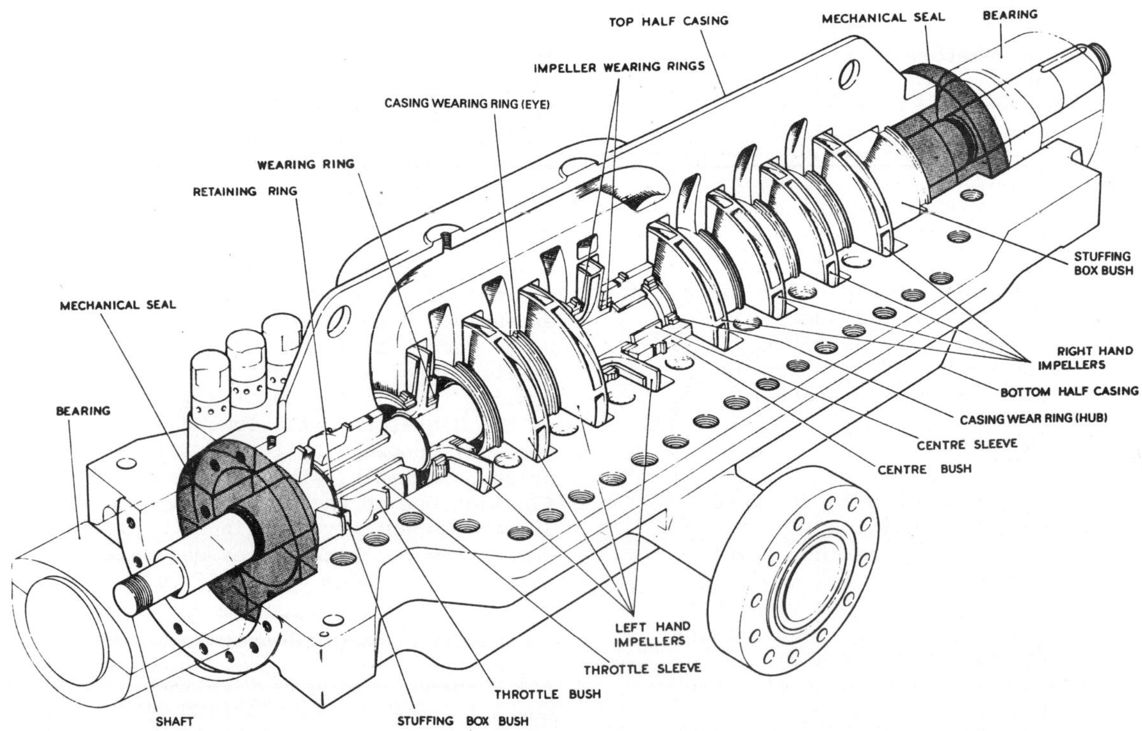

Fig 1 Cut-away of 8 stage axial split (horizontal) casing pump

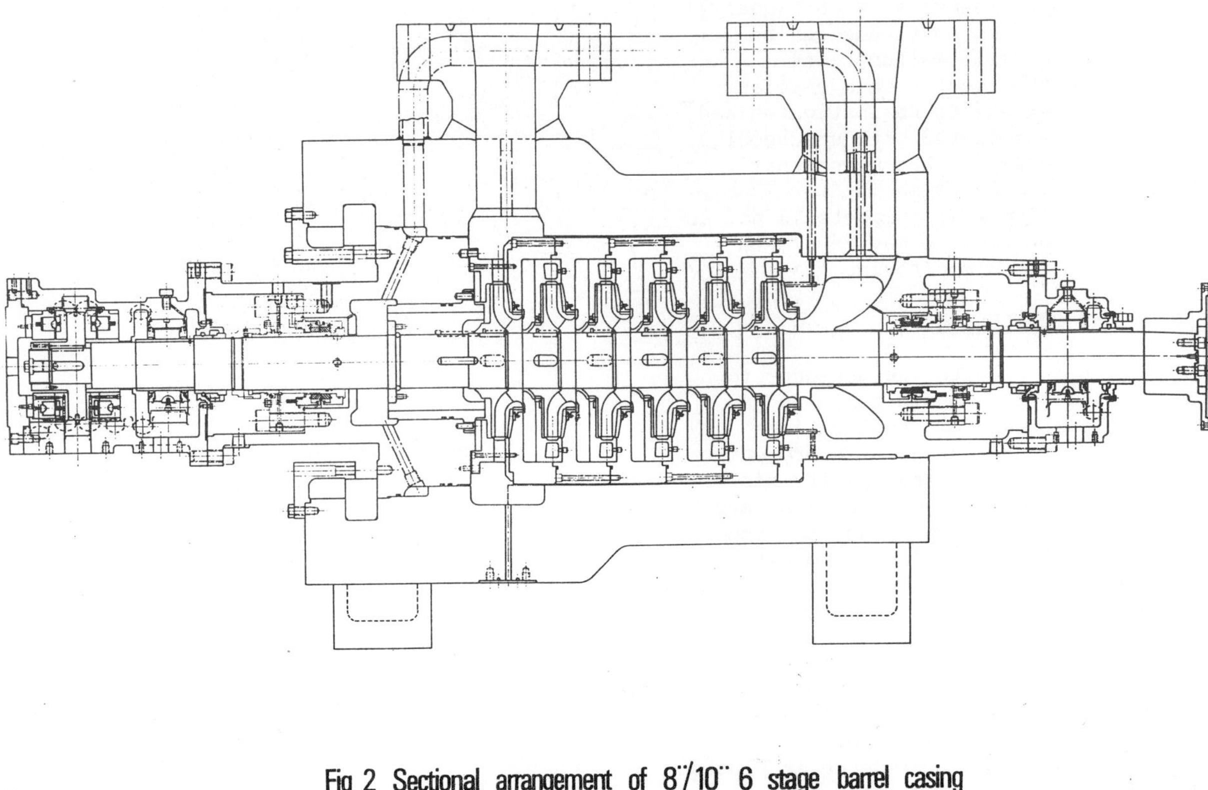

Fig 2 Sectional arrangement of 8"/10" 6 stage barrel casing pump constructed in duplex stainless steel (Ekofisk Field)

Fig 3 8"/10" Vertical oil export pumps 3.6MW at 3600 rpm

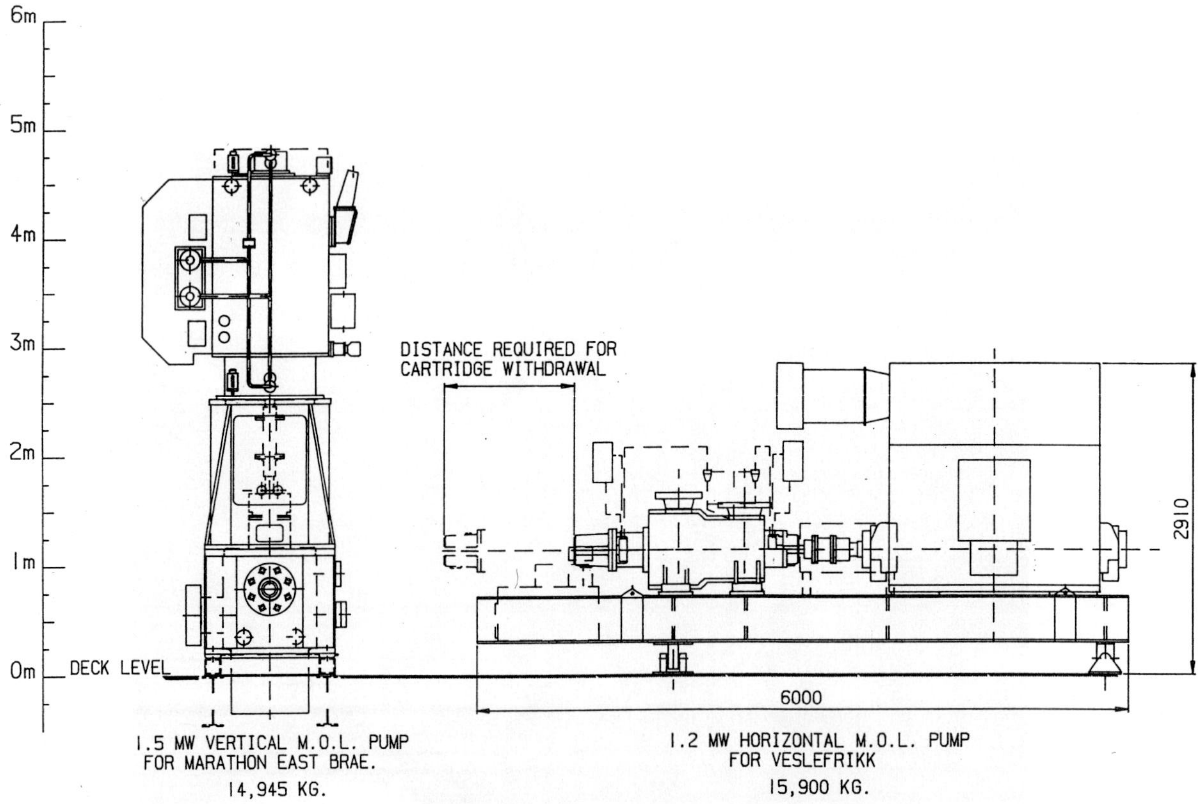

Fig 4 Comparison of horizontal and vertical pumpsets

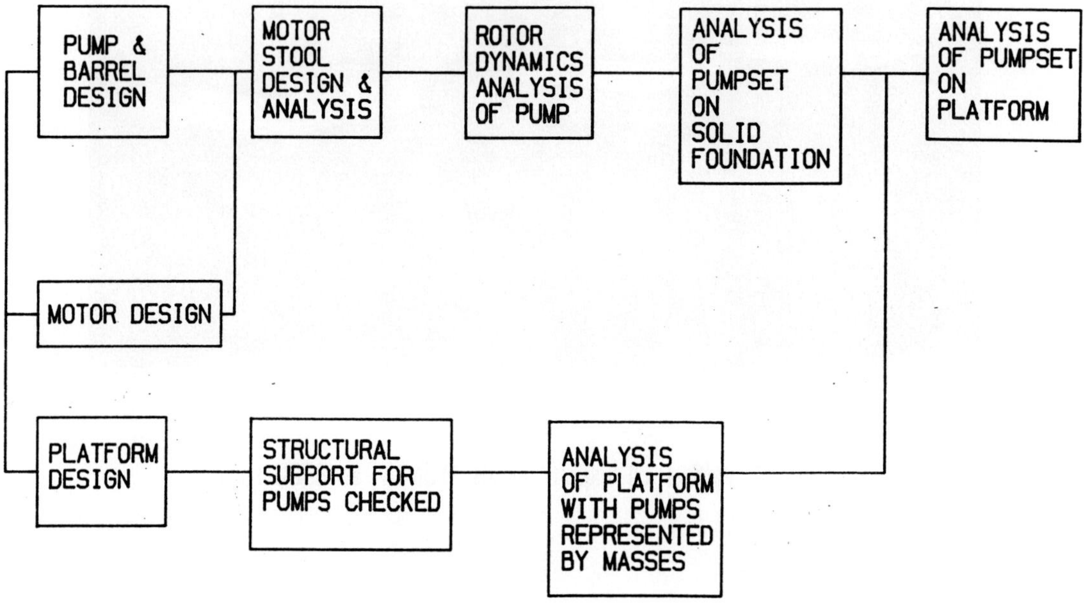

Fig 5 Plan of design/analysis for vertical pumpset

Fig 6 12 200 rpm high speed geared configuration

Capacity 748 m^3/hr., Suction pressure 0.37 bar g., Discharge pressure 235 bar g., Motor rating 6600 kW.

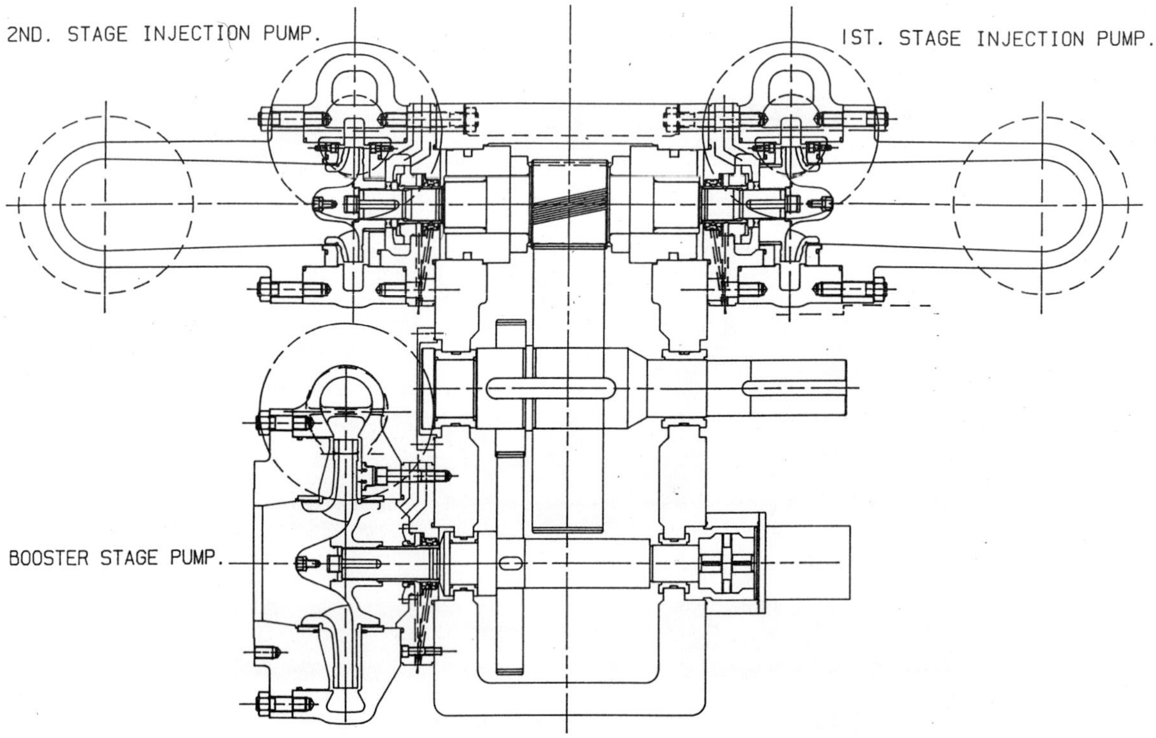

Fig 7 Sectional arrangement of 12 200 rpm high speed geared configuration with integral booster stage

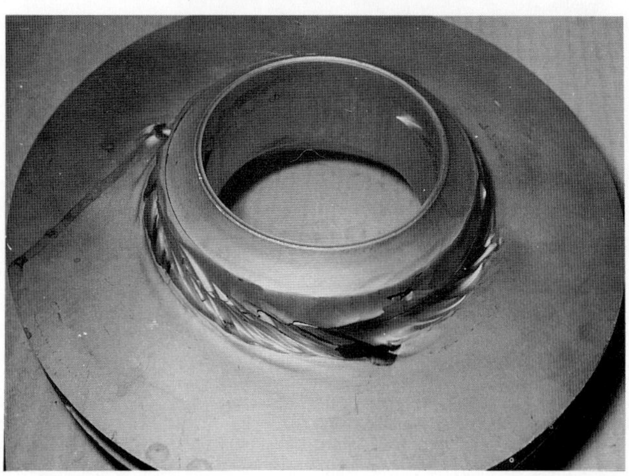

Fig 8 Duplex stainless steel impeller after 6 months in produced water service

Fig 9 Balance drum bush in Sintered Tungsten Carbide after 6935 hours in produced water service

C449/005/93

The water injection system on the Heidrun platform

C CARSTENSEN and T WILK
Satoil, Stavanger, Norway
K Van BROCKLIN
Conoco, Lysaker, Norway

SYNOPSIS Pressure maintenance in an oil reservoir is often a vital part of an oil production system. On an oil platform the water injection system forms one of the largest power consumers. The system key functions: Sea water lift, Filtering, Deaeration, Pressure boosting, and Injection can be fulfilled with a variety of technical distinguishing solutions. For the pumps involved, several different solutions exist as well.

The system of the largest concrete tension leg platform, the HEIDRUN platform, is presented. Selection considerations, the feed back of operational and maintenance experiences are described and the procurement activity is touched. Specific attention is spent on the rotating equipment involved.

This paper briefly describes the alternative bids, which was evaluated and concentrates on the solutions chosen for the Heidrun installation. A description of the system, former operational experience and the planned future use is given.

Special emphasis is placed on the identification of flexibility and avoidance of problem areas.

1 INTRODUCTION

The Heidrun field lies about 175 km off the coast of Mid-Norway about 270 km northwest of Trondheim in 340 - 370 m of waterdepth, see Figure 2. The recoverable field reserves consist of 120 million m^3 of oil and 9,35 billion m^3 of gas associated with oil and an additional 37 billion m^3 in the gas cap. The associated gas will be routed through a 26 inch pipeline to Tjeldbergodden north of Kristiansund to a methanol plant, which is planned. The concrete tension leg platform will have a topside dry weight of around 42000 t and consists of 4 columns with 31 m diameter and 4 tethers per corner. An overall sketch of the installation is given in Figure 2. The distance between the leg centres is 80 m. The platform normal operating staff will be 220 people. However, 350 beds can be made available on peak basis. On the seabed a 2000 t 56 slot template is located. The production start is scheduled in 1995 with a plateau production of 200000 BOPD. The oil production will last at least to the year 2016. After this date, the gas cap will be produced. The water injection capacity is 330000 BWPD and injection will start soon after production start-up.

2 WATER INJECTION SYSTEM

The purpose of the seawater system is to maintain the oil field pressure by delivering high quality, high pressure seawater to the reservoir. This system is vital for the revenue of the entire production system and relies on critical rotating equipment units. The power installed in the complete water injection system is 16,8 MW. This is 39 % of the total operation power for the first phase. The installed power on the platform which is 75 MW, includes contingencies for later phases. The Heidrun overall system design is shown in Figure 3. Water supply to the 6 subsea wells is controlled from the water injection header. A minimum of 5 additional injectors will be drilled from the platform.

The requirements to the system is to obtain a maximum of *reliability*, *flexibility*, optimum *size*, *weight*, and *cost*. The system shall be of proven design, easy to operate and to maintain.

The water injection system consists of the following major components:

2	Injection manifold
1 x 100%	Deaeration tower
1	Chemical injection
2 x 100 %	Vacuum pumps
5 x 25 %	Cartridge-type seawater filters
2 x 100 %	Ejectors for deaeration
3 x 33 %	Booster pumps
3 x 33 %	Water injection pumps

The chosen system with parallel filters and parallel booster and main injection pumps can be regarded as common practice in comparable plants. To increase the availability despite no standby capacity for injection and booster pumps there is ordered one spare cartridge for the injection pump, spare impeller for the booster pump and one complete spare motor/pump unit for seawater lift pump.

3 WATER TREATMENT

The seawater is lifted by the seawater lift pumps, which supply all systems on the platform, except the firewater system. The 3 x 50% seawater lift pumps, each 1,45 MW are located 19 m below LAT in seachest whose water intake is 63 meters below the sea level. The injection water is first used to cool the separated crude oil and the platform's primary cooling water system. This provides an added benefit of increased seawater temperatures, which enhances the deaeration process.

3.1 Filtration

Filtration is necessary to avoid problems in the reservoir and in small clearance areas in the pumps involved. The water is filtered to less than 1000 particles per millilitre larger then 2 microns particle size. Each filter has a capacity of 546 m^3/h at an inlet pressure of 7,1 bar and an operating temperature of 7 - 29 °C. The two-stage filters have an area of 1150 m^2 pr. vessel. The wetted parts of the filter are made of 6MO stainless steel, while the pressure vessel is made of mild steel with a HYPALON liner. The operating weight of each filter is 9,2 t.

3.2 Deaeration tower

Deoxygenation of the seawater is necessary in order to minimise corrosion. Reduction of oxygen level from 10 ppm to 0,02 ppm is done by a 3 stage vacuum unit. The seawater is degassed at a vacuum between 0,018 bara and 0,06 bara dependent on the current water temperature.

Table 1 Deaeration tower data

Type	-	Barrel casing
Height	m	24
Diameter	m	4,8
Capacity	m^3/h	2200
Op. Weight	t	205

3.3 Chemical injection

Chlorination prevents growth of microorganisms. Chlorine injection (sodium hyperchlorite) is done in the seawater system close to the seawater lift pumps. Any free chlorine is removed in the deaeration tower.

The system design includes facilities for injection of oxygen scavenger, biocide and anti-fouling agent.

4 BOOSTER PUMPS

Due to the large height of the deaeration tower, adequate spacing between the tower and the pump inlets was not sufficient to comply with the NPSH requirements of the injection pumps. It was therefore necessary to install booster pumps. The purpose of these pumps is to draw water from the deaeration tower storage sump and boost the pressure to a level which satisfies the NPSH requirements of the main pumps downstream. It is customary to choose slow speed single stage booster pump, which have moderate NPSH requirements. This arrangement ensures long life and reliable operation of the main pump under transient and varying load conditions. The 3 parallel booster pumps are vertical, single stage pumps.

Table 2 Booster pump data

Pump type	-	Vertical, single stage
Head	m	55
Capacity	m^3/h	729
Speed	rpm	1490
No. of Stages	-	1
Driver	-	2 pole- 690V-50Hz-3phase-el. motor
Manufacturer	-	Loher, 185 kW
Rated power	kW	140
Seal code	-	BSTFX, Plan 11
Impeller	mm	438

5 MAIN PUMP

The main injection pumps are high-pressure, barrel-casing diffusor pumps. The horizontal, 7-stage pumps are direct driven by two pole induction motors of 4,9 MW at 50 Hz. The motors are direct started with open valves. The pump is based on a cartridge type rotor/stator configuration, which allows the complete rotating assembly, including bearings and seals, to be withdrawn from the pump as a single item. This withdrawal capability eliminates the need for unnecessary dismantling.

Compared with a non-cartridge design, the time to replace all rotating components is reduced to less than one shift. The cartridge design also eliminates the need for extensive piping-to-pump re-alignment checks after the intervention. The chosen in-line impeller solution offers several advantages compared with other solutions. In the Heidrun case, the choice was between 5 pump alternatives.

The ranking had to be made between two high speed- , three direct driven alternatives. Two of them were horizontal, and one vertical arrangement. All quotations were based on company specifications but the manufacturers were encouraged to propose alternative solutions. The four features : footprint area, skidheight, weight and price are summarised in Figure 4 and Figure 5 in a relative manner. Alternative A in that figure is the chosen alternative. The high speed pumps were characterised by having a somewhat reduced operation window. A number of additional features such as weight penalty, long delivery time, or limited references with respect to certain parameters were taken into consideration.

The design data for the pumps are given in Table 3.

Table 3 Water injection pump data

Pump type	-	Barrel casing
Head	m	1695
Capacity	m^3/h	729
Speed	rpm	2987
No. of stages	-	7
Shaft power	MW	4,4
Coupling	-	Turboflex 2
Impeller	mm	454
$NPSH_A$	m	60
Suction temp.	°C	29
L x B x H	m	8.2 x 3 x 3
Weight	t	33,8

Table 4 Motor data

Type	-	2 pole-induction motor
Manufacturer	-	Holec
Rated power	MW	4,9
Speed	rpm	2987
Weight	t	20
Features		11kV-50Hz-3phase

5.1 Bearings/ bearing forces

Axial thrust balancing is achieved by taking approximately 90 % of the hydraulic thrust by the balance piston and the rest by double acting Stainless steel/white metal tilting pad bearings. The bearings for each pump are lubricated by a separate API 614 - design lube oil system.

5.2 Seals

Single mechanical seals were selected. The API code was BSTHN, Plan 11.

5.3 Rotor dynamics

A lateral critical speed analysis determined that for design clearances and 200 % design clearances, the rotor critical speed are well above the operation speed and exceeds also the requirement of API 610.

5.4 Materials

The major problems in existing seawater injection plants are related to corrosion, erosion in case of high velocity domains. The references (3), (4) and (5) place considerable attention on the discussion of material selection. The use of MONEL for pump shafts has frequently lead to pitting and stress cracking as a result of contact with chlorinated seawater. Therefore this material was avoided in the Heidrun plant. The piping material is GRE for the system upstream of the injection pumps and carbon steel for the remaining part. Carbon steel is considered adequate for the service because the seawater has been deoxygenated. The pumps however are also in the deoxygenized part but have been made of alloys due to the significantly higher local velocities.

Injection pump: DUPLEX ZERON 25
Injection piping: CS
Booster pump: DUPLEX
Seawater distribution piping: CS

5.5 Condition monitoring

The main quantities, which are monitored, are illustrated in the P&ID in Figure 6.

5.6 Control

With a constant speed drive and no discharge throttling valve, pump capacity is solely dependent on the shape of its head - capacity performance curve and the downstream reservoir pressure. An overboard dump valve is provided for pump minimum flow protection.

6 TEST REQUIREMENTS

The usual API 610 requirements: hydrostatic test, performance test and NPSH test were specified. Any tests are witnessed by purchaser.

7 CONCLUSIONS

Several types of pumps are possible for water injection services offshore. The final choice for the Heidrun installation was based on the following considerations: The solution is proven, the majority of seawater injection pumps in the North Sea have in fact a similar architecture. A combination of weight and space advantages in combination with existing experience and acceptable price finally lead to the chosen alternative.

8 Acknowledgement

The authors thank the Heidrun Unit owners for permission to publish this paper. Thanks are expressed to Frank Mohn A/S - Weir pumps for their contribution with drawings.

9 References

(1) The present Heidrun unit consists of Den norske stats oljeselskap a.s (STATOIL) 75,000 %, Conoco Norway Inc. 18,125 %, Neste Petroleum A/S 5,000 %, Norsk Hydro a.s 1,250 %, and Det Norske Oljeselskap a.s (DNO) 0,625 %. The license 095 (block 6507/7) is operated by Conoco, while the license 124 (block 6507/8) is operated by Statoil. An agreement has been reached under which the total operatorship will be transferred to Statoil at a time to coincide with the start of production. Conoco will be the operator during the field development phase.

(2) Edwards, R. M.: "Evaluation criteria for high energy centrifugal pumps", 10th International Conference of the British Pump Manufacturers Association, Cambridge, UK, 1987

(3) Goodacre, G. E.: "High speed pump reliability: Development for offshore water injection duty", 9th International Conference of the British Pump Manufacturers Association, Coventry, UK, 1985

(4) Hancock, W. P.: "Operating experience and extension of water injection system on the Statfjord field to over 1 Million BWPD", OTC 5839, 1988

(5) Ryall, M. L.: "Some recent developments in high-speed multi-stage pumps for secondary recovery in oil fields", IMechE Conference, paper C31/84, 1984

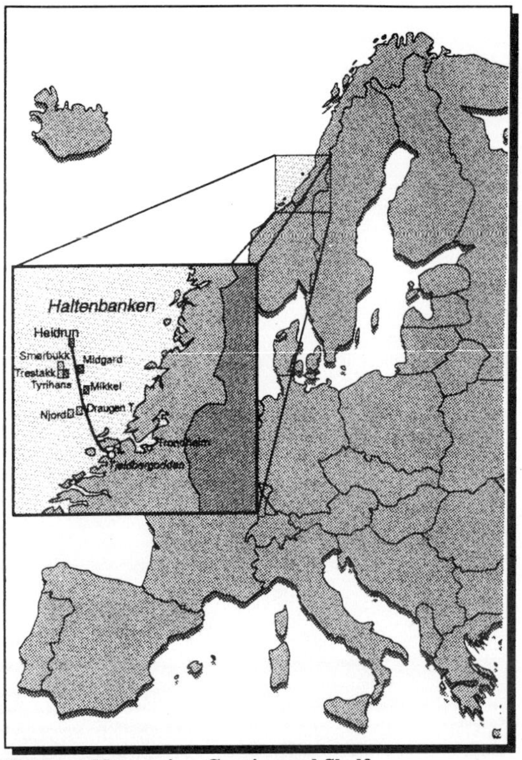

Figure 1 Norwegian Continental Shelf

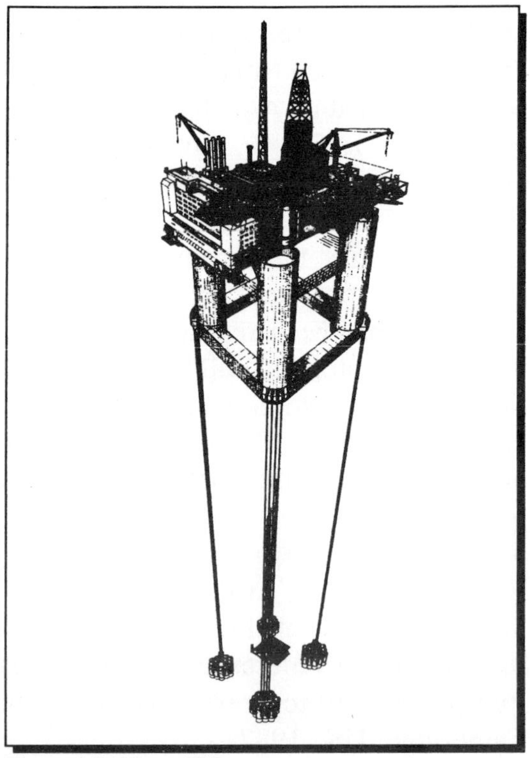

Figure 2 Heidrun TLP

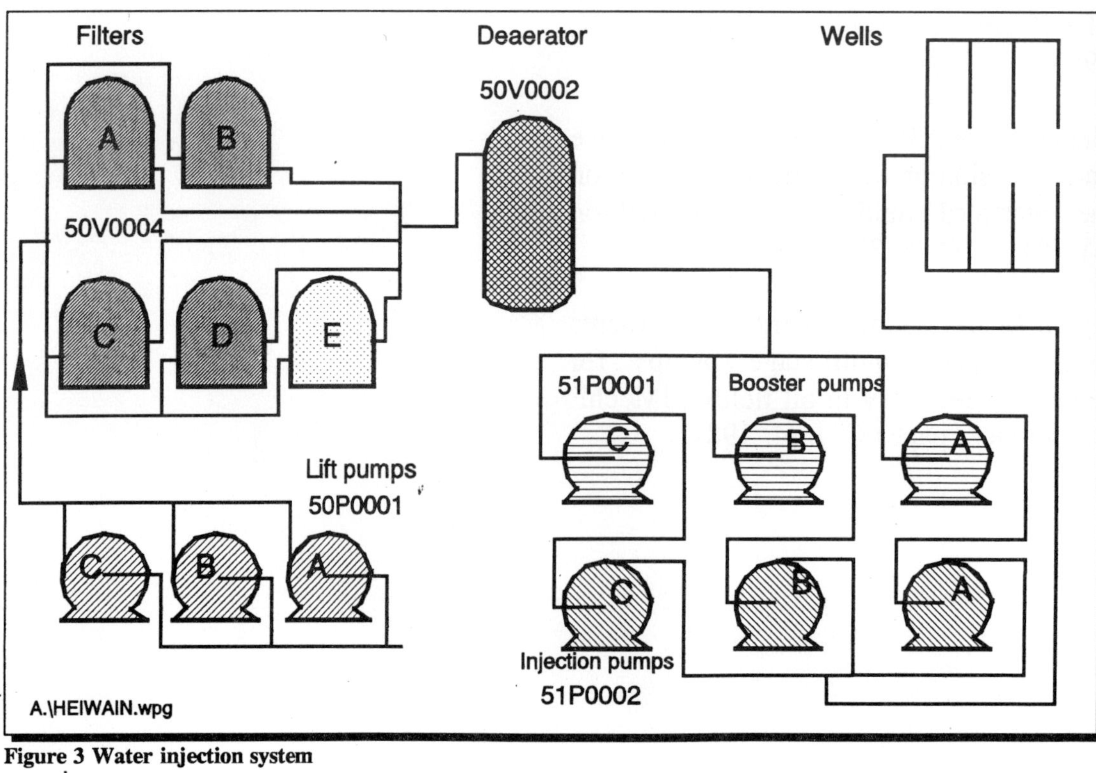

Figure 3 Water injection system

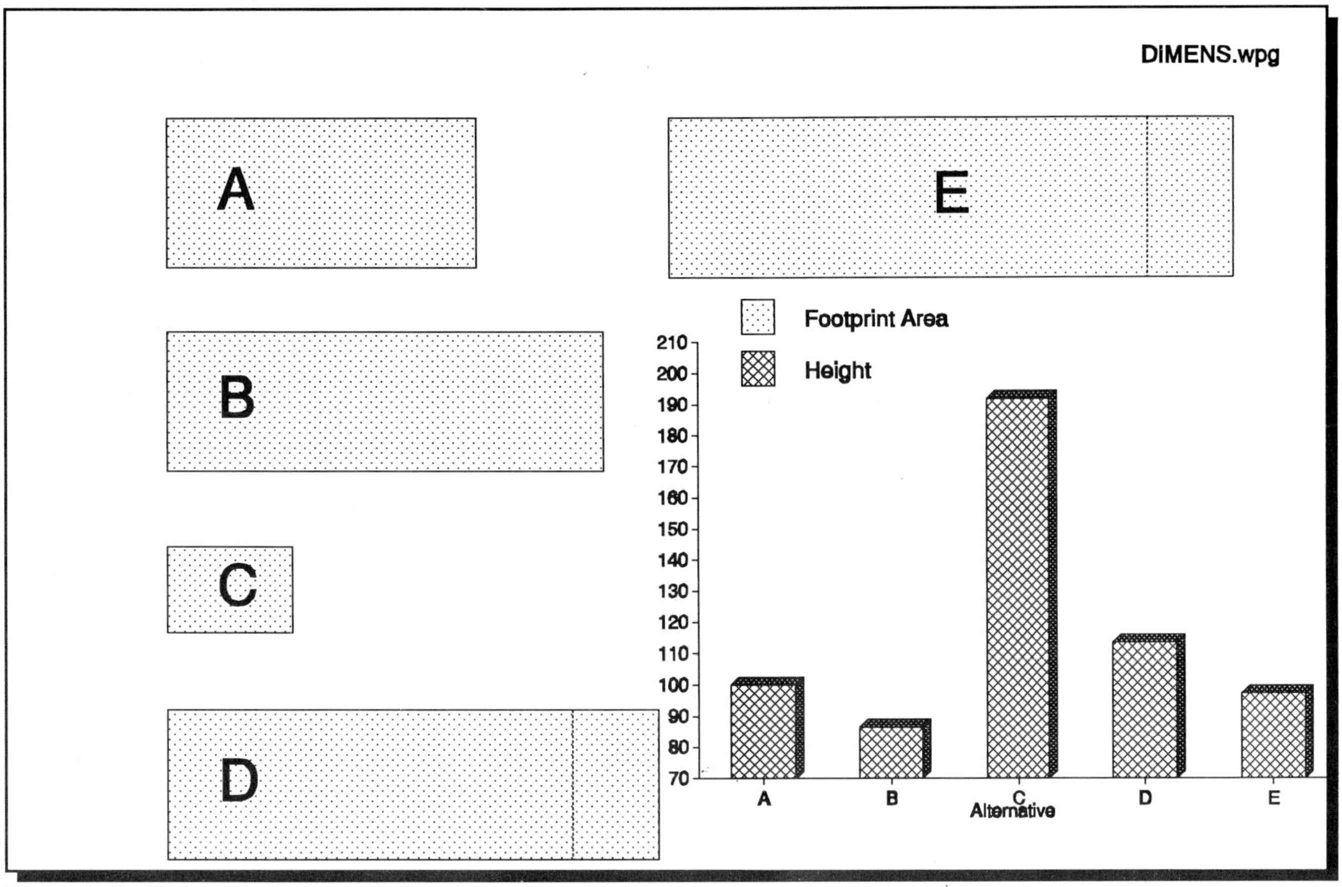

Figure 4 Bid comparison

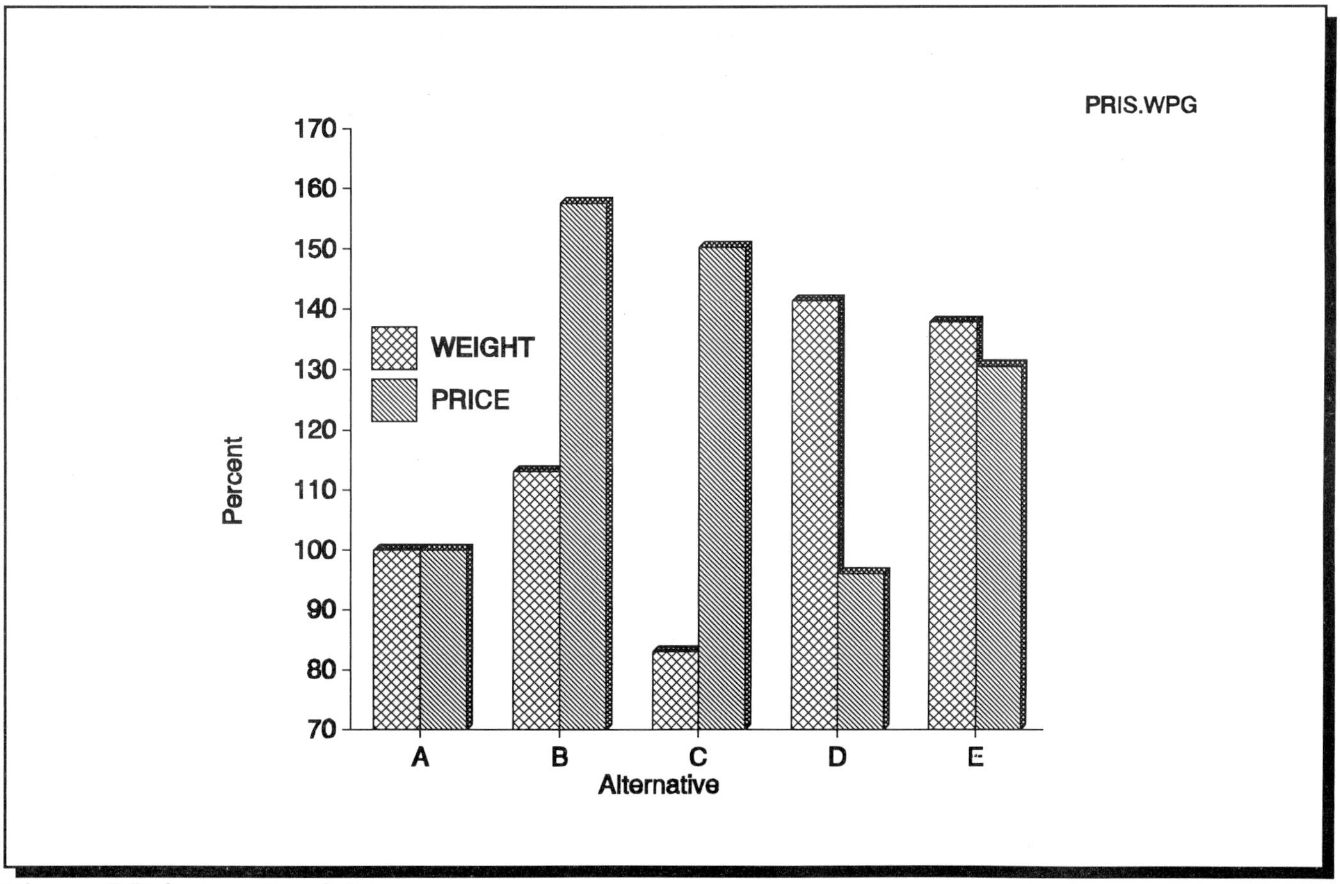

Figure 5 Prices and weights

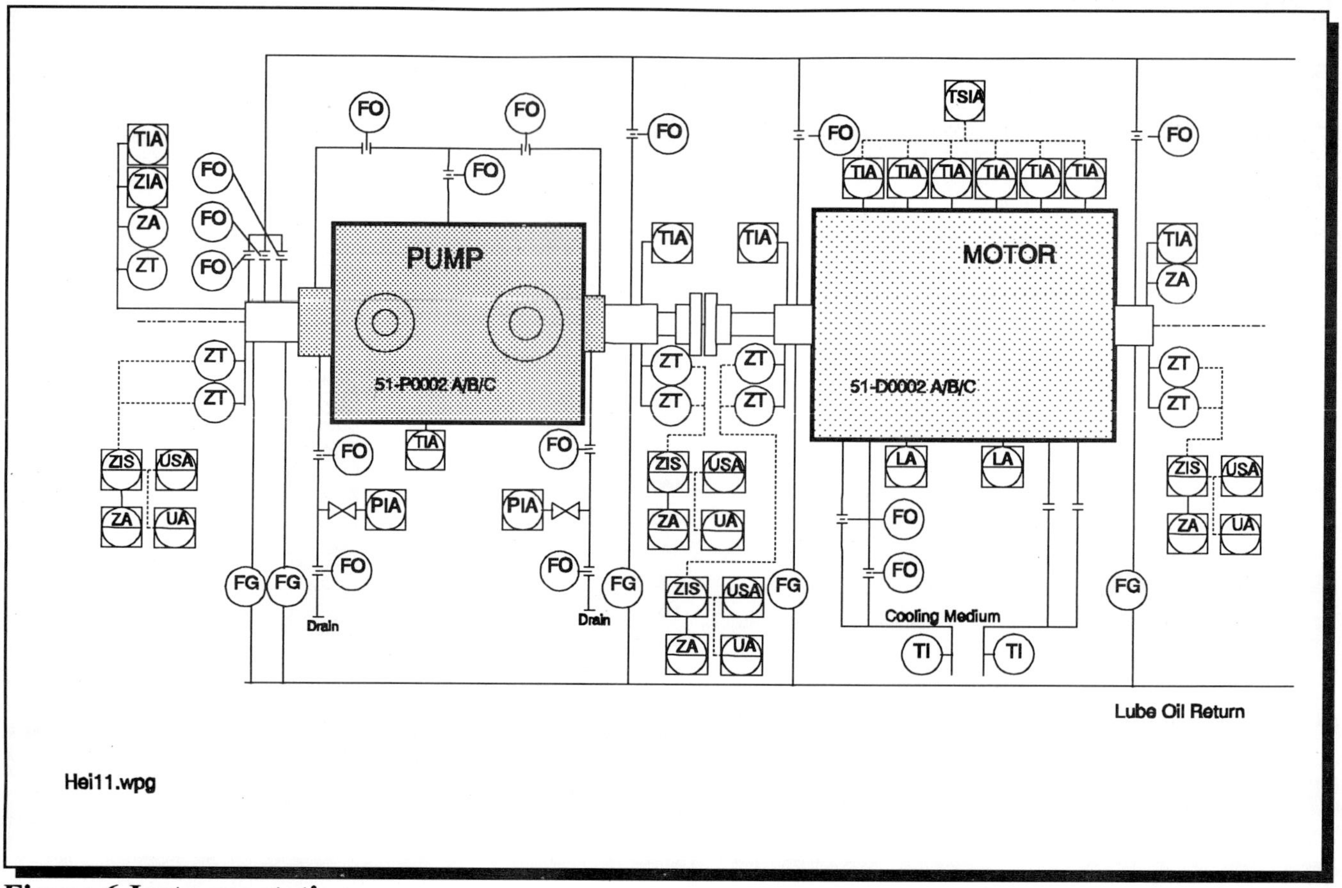

Figure 6 Instrumentation

C449/022/93

Upgrading of the seawater injection facilities on a North Sea offshore production platform

Eur Ing A K CROZIER, BSc, CEng, MIMechE
BP Exploration Operating Company Limited, Aberdeen

SYNOPSIS This paper addresses the requirement for the Seawater Injection System upgrading, the limitations of the original equipment, new equipment specification, supply, installation and operation.

Particular mention is made of how performance from the High Pressure Pumps was maximised within the existing motor power and casing pressure limitations and how specification of the new Booster Pumps provided improved system availability and flexibility.

The modified system satisfied significant space, weight, power and access constraints and minimised disruption to the platform production during installation and commissioning.

NOTATION

m^3/hr	cubic metres per hour
bpd	barrels per day
ppb	parts per billion
NB	nominal bore
D/A	deaerator
CCR	central control room
ESD	emergency shut down
CMS	computer monitoring system
HAZOP	hazard and operability study
PLC	programmable logic controller

1. INTRODUCTION

As North Sea oil fields mature and deplete, reservoir characteristics alter and differing enhancement techniques are used to maintain and maximise production for as long as possible. Reservoir Engineering is a very complex science which defines the particular techniques best suited to optimising production from specific installations and how these should be applied throughout the producing life of the field.

BP's Magnus Field is the most northerly producing field in the UK sector of the North Sea, the platform being located 120 miles North East of Shetland. Output from the field is particularly sensitive to seawater injection into the reservoir, and after approximately six years of operation, studies showed that plateau production could best be maintained by a significant upgrade of the original seawater injection facilities.

A dedicated Project Team was set up by BP to handle the project on behalf of the client (Magnus Asset Group). Detailed design was carried out by a Design Contractor (Foster Wheeler Wood Group), and the project team were responsible for complete project management including all interfaces between client, design contractor and vendors/sub-contractors. They also supervised all installation, hook-up and commissioning activities.

2. ORIGINAL SYSTEM DESIGN

The original water injection system was rated for the following design conditions:

(a) Water injection capacity

1324 m^3/hr (200 000 bpd)

(b) Injection manifold operating pressure

200 barg

(c) HP pump discharge header rating

1500 LB (design pressure 245 barg @ 16deg.C)

(d) Water filter quality

98% of particles greater than 2 micron removed at rated flow

(e) <u>Residual oxygen level</u>

50 ppb

A schematic diagram of the original system is shown in Figure 1.

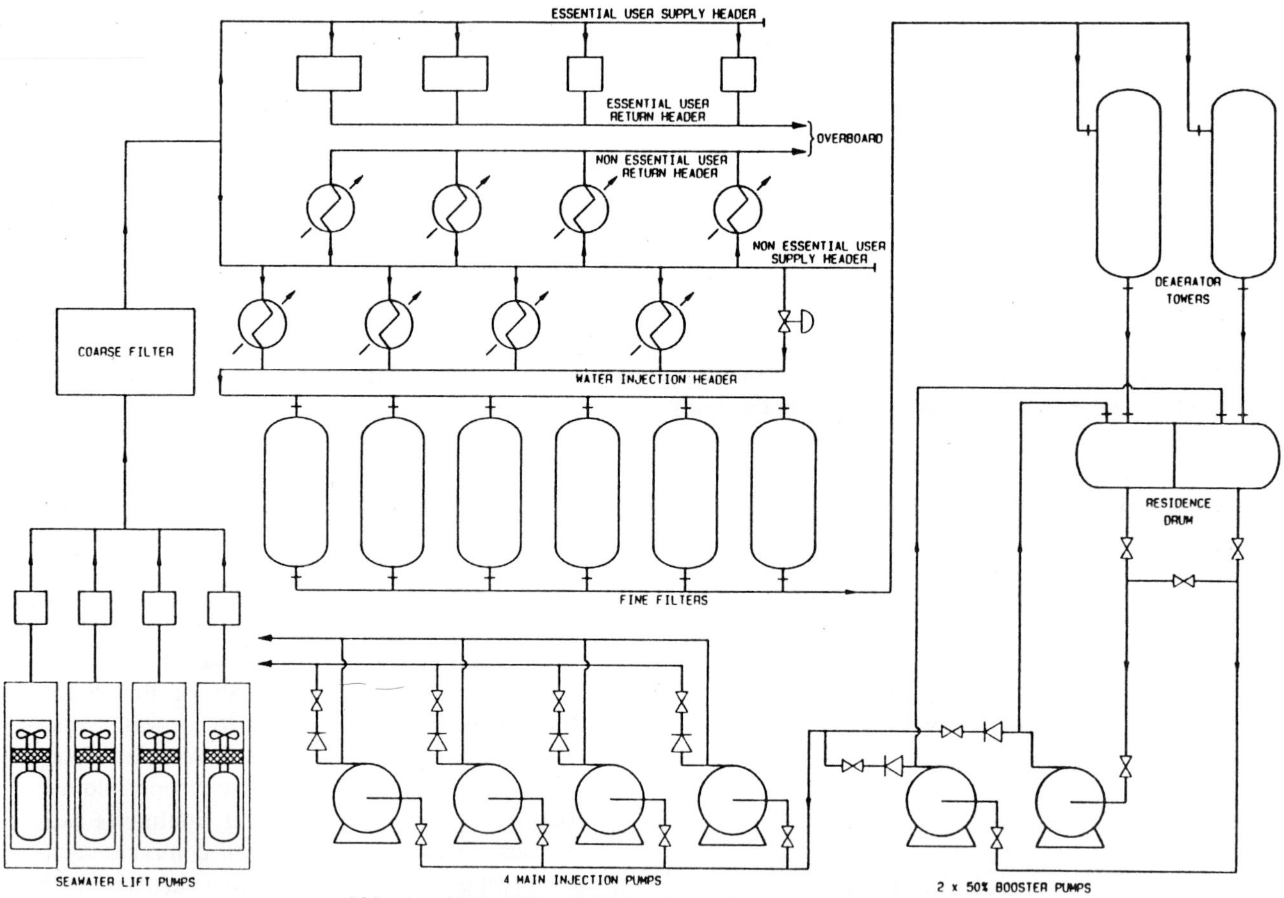

FIG. 1 SCHEMATIC DIAGRAM OF ORIGINAL SYSTEM

Raw, chlorinated seawater was brought onto the platform by four seawater lift pumps and after passing through a coarse filter package, was utilised for direct process cooling and other services, the warm cooling water being collected in three return headers. A portion of the warm seawater from one return header was processed for water injection, the excess water from the other headers being dumped overboard. If water injection demand ever exceeded the return flow to the dedicated injection header, then the balance was provided by cold water make-up from the "non essential user supply header". Injection water was then fine filtered in six dual media filter units and deaerated in two 50 per cent towers with dedicated residence drums.

Originally the towers were 'gas stripped' and oxygen concentration was further reduced by adding oxygen scavenging chemicals to the residence drums. From these drums two dedicated booster pumps fed the four main seawater injection pumps.

These booster pumps also provided water for low pressure desanding and fuel oil displacement. The main injection pumps fed to a 16" NB discharge header which in turn fed 16" NB injection manifolds. High pressure water was also supplied for desanding to the pig launchers and for service water to the sub-sea wells.

Although the original seawater injection system satisfied the requirements of the initial specification, as the field matured additional injection capacity and pressure were required. High corrosion rates had also been observed in the pipework downstream of the deaerators. Contributory factors were considered to be the high concentration of carbon dioxide in the fuel gas used for stripping the deaerators becoming dissolved in the seawater and the high residual oxygen levels. An improvement in residual oxygen levels was therefore also required.

3. CONSTRAINTS ON UPGRADING

A detailed review was carried out to determine how much design margin existed within the original equipment and which items would bottleneck upgrading. This showed that very few items had available spare capacity and replacement or modification would be required of the following:

(a) Fine Filters

(b) D/A Towers

(c) Booster Pumps

(d) Main Injection Pumps

(e) Pipework

(f) Utilities

(g) Support Structures

4. REVISED SYSTEM PERFORMANCE REQUIREMENTS

In depth review of reservoir characteristics plus analysis of plant modifications required to produce varying degrees of enhancement showed the following system specification would give the best cost/benefit ratio.

(a) Water Injection Capacity

1855 m^3/hr (280 000 bpd).

This will give an annualised average figure of 1656m^3/hr (250 000 bpd) when all equipment and system outages are considered.

(b) Injection Manifold Operating Pressure

225 barg

(c) Pump Discharge Header Rating

2500 LB (design pressure 300 barg at 38°C)

(d) Water filter quality

98% of particles greater than 5 microns removed at rated flow

(e) Residual Oxygen Level

10 ppb

(f) Pipework Design Life

15 years

(g) Pipework Corrosion Allowance

6 mm

5. SYSTEM/EQUIPMENT MODIFICATIONS

5.1 Overview

An additional cross-over line ensured adequate warm seawater was supplied to the system and suction pipework to the fine filters was replaced to limit velocity and pressure losses at the increased flow rate. Modifications were also carried out to the filter internals.

The original 'gas stripped' Deaerator Towers were replaced by larger vessels suitable for vacuum operation and the diaphragm in the residence drum into which the D/A towers discharged was breached to form a single vessel.

The original 2 x 50% Booster Pumps were replaced by 3 x 33 1/3% pumps of increased capacity and the discharge pipework through to the main injection pumps was replaced with pipework suitable for the increased pressure and flow.

The original four Main Injection Pumps were modified to give increased performance and a fifth identical pump is to be installed at a future date in the space vacated by the original booster pumps. This necessitated replacement of the downstream header and pipework to accommodate the upgraded conditions. Modifications were also required to the power distribution and controls systems and to access/support structures.

A schematic diagram of the upgraded system is shown in Figure 2.

5.2 Impact on Original Facilities

5.2.1 New Equipment Supplied

3 x 33 $^1/_3$ % Booster Pumps
1 x Main Seawater Injection Pumpset
2 x 50% (Vacuum) Deaerator Towers
3 x 50% Vacuum Pumps
Extended 11KV switchboard
Extended 3.3KV switchboard
Significant quantities of pipework suitable for the revised system operating conditions

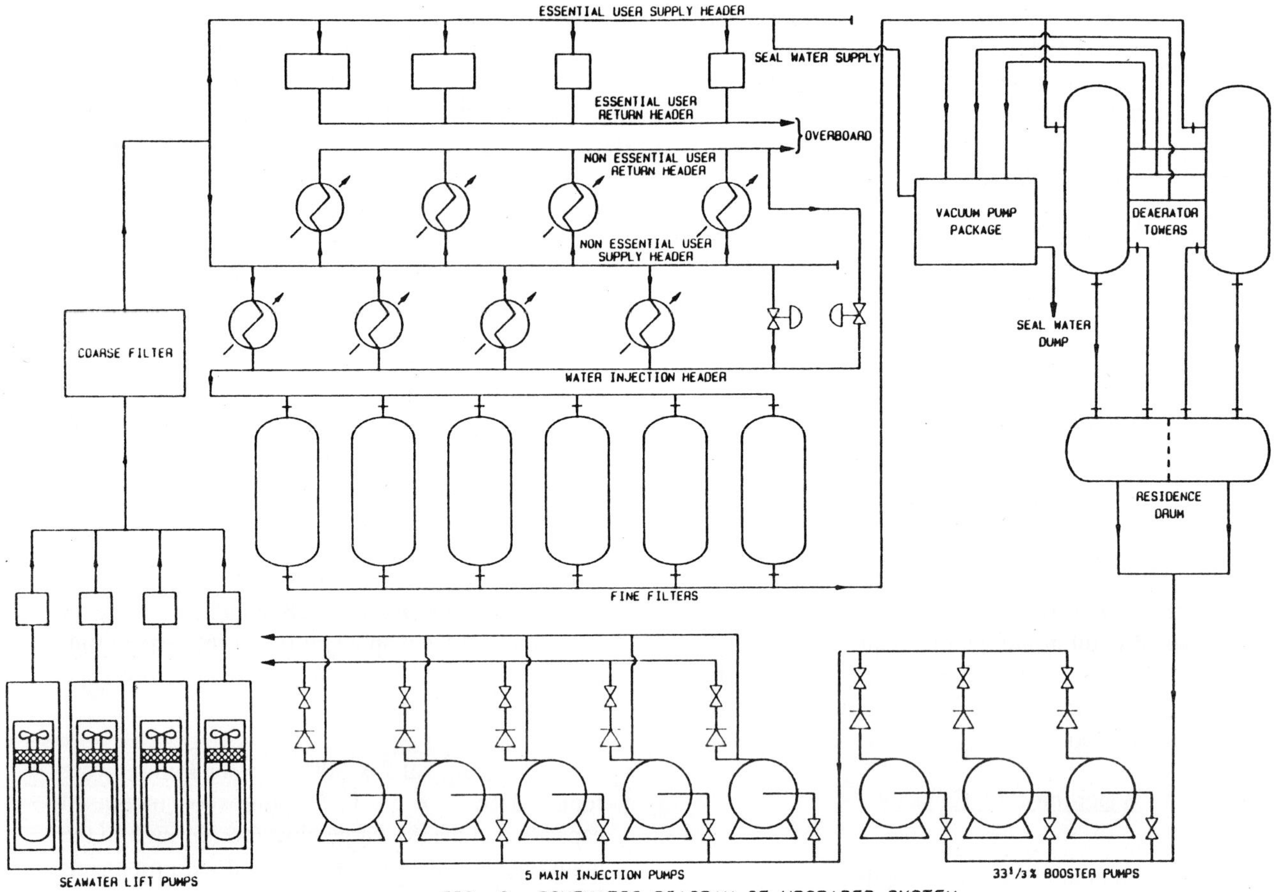

FIG. 2 SCHEMATIC DIAGRAM OF UPGRADED SYSTEM

5.2.2 Original Equipment Modified

4 x Main Seawater Injection Pumps
Central Control Room Panel Section 32
415V Switchboard
Emergency Shut Down Panel
Fine Filter Package (including piping)
CMS Equipment
Support Structures

5.2.3 Original Equipment Removed

2 x 50% Booster Pumps
2 x (Gas Stripped) Deaerator Towers
Pipework replaced due to uprating

5.2.4 Original Equipment Relocated

Fire Pump G4403 (Reorientation)
Fire Monitor - Module 2
Foam Tank - Module 4
Dirivent Fan - Module 6

5.3 Modification Details:

5.3.1 Seawater Supply

A 12 inch nominal bore crossover was installed between the "Non Essential User Return Header" and the "Return Header to Water Injection". This ensured that the increased water injection rate could be supplied without drawing additional cold water from the "Non Essential User Supply Header". The new crossover was provided with a pressure control valve.

The 16 inch NB section of pipework upstream of the Fine Filters was replaced by 20 inch NB.

A 3 inch NB line was tied in to the "Essential User Supply Header" to provide a supply of cold, filtered water to the Vacuum Pump liquid seal system.

5.3.2 Fine Filters

To accommmodate the increased throughput without incurring an unacceptable pressure drop, the filter media was replaced with a coarser grade. This resulted in a slightly reduced filter efficiency but was not considered to be detrimental to the long term operation of the reservoir. The backwash cycle was also made more frequent and the air scour rate increased during the backwash period. New internal collection headers were installed. Trim in flow control valves was changed to handle increased flow and PLC software altered to suit the above changes.

5.3.3 Vacuum Deaeration

The new Deaerator Towers were three stage 'Vacuum' units, having the same diameter but being taller than the originals. Design parameters of the new towers were as follows:

(a) Maximum throughput

2120 m^3/hr l(320 000 bpd)

(b) Optimum performance capacity with inlet distributors originally supplied

1855 m^3/hr (280 000 bpd).

(c) Design Pressure

3.75 barg/full vacuum

(d) Design Temperature

-15/+120°C

(e) Internal Lining

Glass Flake

The new towers were supplied complete with internals suitable for vacuum operation and with a compatible vacuum pump package. Although taller and heavier than the original towers, the additional static head loss for the feed system was offset by the towers' low internal pressure.

The inlet pipework local to the towers was modified to suit and inlet flow control valves replaced. Vacuum take-offs from each deaeration stage of the two vessels were commoned and run to the vacuum pump skid.

The vessels were designed to mitigate foaming and as a further precaution, a chemical injection point was provided upstream for anti-foam dosing. Vessel protection consisted of relief valves and bursting discs connected to the LP flare system. Access ladders and platforms were provided for maintenance purposes and inspection windows permitted foaming checks to be carried out. Sample points with barometric legs to allow sampling under vacuum conditions were provided between each stage of deaeration and after the final stage prior to entering the Residence Drum for performance guarantee verification.

The new vacuum package comprised 3 x 50% pump sets on common manifolds. Each set was on a separate skid for ease of installation and comprised Vacuum Pump, Motor and Coupling, Atmospheric Air Ejectors, Seal Water Separator Tank and all associated Valves and Instrumentation. Control of the Vacuum Pumps was by a local panel. Pump status indication was also repeated at a Central Control Panel. The Liquid Ring Vacuum Pumps were supplied with cold `Seal Water' filtered to remove particles greater than 80 micron. After use the warm seal water discharged to dedicated `Seal Water Separators' prior to dumping.

5.3.4 Residence Drum

The original Residence Drum was retained, being suitable for vacuum service. Originally it had a mid position diaphragm plate installed to effectively create dedicated residence vessels for each deaerator tower. The original 2 x 50% Booster Pumps were also dedicated to one D/A Tower/Residence Drum. In order to simplify the pipework and controls for the new 33 1/3% Booster Pumps, the diaphragm plate was breached to create a single vessel. The original booster pump minimum flow recycle lines were removed and the redundant 8" NB connections on the residence drum utilised to provide an increased transfer capacity from the D/A towers. 12" NB nozzles initially used for gravity flow level controlled dumps were unsuitable for vacuum operation and were utilised as additional water outlets to the new booster pumps. A new level controlled dump was provided on the booster pump discharge.

The Chemical Injection dilution lines were retained to ensure good mixing of the chemicals across the horizontal vessel. Level instrumentation was modified as described in Section 5.9.2.

5.3.5 Booster Pumps

The original two x 50% Booster Pumps were initially rated at a head of 44 m and flow of 696 m^3/hr (106 000 bpd) and had previously been upgraded to give a flow increase of 15%. They were unsuitable for further upgrading for the following reasons:

(i) NPSHR at increased flows is in excess of that available.

(ii) Insufficient generated head available.

(iii) Driver under-rated for new service.

The Main Injection Pumps had a limited capacity for upgrading due to Motor Power limitations and even when all of the available potential was utilised, this did not provide adequate discharge pressure to satisfy the new system requirements. Additional pressure was therefore required from the Booster Pumps. Ideally, 3 x 50% pumps were preferred which would give an installed spare pump. However, pumps satisfying that specification would have been too large to be accommodated without very extensive structural modifications and would also have required extensive expansion of the power distribution system to accommodate the required motor power.

A balance was reached whereby 3 x 33 1/3% pumps were supplied with motor ratings which fully utilised the potential of the existing power distribution system. The resulting pump specification gave a large margin of 'run-out' capacity on the pump duty points such that even with one pump out of service, the two remaining pumps would provide adequate NPSH to the Main Injection Pumps (including the additional 5th pump). The Main Pumps would then however have a slightly reduced discharge pressure.

The duty of each pump was specified as follows:

(a) Duty flow

641 m^3/hr (96 800 bpd)

(b) Generated Pressure

181 m

(c) Suction Pressure

-0.4 barg

The above duty was based on a total system throughput of 1855 m^3/hr (280 000 bpd) plus chemical injection recycle for the two deaerators. The chemical injection recycle also serves as the pump minimum flow recycle thus simplifying the control system.

To satisfy the above conditions and the requirement for a minimum space 'footprint', vertical 'Canister' type pumps were selected. They were located between Modules 4 and 6 at a level between the main and mezzanine decks as shown in Figure 3.

To cater for the vacuum suction conditions, a continuous flush of deoxygenated seawater was provided to the mechanical seals to prevent possible air ingress when a pump was on standby. Provision was also made for venting the suction side high points back to the deaerators.

Unplanned outages were covered by procurement of a spare set of pump internals. A spare motor was also purchased and is held in a preserved condition Onshore.

Performance Curves for the new and original booster pumps are shown in Figure 4.

5.3.6 Main Injection Pumps

The four original Main Injection Pumps were specified at a rated flow of 332 m^3/hr (50 000 bpd) at a generated head of 2000m.

To satisfy the new system requirements of 280 000 bpd throughput at 225 barg presssure, the original pumps

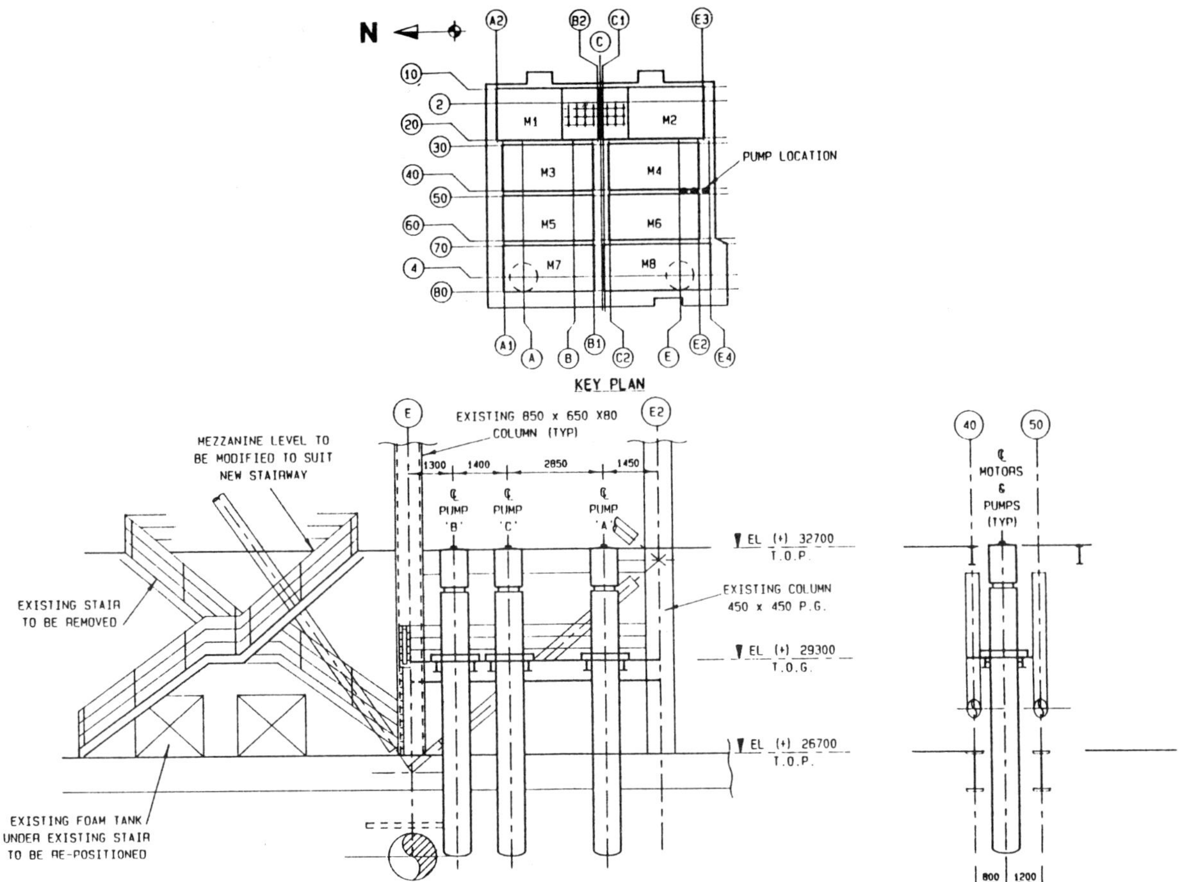

FIG 3 LOCATION OF NEW BOOSTER PUMPS

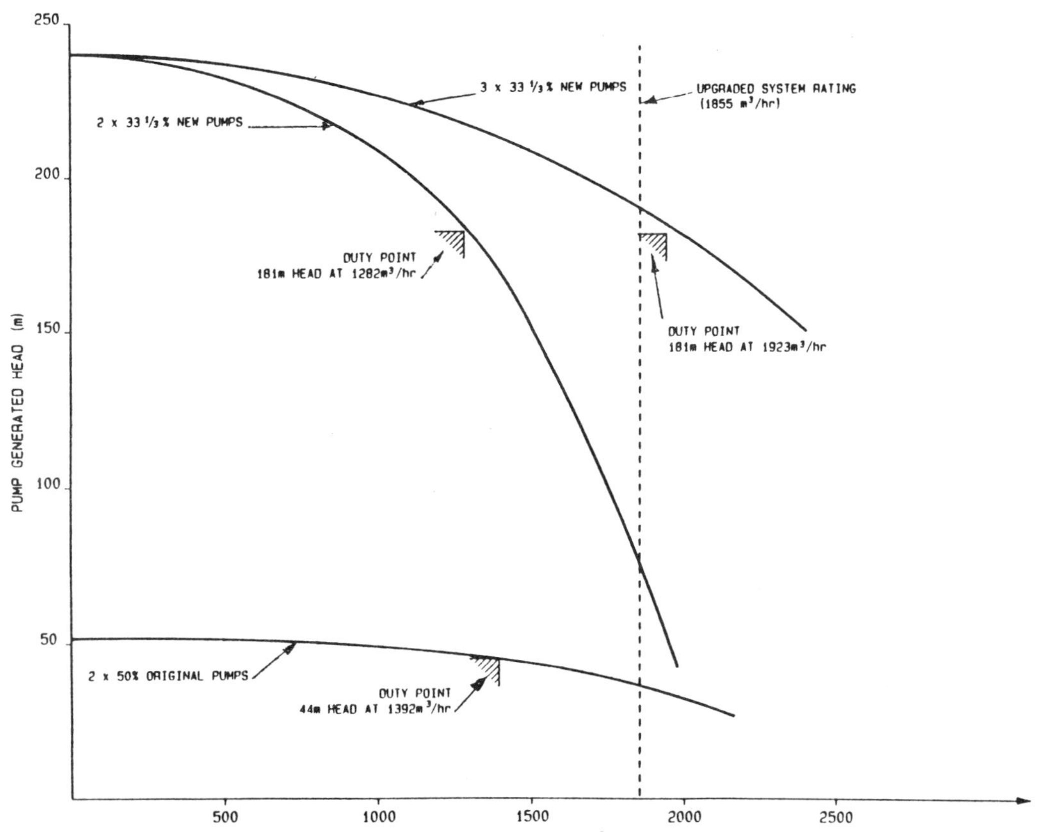

FIG 4 BOOSTER PUMP PERFORMANCE CURVES

required upgrading and an additional 5th pump of identical 'upgraded' specification will also have to be installed. The 225 barg system pressure requirement is equivalent to 227.5 barg at the pump discharge.

The pump vendor had conducted considerable hydraulic design optimisation work and associated performance testing on the existing pumps and was able to provide guaranteed performance improvements by fitting impellers of modified configuration and larger diameter within the original casings. When added to the upgraded performance from the new Booster Pumps, the overall system performance requirements could be satisfied. The new Main Pump duty specification was :

(a) Duty flow

371 m^3/hr

(b) Generated pressured

2100m

(c) Suction pressure

15.9 barg

Main Pump casing Design Pressure was 265 barg and this would accommodate the proposed increase in discharge pressure to 227.5 barg. Similarly, the original motor power rating of 3055 Kw would accommodate the increased pump power rating of 2969 Kw. Gearbox and Couplings also had sufficient capacity to accommodate the power increase.

The casings of the original four pumps were supplied in 318 Stainless Steel. These suffered from excessive corrosion and consequent leakage along the joint faces and through the stud holes. A remedial Inconel 625 inlay on the joint faces was incorporated in these units. An existing 'spare' pump had previously been purchased having a duplex alloy casing. This material had showed much improved corrosion resistance and was also specified for the new pump.

The location for the new pumpset is on the lower production level of Module 6 in the space originally occupied by the 2 x 50% Booster Pumps.

This location was selected because:

(i) The pipe connections to thc pump arc relatively straightforward.

(ii) All Main Injection Pumps will be in Module 6.

(iii) Good maintenance access is availablc.

Performance curves for the original and upgraded main pumps are shown in Figure 5.

Combined system performance curves from the new Booster Pumps and upgraded Main Pumps are shown in Figure 6.

Proposed Module 6 layout before and after installation of the fifth Main Injection Pump is shown in Figure 7.

5.3.7 Control Philosophy

(i) Seawater Supply

With the exception of the crossover between the seawater return headers, the control philosophy upstream of the deaerators remained unchanged.

(ii) Deaerator Flow/Level Control

The flow into the Deaerators was balanced by individual flow controls on each tower inlet, the inlet flow controllers deriving their set point from the common level control on the residence drum and deaerator outlet flow transmitter. The two deaerators therefore responded as one thus eliminating interaction between the two inlet flows. The inlet flow control also responded quickly to outlet flow conditions, including Injection Pump start up.

Under normal operation both towers are run simultaneously. To allow single tower operation, a selector was provided that adjusts the bias to the inlet flow controller set points.

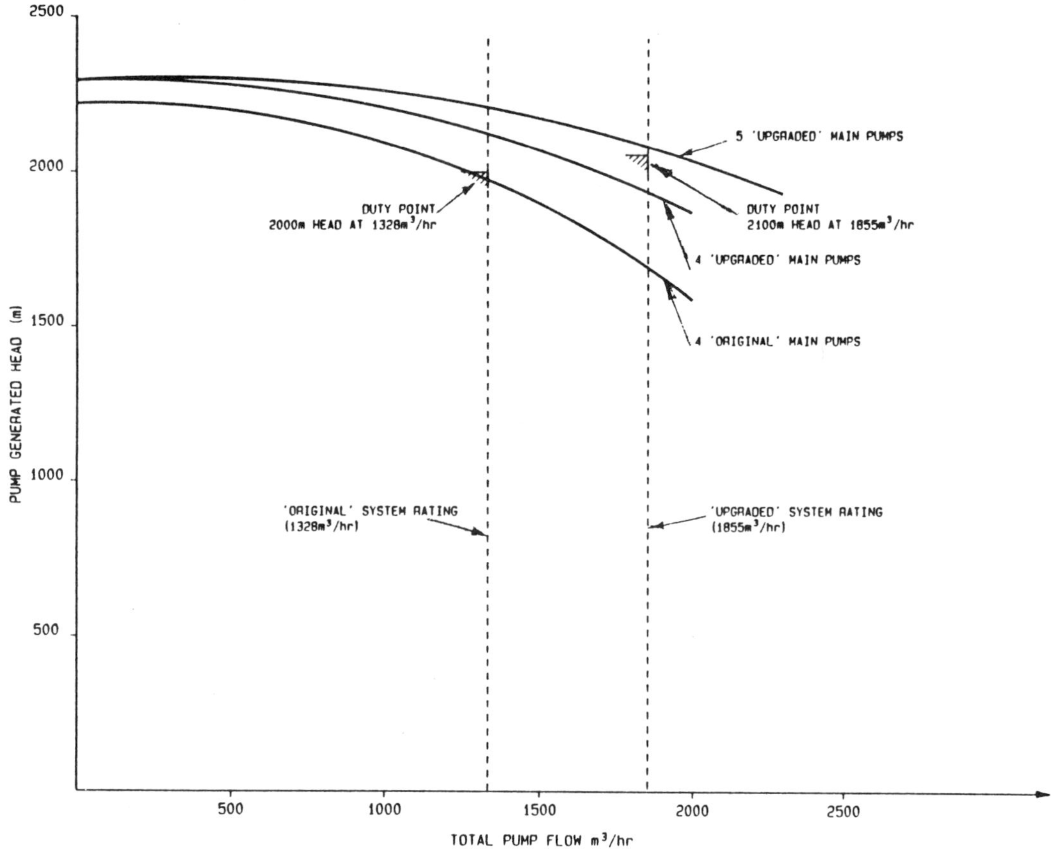

FIG 5 MAIN PUMP PERFORMANCE CURVES

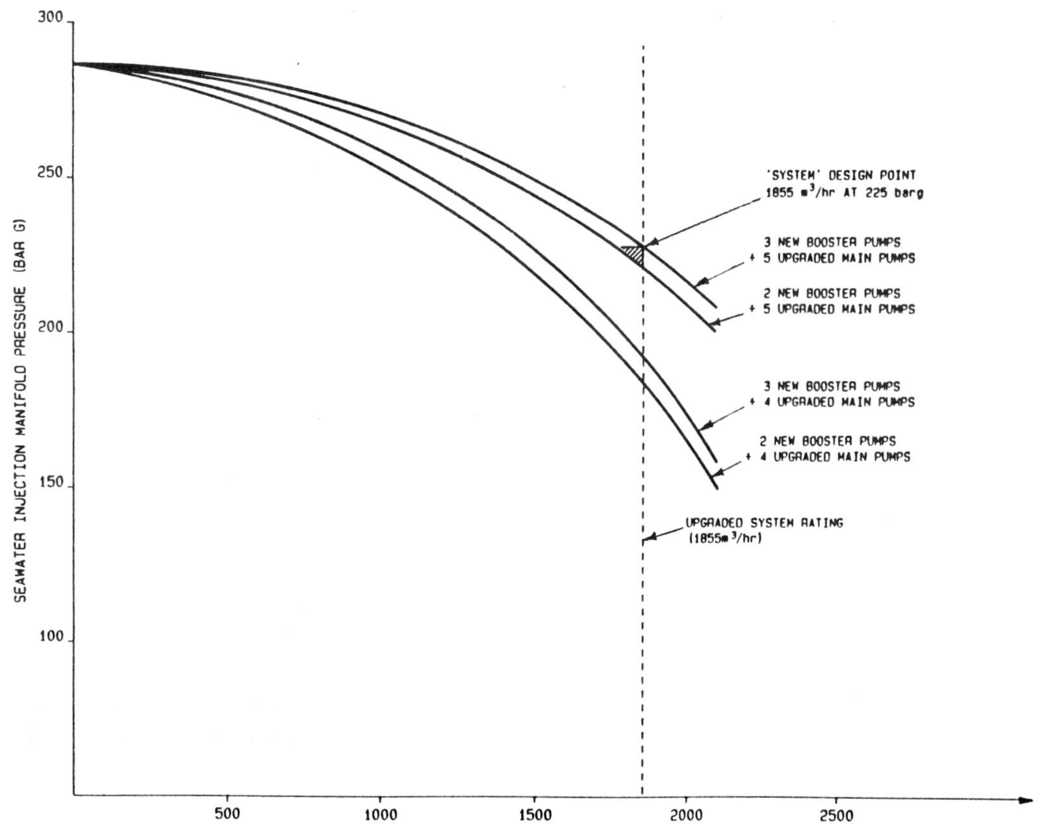

FIG 6 COMBINED SYSTEM PERFORMANCE CURVES

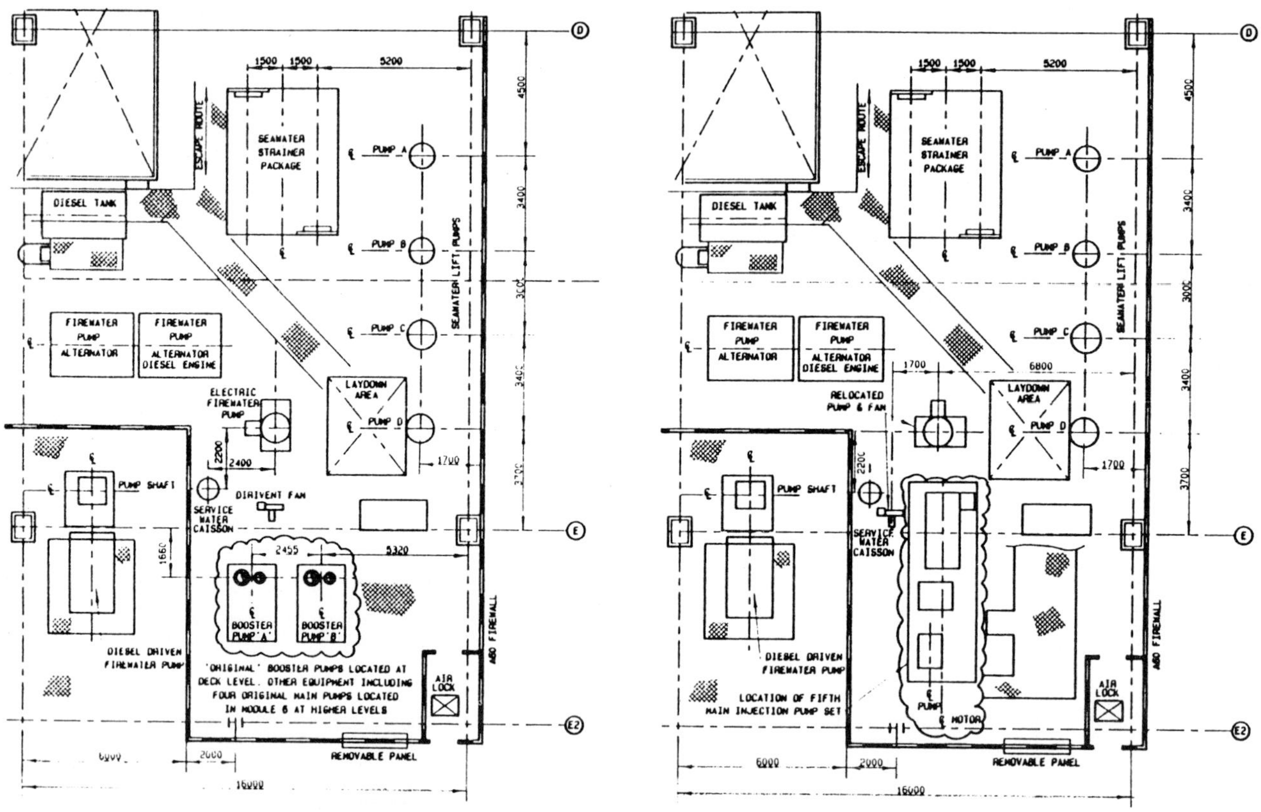

EXISTING LAYOUT

PROPOSED LAYOUT WITH 5TH INJECTION PUMP INSTALLED

FIG 7 EQUIPMENT LAYOUT IN MODULE 6

The residence drum was provided with low and high level pre alarms and trips. A low trip trips both Booster and Main Injection Pumps. A high level trip closes inlet Flow Control Valves and Shutdown Valves.

(iii) Booster Pumps

These pumps were not provided with individual minimum flow protection. The pumps are started against against a closed discharge and once the discharge is opened, sufficient short term minimum flow protection is provided by the chemical injection dilution recycle.

(iv) Main Injection Pumps

The control philosophy of the original four pumps remained unchanged and that of the new pump was identical to these.

5.3.8 Piping Modifications:

Extensive piping modifications were required to many parts of the Water Injection and support systems. In addition to the crossover line and fine filter suction header already mentioned in Section 5.3.1, and the deaerator/residence drum transfer lines mentioned in Section 5.3.4, the following principal alterations were also required:

(i) Booster Pump Suction, Discharge and Recycle Pipework

The recycle lines from the original Booster Pumps were redundant and were removed. New Suction and Discharge Pipework was supplied to suit the higher flows, new location and additional Booster Pump.

Discharge Pipework rating was also increased to 300 LB to cater for the increased Booster Pump pressure and give an additional corrosion allowance.

Suction Pipework routing was constrained by NPSH considerations and the requirement to allow the original pumps to remain in service while the new pumps and pipework were installed.

(ii) Main Injection Pump Discharge and Injection Manifolds

The Design Conditions selected for pipework downstream of the Main Injection Pumps were:

(a) Pressure Rating

300 Barg

(b) Temperature Rating

30 deg C

(c) Corrosion Allowance

6mm min.

(d) Max Fluid Velocity

5m/s

The original pump discharge header was rated at 248 barg at 38 deg C and the original injection manifold extruded header sections were rated at 260 barg at 38 deg.C.

This only gave a 3.2mm corrosion allowance, most of which had already been utilised giving no effective allowable increase in rating. The complete HP Injection Pipework System from the Main Injection Pump block valves to the Injection Flowlines, including the Flowline Isolation Valves was replaced. The following material was specified to suit the new operating conditions:

(a) 20" NB Carbon Steel x 60mm min wall thickness, API Grade X 52

(b) 6" NB Carbon Steel x 28mm min wall thickness, API Grade X 52

In order to minimize downtime, the new header and manifold sections were supplied and installed prior to destructing of the original assemblies.

The predicted corrosion rate for the uprated injection system is 0.4 mm/year for a water velocity of 5m/sec and the specified maximum oxygen concentration of 10 ppb. To achieve a 15 year life then a minimum corrosion allowance of 6mm is essential.

5.3.9 Utilities:

(i) Power Distribution

Power supplies to the following additional equipment were required:

(a) 11 kV seawater injection pump motor

(b) 3.3 kV additional injection booster pump motor with uprating of starters on the two other original pump drives

(c) 415V deaerator vacuum pump motors

(d) 415V lube oil heater and lube oil pump motor for additional main seawater injection pump

(e) 240V trace heating for vacuum pump skid

(ii) Central Control Room Modifications

Significant modifications were required to section 32 of the central control panel to incorporate the additional controls for the new seawater system equipment. All major controls associated with the upgrade with the exception of the vacuum pump controls were located on the section 32 panel.

The mimic panel at the top of section 32 and the CMS were modified to include the additional equipment.

5.3.10 Structural Modifications:

The overall weight effect of the Water Injection System upgrade on the platform was:

Weight Added:	+303.8 Tonnes
Weight Removed:	-115.5 Tonnes
Balance:	+188.3 Tonnes

Although a significant weight increase, most of this was positioned on substantial existing steelwork and only relatively minor modifications were required including adding trimmers, maintenance runway beams and strengthening to plate girders under the D/A towers.

Several walkways, stairs and access platforms required modification or reorientation.

5.3.11 Safety/Loss Prevention Considerations:

The full safety implications of the upgrading of the seawater injection system are shown on the Operability and Safety Assessment Sheet (Figure 8). All of the implications mentioned therein were addressed by suitable HAZOP and other design reviews.

5.3.12 External Approvals:

Certifying Authority approvals were obtained where necessary at all stages of Design, Fabrication, Installation and Commissioning.

6. PROJECT PROGRAMME/ CO-ORDINATION:

During all work on the project, production losses from the platform had to be kept to an absolute minimum. This necessitated a large degree of supply, fabrication and installation of new equipment prior to destructing of original items. It also required very close control of material movements since lay down and storage space on a fully operational installation were extremely limited. Tie-ins to existing systems and facilities were phased to coincide with planned shutdowns where possible and to further minimise disruption, three phases of implementation were defined as follows:

6.1 Phase 1

During this phase, all of the tie-in connections to the original system to which new equipment would be connected were installed and blanked off. This then provided terminal points from which construction or installation of all new equipment could be run while the original system was still live.

By-pass lines were also installed around the Fine Filter package and Deaerator Towers which would enable injection to be continued via these routes for short periods while those items were replaced or modified. By-passing the fine filters for brief periods was not considered to be detrimental to the reservoir in the long term and oxygen content with the D/A towers by-passed was contained by chemical injection.

6.2 Phase 2

This phase included installation of the new 3 x 33 1/3 % Booster Pumps and the Vacuum Deaerator Towers and associated Vacuum Package. All associated pipework, the majority of which had been installed previously, was connected. The D/A tower residence drum diaphragm was breached and all associated control and ancillary systems modifications carried out. The new high pressure injection header and individual Well flowlines were connected and all original, redundant equipment removed.

6.3 Phase 3

This phase will involve re-location of the equipment mentioned in Section 5.2.4 and the installation and hook-up of the 5th Main Seawater Injection Pump.

Although this has yet to be carried out, the fifth Main Injection Pumpset has been purchased and successfully works string tested and is presently in controlled storage at the Vendor's works.

7. SUPPLY AND INSTALLATION:

Ordering of new equipment was geared to achieving delivery dates such that the programme defined above could be maintained. Tight deadlines had to be met on system design to finalise details of principal long lead items (Booster Pumps and D/A Towers with Vacuum Package). The High Pressure Injection Header also required concentrated design input to finalise details of the non standard components which were on protracted delivery.

Installation work by all disciplines was very strictly controlled by working to detailed 'Workpacks' for small sections of the overall workscope. This kept the construction workforce to the minimum required to meet the stringent accomodation allowance which was available on the platform. This allowance depended on the number of staff engaged on other essential platform activities.

To eliminate site welding of the high pressure injection pipework spools, sophisticated measuring techniques were used to enable these to be fully welded and heat treated onshore. The size of each of these heavy spools was limited to enable them to be installed in the congested wellhead area using existing lifting facilities.

The limited load/radius capacity of the platform cranes led to a technique being used during offloading of the D/A towers from the supply vessel whereby their transit platform collapsed on first release of load. This gave sufficient margin to accommodate deck movements of the vessel at the prevailing sea states.

A two week shutdown of the Seawater Injection System was necessary to implement all of the transfer and hook-up work associated with Phase 2 of the programme when in addition to the mechanical and piping work, significant input to control and utility systems was required.

8. OPERATIONAL EXPERIENCE:

On first commissioning of the new system, some problems were experienced with the flow controls on the Deaerator Towers and foaming within the towers was also evident. These were overcome by minor modifications to the controls and by optimising the specification of the anti-foam chemical used.

The system in its present Phase 2 condition fully meets the requirements of the specification.

All new equipment has proved reliable in service and is providing the required improvement in reservoir performance. This is being continuously monitored to determine the optimum time at which to install the additional Main Injection Pump.

9. CONCLUSION

The project was implemented in a condensed timescale within demanding constraints on a fully operational major North Sea installation.

The high capital value of the project (£17 000 000) will be quickly amortised by the extended production plateau of the field.

The Seawater Injection System is now more flexible in operation, has increased reserve capacity and is less sensitive to changes in demand.

ACKNOWLEDGEMENT:

The author wishes to thank the management and colleagues of BP Exploration Operating Co.Ltd. for permission to publish this paper and their assistance in its preparation.

PROJECT TITLE : MAGNUS WATER INJECTION UPGRADE   CLIENT : MAGNUS   PROJECT No : M410   DATE : 7/2/90

CIRCLE THOSE TOPICS WHICH MAY BE CHANGED BY THIS PROJECT

PROCESS PARAMETERS

TEMPERATURE
PRESSURE
FLOW
LEVEL
COMPOSITION/FLASH POINT/HYDRATE
TOXICITY

OPERATING METHODS

START UP
SHUT DOWN
ROUTINE OPERATION
UNUSAL OPERATION
CHANGE IN INSTRUMENT LOCATION
CHANGE IN INSTRUMENT TAPPING
PREPARATION FOR MAINTENANCE

SAFETY EQUIPMENT

E.S.D.
FIRE DETECTION SYSTEM (P.O.C./U.V. FRANGIBLE BULBS)
GAS DETECTION SYSTEM
WATER DELUGE SYSTEM
PERSONNEL SAFETY EQUIPMENT
MEANS OF ESCAPE
LIFEBOATS

ENVIRONMENT

OIL SLICKS
LIQUID EFFLUENT
SLID EFFLUENT
GASEOUS EFFLUENT

ENGINEERING HARDWARE EQUIPMENT AND DESIGN

ENGINEERING LINE DIAGRAM
WIRING/SINGLE LINE/SCHEMATIC DIAGRAMS
CAUSE AND EFFECT DIAGRAMS
PLANT LAYOUT (INCLUDING PIPING)
DESIGN TEMPERATURES
DESIGN PRESSURES
MATERIALS OF CONSTRUCTION
LOADS ON OR STRENGTH OF :-
STRUCTURES
VESSELS
PIPING/SUPPORTS
TEMPORARY OR PERMANENT :-
PIPING/SUPPORT
VALVES/NON RETURN VALVES
SPEC. PLATES
FILTERS
INSTRUMENTATION SYSTEMS
TRIPS AND ALARMS
STATIC ELECTRICITY
RADIOACTIVITY
RATE OF CORROSION
LOW TEMPERATURE EMBRITTLEMENT
ISOLATION FOR MAINTENANCE
ELECTRICAL OR MECHANICAL
HANDRAILS
LADDERS
PLATFORMS
WALKWAYS
TRIP/COLLISION HAZARDS
ACCESS FOR :-
OPERATION
MAINTENANCE
FIRE FIGHTING

NOTE :-
THE "REPORT" SECTIONS AND APPENDICES THEREOF REFERRED TO ABOVE IS THE MASTER "BASIS OF DESIGN" DOCUMENT FOR THE PROJECT

LEGEND :-
+ UPS - ININTERRUPTED POWER SUPPLY
++ ESD - EMERGENCY SHUTDOWN

HOW DOES THE PROPOSAL AFFECT THE FOLLOWING CATEGORIES ?		WHAT IS THE AFFECT AND HOW WILL SAFETY AND OPERABILITY BE MAINTAINED ?
1. RELIEF AND BLOWDOWN 1.1 INTRODUCE OR ALTER IN ANY WAY ANY POTENTIAL CAUSE OF OVER PRESSURE/UNDER PRESSURE IN THE SYSTEM OR PART OF IT	YES	OVERPRESSURE - NEW BOOSTER PUMPS AND MODIFIED/NEW WATER INJECTION PUMPS GENERATE SUFFICIENT PRESSURE TO EXCEED CURRENT DOWNSTREAM LINE RATINGS. NEW SUITABLY RATED PIPEWORK WILL BE INSTALLED TOGETHER WITH PCV's AND PSV's ON WATER TAKE-OFFS FROM MAIN LINES UNDERPRESSURE - NEW VACUUM PUMPS GENERATE CLOSE TO FULL VACUUM. ALL AFFECTED VESSELS, PIPEWORK AND EQUIVALENT IS DESIGNED FOR FULL VACUUM
1.2 INTRODUCE OR ALTER IN ANY WAY ANY POTENTIAL CAUSE OF VIOLATING THE UPPER OR LOWER TEMPERATURE LIMITS OF THE SYSTEM OR PART OF IT	NO	
1.3 IN ANY WAY INVALIDATE EXISTING PROTECTIVE SYSTEMS	NO	BUT ADDITIONAL PROTECTIVE SYSTEMS ARE REQUIRED FOR NEW EQUIPMENT
2. AREA CLASSIFICATION 2.1 INTRODUCE NEW OR ALTER EXISTING ELECTRICAL EQUIPMENT	YES	MOTORS * 3 x NEW BOOSTER PUMP MOTORS MOD 4 *REMOVE 2 x BOOSTER PUMP MOTORS MOD 6 *1 x NEW INJECTION PUMP MOTOR MOD 6 1 x NEW LUBE OIL PUMP MOTOR MOD 6 *3 x NEW VACUUM PUMP MOTORS MOD 12 RELOCATE 1 x DIR VENT FAN MOTOR MOD 6 UPRATE EXISTING INJECTION PUMP MOTORS OTHER - EXTEND 11KV SWITCHBOARD MOD 7 - EXTEND 3.3KV SWITHBOARD MOD 7 - MODIFY 415V SWITCHBOARD MOD 6 *- HEAT TRACING & CONT SUPP TO VACUUM PKG MOD 12 - ADDITIONAL SUPPLY TO UPS IN CCR - RELOCATE LIGHTS MODS 4, 6, 12 - TEMPORARILY DISCONNECT REFRIGERATION UNIT MOD 12 - RE-TERMINATE SUPPLIES TO FIREPUMP G4403
2.2 INTRODUCE OR ALTER LOCATION OF POTENTIAL LEAKS OF FLAMMABLE FLUID	YES	SEE MARKED UP HAZARDOUS AREA DRAWINGS APPENDIX 19 NOTE : ITEMS MARKED * AFFECT EQUIPMENT LAYOUT ON HAZARDOUS AREA DRAWINGS
3. SAFETY EQUIPMENT 3.1 REQUIRE NEW SAFETY EQUIPMENT	NO	
3.2 AFFECT ANY OF THE EXISTING SAFETY EQUIPMENT	YES	- FIREPUMP G4403 TO BE TURNED THROUGH 90° AND PIPEWORK RE-ROUTED + - UPS MODS - SEE SECTION 4.2.1 DELUGE SYSTEM - SEE SECTION 5.14 ++ - ESD SYSTEM MODS - SEE SECTION 7.4
4. OPERATIONS AND DESIGN 4.1 INTRODUCE NEW OR ALTER EXISTING EQUIPMENT	YES	MAIN ITEMS 1 x NEW INJECTION PUMP SET 3 x NEW BOOSTER PUMP SETS 3 x NEW VACUUM PUMP SETS 2 x NEW DEAERATOR TOWERS 11KV + 3.3KV SWITCHBOARD EXTENSIONS MODIFIED CONTROL SUITE IN CCR EXTENSIVE PIPING AND CONTROL MODS LV SWITCHGEAR MODS SEE REPORT FOR DETAILS
4.2 REQUIRE CONSIDERATION OF RELEVANT API/ASME CODES OR B.S.	YES	API 520/1 FOR PSV's ALSO USED EXTENSIVELY IN EQUIPMENT SPECS (SECTION 14.0 REFERS) MAIN ITEMS COVERED BY API STANDARDS 610, 613, 614, AND BRITISH STANDARDS 4999, 5000, 5500
4.3 REQUIRE NEW STRESS CALCULATIONS ON PIPING, SUPPORTS OR VESSELS	YES	FOR ALL NEW CRITICAL PIPEWORK & SUPPORTS FOR ALL NEW TAG NUMBERED EQUIPMENT ITEMS
4.4 AFFECT THE PROCESS UPSTREAM OR DOWN STREAM OF THE CHANGE	YES	SEE SECTION 13.0 OF REPORT
4.5 AFFECT THE SAFE ACCESS FOR PERSONNEL, SAFE PLACES OF WORK OR SAFE LAYOUT	YES	WALKWAYS/ACCESS/ESCAPE ROUTES MODIFIED IN MODULES 4 & 6. SEE SECTION 7.7 OF REPORT
4.6 AFFECT ANY PROTECTIVE SYSTEM WHATEVER OR REQUIRE ADDITIONAL PROTECTIVE SYSTEMS	YES	SEE 3.2 ABOVE
4.7 AFFECT EXISTING OPERATING OR MAINTENANCE PROCEDURES OR REQUIRE NEW PROCEDURES	YES	NEW OPERATING AND MAINTENANCE PROCEDURES WILL BE REQUIRED FOR NEW EQUIPMENT AND MODIFIED CONTROL SYSTEMS
4.8 ALTER NOISE LEVELS	YES	SEE SECTION 7.8 OF REPORT
SPECIAL TOPICS HAS THIRD PARTY CERTIFICATION BEEN SOUGHT ?	YES	CERTIFYING AUTHORITY REQUIRED
ARE ANY SPECIAL INSTALLATION/TESTING PROCEDURES REQUIRED ?	YES	SEE SECTION 9.6 OF REPORT
IS A PLANT/SYSTEMS SHUTDOWN REQUIRED ?	YES	SEE APPENDIX 7

FIG 8 OPERABILITY AND SAFETY ASSESSMENT SHEET

C449/015/93

LNG power recovery

Eur Ing W WONG, CEng, FIMechE, FIMarE
Bechtel Limited, London

SYNOPSIS

The thermal efficiency of a Carnot cycle is limited by the maximum and minimum temperatures available. The construction of LNG terminals and the need to vaporise LNG offers a cooling source at a very much lower temperature than sea water.

By using the cold sink available and by adapting the waste heat available from a combined cycle plant, it is possible to recover power from the vaporisation of LNG. This is done by the use of a reverse cascade refrigeration process where energy is extracted by expanders instead of energy input by compressors.

The paper explains how the hot and cold processes can be integrated to produce a compound cycle with an overall efficiency of some 55% or more depending on the quantity of LNG to be vaporised.

Special reference is be given to the features needed for the rotating equipment and how equipment selection is critical to the realisation of a practical working power station.

1 INTRODUCTION

Natural gas is often found in remote locations far from developed industrial nations. Large reserves have typically been found in Siberia, Sumatra, Alaska, the Middle East, the Sahara Desert and the North Sea. Where possible the gas can be transported by pipeline to the end user. Currently, where the gas source and the user are separated by oceans and continents, the only viable transportation is by conversion to liquid natural gas (LNG) and carriage by insulated LNG tankers.

To produce LNG the natural gas is piped to a coastal location where the gas is processed to remove impurities and inerts, and to extract the heavy ends which are valuable for separate sale. The processed gas is finally refrigerated to make LNG and sent to storage at a temperature of approximately -160°C.

After loading the LNG into a tanker and transportation it is then off-loaded into storage tanks at the receiving terminal. The LNG is pumped from storage to the required pressure and vaporised for final transmission to the consumers.

At the receiving terminal it is cheap and convenient to vaporise the LNG by using sea water. However, this results in thermal pollution of the sea, the ocean near the terminal becomes colder with the resulting effect on the ecology of the area. In contrast, thermal power and processing plants usually have the effect of heating the sea near its location.

The energy consumed and the possible energy recovery based on existing technology is given in Table I. From these figures the efficiency of the LNG and LNG vaporisation processes can be calculated,

TABLE I Energy Per Tonne of LNG

Liquefaction	
Polytropic conversion	300 kWh
Driver power	412 kWh
Plant power	850 kWh
Vaporisation	
Isothermal Vaporisation	213 kWh
Sea Water Pumps (for heating)	3 kWh
Possible energy recovery	50 kWh
Heat of Combustion LHV	12500 kWh

Less than 25% of the energy used is reversible and only 25% of which is recoverable. About 7% of the natural gas is used as fuel, and when other losses are taken into account the gross conversion loss is about 10%.

In the production and transportation of LNG, every effort has been made by the suppliers to conserve energy and to eliminate inefficiency. Except in the case of Japan, users seem to be content with conserving LNG loss with no attempt to recover the available energy. In Japan the LNG is vaporised to supply fuel for conventional steam cycle power stations. A comparison of the condenser cooling requirements of a 1000 MW power station and the thermal energy in vaporisation of the LNG shows that the thermal content of the LNG used is insignificant in comparison. The vaporisation process in this situation is independent and it is convenient to use any hot water available for LNG energy recovery purposes. A typical cycle as used in Japan is shown in Figure 1. There are now over 11 such plants in operation in Japan.

The natural gas output pressure required from LNG vaporising terminals varies depending on the use required for the natural gas. It should be noted that in the next decade, it will be unlikely that natural gas will continue to be burnt in conventional steam cycle power plant. A 1000 MW combined cycle power station will become common and the fuel pressure required will be much higher. In the USA, for example, the vaporised LNG is typically required for onward transmission in a pipeline and an even higher pressure is required [Table II].

TABLE II Required Fuel Gas Pressure at Battery Limit

PURPOSE	REQ. PRESSURE BAR A
Steam Power Stations	6
Combined Cycle Stations	25
Local Distribution	30
Long Distance Transmission	70
Power Station Consumption 0.16 to 0.25 tonne/MWh	

In the case where LNG has to be vaporised to provide fuel gas for a combined cycle, the cold sink can be integrated with the cycle if the plant is located in a arid location with continuous high ambient temperatures. It can be shown that the cold sink available from vaporisation of the fuel needed will provide a temperature drop to the inlet air to the gas turbine of some 16°C, so that an inlet ambient of 30°C can be cooled to 14°C. This is a special case which is only applicable if the power station is close enough to the LNG terminal to allow integration of the vaporisation process with the inlet cooling of the gas turbine.

For other cases, energy recovery is still possible and can be based on the use of hot water waste heat if available, or the use of a compound cycle as discussed in this paper.

2 THEORY

Carnot cycle efficiency is based on the maximum and minimum temperatures available. Although the Carnot cycle efficiency is theoretical and impossible to obtain in practice, the theoretical gain in efficiency will indicate the possible gain in practice.

$$\text{Carnot efficiency } E = \frac{Tmax - Tmin}{Tmax}$$

For a combined cycle plant with a maximum firing temperature of 1200°C and condensing temperature of 35°C, the Carnot efficiency will be 79%. With a minimum temperature of -160°C the Carnot efficiency will be 92.32%. The Carnot gain in efficiency will be 13.3%.

3 THE BASIS FOR ENERGY RECOVERY

It is impossible to find a single working fluid that will operate within the full range created by the maximum practical temperature and the temperature of the LNG. Lower maximum temperatures however are possible such as the gas turbine exhaust temperature or the steam condensing temperature of a combined cycle. However, even these lower maximum temperatures will not enable a single fluid to be used.

In practice, therefore, a series of cascading cycles (ie. a compound cycle) must be adopted [Figure 2].

In the design of the cycle, for true integration, the heat rejected at each level must be the heat input into the next. The overall thermal efficiency (E) will be:

$$\text{Overall } E = E_1 + (1-E_1)E_2 + (1-E_2)E_3$$
$$\text{where } E = \frac{Tmax - Tmin}{Tmax}$$

For an ideal Carnot cycle, it can be shown that the overall efficiency is the same for any number of cycles between the maximum and minimum temperatures and that the temperature levels of the intervening cycles will have no effect on the overall efficiency.

These theoretical findings are important because it means that the temperature levels can be solely dictated by practical convenience and economic choice.

In practice the processes are not reversible and a temperature difference is required between each cycle. The actual power recovered will depend on the cycle design and the operational needs of the terminal.

4 THE COMPOUND CYCLE

In a combined cycle gas turbine power station, the exhaust gas waste heat is used to generate steam. The steam is then used to drive a steam turbine. The exhaust from the steam turbine, which is waste heat, is cooled to condense the steam before being returned to the steam raising plant.

Due to the availability of lower temperature cooling streams, the steam turbine waste heat can be used to boil a suitable fluid instead of being dumped into the environment. This fluid, boiling at high pressure, is then used to drive a gas expansion turbine so that the exhaust, in order to complete the cycle, is condensed by heat exchange with the LNG before being returned to the vaporisation section. The waste heat from the original fluid is in effect being used to heat the LNG. Therefore, the use of a compound cycle prevents the disposal of waste heat into the environment but instead uses it for vaporising LNG.

5 THE GAS TURBINE

The availability of LNG which needs to be vaporised in reality provides a cold source which is provided free of charge in the supply of the LNG to the user. There is a thermal capacity of some 213 KWh/t which can be exploited.

The first element in the compound cycle is the gas turbine. Lowering of the gas turbine inlet temperature has the effect of increasing output and reducing heat rate. Table III shows the effect of this for a Frame 6 gas turbine.

TABLE III Effect of Gas Turbine Inlet Cooling

	Penalty	Gain per °C
Output	0.375%	+0.54%
Heat rate	0.125%	-0.18%

There are various methods of inlet cooling, evaporative, spray and heat exchange. The former two have direct contact between the air and the cooling water and consumes water. The latter allows process integration of the inlet cooling duty with the compound cycle. The penalty imposed for inlet cooling is due to the pressure drop required by the exchanger and the need to install water removal vanes into the intake system.

A further energy loss will be the need to provide fuel gas at a higher pressure due to the increased compression ratio of the gas turbine compressor caused by the lower inlet temperature. Based on an economic approach temperature, and a cooling liquid at 5°C, cooling of the intake much below 15°C is not practicable without lowering the cooling liquid temperature. Intake cooling therefore will only provide a continuous gain in energy for locations which are always above 15°C in temperature.

It is possible to consider a 2°C cooling liquid, and say, an 8°C approach so that the inlet air is cooled to 10°C. This will give a gain of 2.3% in output for 15°C ambient and a reduction of heat rate of 0.77%. However, accurate temperature control is needed to ensure that the cooling liquid does not drop too low, causing the water in the inlet air to condense and freeze on the cooler surfaces.

The heat load required for cooling will depend on the relative humidity of the air. Cooling of low humidity air to reach 100% relative humidity requires a much lower cooling heat load, than when air has high humidity so that water vapour is condensed in the cooling process. This is illustrated in Table IV.

The gas turbine increase in rating shown is due to a larger mass flow of air being made available as a result of the lower temperature. The increased output, of course, requires more fuel to be burnt.

The heat to be removed from the air can be more than the gain in output, and although the use of waste heat from a cold sink will be inefficient, the integration of inlet cooling with the vaporisation of LNG fuel is of interest. In some cases, where the increase in capital cost is justified, the use of a refrigeration plant could also be of interest. Inlet cooling is marginal, and the relative merits of the different options will need to be fully evaluated for each situation.

The gas turbine itself has a lot of waste heat, the most important of which is the exhaust gas. The complete heat recovery from this exhaust gas means cooling down to the ambient temperature which at ISO conditions is 15°C. It is possible to use this heat but cooling of the exhaust to these low temperatures will not be acceptable because of condensation and corrosion. Burning natural gas with low/no corrosive elements should allow temperatures of 90 to 100°C stack outlet temperatures to be adopted. Every 10°C of exhaust temperature represents a large energy loss and there is an incentive to reduce these to a minimum. The lubrication oil cooling system also represents a considerable heat source. The system will accept up to 54°C inlet cooling water and a return temperature of up to 80°C can be obtained. The waste heat available from a PG6 is shown in Table V.

TABLE IV Inlet Cooling Heat Load: 30°C down to 14°C

Ambient °C	Inlet °C	Outlet MW	Fuel MW/MW	Effect of Humidity % RH	
30 30	30 14	34.47 38.34	3.265 3.182	36.3	75
Inlet cooling heat load MW Chilled water plant power MW (Ammonia)				2.21 0.46	5.8 1.2

TABLE V Waste Heat from a PG6

Heat Sources	Stack Outlet Temp	Heat Available MW
Exhaust gas	120°C	65.0
Exhaust gas	100°C	68.0
Lube oil cooling system		1.3188

6 THE ALTERNATOR

It should be recognised that the efficiency and allowable output of an alternator is also affected by the ambient temperature.

Increase in ambient temperature or the temperature of the available cooling medium will reduce the temperature difference between the maximum allowable working temperature of the windings and the cooling air. The resulting loss in heat transfer capacity requires the alternator output to be proportionally reduced. A reduction in alternator output will reduce the copper losses in proportion, but the iron losses remain constant so that the nett effect will be a small reduction of electrical efficiency. These characteristics allow gas turbine package suppliers to provide standard electrical generation sets suitable for all ambient conditions.

If gas turbine inlet cooling is adopted then the use of the same chilled water cooling circuit can be applied to a closed air cooled water cooled alternator to ensure that the resulting increase in gas turbine power output is matched by the alternator electrical generating capacity. The alternator heat load will be in the order of 1% which can be easily accommodated. The loss in allowable output is 0.66% per degree C (Figure 3) for a typical alternator as compared to 0.785% per degree C for a PG9 gas turbine.

Independent of gas turbine inlet cooling applications, chilling of closed air circuit water cooled alternators will have capital cost advantages, and a very small gain in efficiency. About 0.05% efficiency improvement can be gained for 10°C drop in cooling water temperature. In some circumstances, these points will be worth consideration.

7 THE STEAM TURBINE

The integration of the Steam Rankine cycle into the compound cycle needs careful consideration. Theoretical analysis has shown that the operational temperature levels within the compound cycle will not affect the overall efficiency. A high condensing temperature will enable higher energy from the following propane cycle but with loss of power from the steam cycle. However, dropping the steam condensing temperature will extract more power for the least cost. There is a limit to lowering the steam temperature which is dependent on the lowest practical vacuum, available steam turbine blade designs, and adequate gland sealing arrangements. Table VI shows the effect of lowering steam exhaust temperature on the steam conditions for saturated steam.

From Table VI, it will be seen that increasing the vacuum from 0.06 to 0.03 bar a will double the exhaust volume flow. The last blade rows will need an increase in length by 1.4 to maintain the same exit velocity.

A survey of leading steam turbine manufacturers shows that steam turbine designs suitable for 22°C exhaust conditions are readily available. The configuration will depend on the size required. For outputs in the region of 40MW, a divided flow exhaust configuration may be needed. One manufacturer is able to design steam turbines to operate down to 0.01 bar a. For this, the cooling liquid would need to be supplied well below freezing and it will be necessary to ensure that ice formation of the condenser surfaces will not take place.

High vacua with the need for long blades will need special attention with regard to avoiding excitation. For constant speed machines for power generation, it should be possible to tune the system for one set of possible forcing frequencies. Different types of blade assemblies such as, axial fir tree roots blades with anti-rotation Z-lock covers, or tangential female dovetail roots with long cover bands to lock groups of blades together, are available with good damping characteristics and tunable response frequencies. Modern techniques which provide accurate identification of all possible sources of excitation, allow the design of low vacuum steam turbines with a high degree of confidence.

High vacuum can also give rise to excessive moisture level in the exhaust steam. The cycle design must keep the moisture within acceptable limits to ensure reliable operation of the steam turbine but without the need for reheating. A standard dual pressure cycle design may need adjustment to the levels of pressure and superheat to restrict the expected moisture level. The use of special metallurgy blading in the wet zone of the turbine can allow operation up to 15% moisture with confidence. Consideration may need to be given for the need for stainless steel inlays and moisture separators for higher levels of moisture content. However, operation well below 15% will avoid the need for this sophistication [Fig. 4 from Reference 6].

Due to the high vacuum, some attention will need to be paid to the gland sealing system but no special design would appear to be required and the use of low clearance staggered labyrinth designs normally available is satisfactory. However, in the future some improvement may be possible as the application of dry gas seals become available for larger shaft diameters.

8 VACUUM DEAERATION

The combined cycle plant lends itself to the use of vacuum deaeration. This results in the supply of low temperature boiler feed water and generally will improve the cycle efficiency. Steam, instead of being bled from the turbine or taken from LP steam, is used from the steam turbine exhaust. Heating of the boiler feed water is from the exhaust gas, where heat extraction is only limited by the need to keep the exhaust gas stack temperature above the dew point with an acceptable margin, as discussed previously.

The use of vacuum deaeration also has some useful impact on reducing capital cost and operational cost due to the ability to deaerate without chemical oxygen scavengers (hydrazine) [Fig. 5 from Reference 4].

9 CRYOGENIC CYCLES

The heart of the methane and propane cycles will be the expansion turbines. Traditionally, these will be radial flow machines and typical data for the methane and propane machines are given in Table VII. The typical construction of such machines is shown in Figure 6. This shows the cross-section drawing of a 12MW Isobutane expander fitted with tandem mechanical oil seals mounted at the end of the pinion drive shaft of a speed reducer.

Dry seal technology is now well established and the need for a pressurised seal oil system is no longer required. The expander is used to drive an induction generator through a conventional gear box. The use of an induction generator has the advantage of being self-regulated as the frequency is controlled by the electrical power supply. This is, of course, assuming that the electrical supply to which it is connected will be very large in capacity compared to the power being generated.

In the case of a compound cycle, advantage can be taken of the presence of a gas turbine driven generator and a steam turbine generator. Both these generators can be arranged for dual drive. The methane expander can be connected to the gas turbine generator and the propane expander connected to the steam turbine driven generator. An automatic self engaging clutch can be used in both cases. These couplings are mechanical clutches which sense the difference in speed between the driver and the load, in this case, the expander and the generator. The clutch will remain disengaged so long as the speed of the expander is less than the generator.

This arrangement allows the start up and operation of the combined cycle plant on its own with the cold cycle plant at standstill. The methane circuit is cooled down by the vaporisation and flow of LNG without the use of expanders. The flow of LNG will cool down the propane circuit and once the propane circuit is down to temperature, the propane expander can be started up. As the propane expander speed increases and comes up to generator running speed, the self synchronising speed clutch engages and the generator takes up the power output from the expander. As LNG throughput is increased further with the assistance of heat input from the propane circuit, the methane expander can be started up and as it gets up to speed, the clutch will engage for power transmission into the generator. An arrangement of such a clutch where the clutch weight is shared between the expander gear box bearing and a bearing within the clutch housing is shown in Figure 7. The drive to the generator is provided by a solidly coupled extension to the generator rotor.

This arrangement will also provide a gain in the cycle efficiency. Induction generators between 5MW and 10MW will have an efficiency of 96% to 96.5% whereas synchronous generators of 20MW or more will have an efficiency of over 98%. Furthermore, each induction generator will require its own switch gear and protective system. The capital expense of this plus the associated foundations will be greater than that of a self-engaging clutch. Lower cost and increased output gives this arrangement a distinct advantage.

10 COMPOUND CYCLE PERFORMANCE

Gas turbines are not tailor-made but are available in a range of sizes, perceived to be marketable by the manufacturers concerned. While it is possible to precisely size the Steam Rankine Cycle to match the heat input required for the LNG cycle, it is not,in the same way, possible to select a gas turbine with the precise waste heat that is required by the Steam Rankine cycle. For these reasons, it is necessary to find the nearest size combined cycle unit available or a discrete number of units to match the heat input needed.

In determining the optimum size of the combined cycle plant, the price of electricity and applicable interest rates are important factors. These factors will determine the size of the combined cycle plant to be integrated with the cryogenic power recovery plant. Because the cryogenic power recovery is fixed by the LNG throughput, the total compound cycle output and fuel consumption will be determined by the size of the combined cycle power plant to be integrated with it. The smallest possible combined cycle plant, as limited by the heating needs of the cryogenic cycle, will produce the highest thermal efficiency. The use of a larger combined cycle plant will produce a greater output of electricity with a lower capital cost per KW installed.

The best overall efficiency is achieved when close integration is made between the heat required for heating the intermediate fluid with the heat obtained from the combined cycle. Heating of the streams which do not contribute to enhancing expander power output can be by use of sea water instead of waste heat.

Table VIII shows a typical combined cycle output for 30°C ambient compared to equivalent 15°C cooled ambient with and without reduced temperature steam turbine exhaust conditions, and Figure 8 shows the performance to be expected from a typical compound cycle power station. From these results, based on the gross electrical output, the progression in thermal efficiency improvement expected is given below:-

Gas turbine simple cycle	31.87%
Gas turbine combined cycle	47.95%
LNG compound cycle	58.61%

The extra output due to the use of the LNG cold sink is calculated to be:-

Steam cycle	1.306 MW
Propane cycle	4.735 MW
Methane cycle	7.325 MW
Total extra power	13.366 MW

The efficiency of cold energy recovery shown is:-

Power recovery	13.366 MW
	ie 25% efficiency
LNG vaporisation energy	53.25 MW

For nett output the following utilities would need to be accounted for:-

BFW and CW pumps	2.0 MW
Propane pump	0.153 MW
Methane pump	1.39 MW

These results are perfectly achievable, based on existing technologies well proven individually, but yet to be amalgamated to demonstrate the whole.

11 ACKNOWLEDGEMENTS

Thanks are due to Bechtel for permission to publish this paper and to colleagues in the Process Department, P Kauders and Dr K A Sheikh who carried out the process simulation of the cold and hot cycles.

Special thanks are also due to ABB, GE, GEC-Alsthom, European Gas Turbines, Siemens, Mafi Trench, SSS Gear Ltd and Dresser Rand, without whose assistance this paper could not have been written.

References

1 A Perspective Look at Japan's Power Generation.
Author: Katsuhei Tanemura, Tokyo Electric Power Co.
Cranfield Institute of Technology 1990.

2 An Inlet Air Washer/Chiller System for Combined Cycle Plant Repowering. U Sengupta and G E Soroka, Bechtel Power Corp, Gaithersburg, Maryland. Power Gen 1988.

3 Turbo Expander for Waste Heat Recovery
W Wong, Bechtel London.
European Turbomachinery Symposium, Brunel University 1988.

4 Case Study - Real Life Optimisation of a 120MW Combined Cycle Co-Generation Plant.
Richard B Stanley, Bechtel, San Francisco.
ASME/IEEE Power Generation Conference 1987.

5 The Status of Liquified Natural Gas Worldwide
Kay McKeough. Energy Series Paper No 25.
The World Bank, March 1990.

6 Recent Advances in Mechanical Drive Turbine Technology
E.T. Vitone, W.G. Catlow, J.I. Cofer IV and B.E. Gans,
G.E. Co. Fitchburg Mass. USA.

7. The Integration of LNG Vapourisation with Power Generation
W. Wong, Bechtel London, Gas producers Association, European Chapter, Poole Dorset, 1992.

TABLE VI Vacuum Saturated Steam Properties (Source: Schmidt, E - Steam Tables)

Pressure bar a	Temp °C	Specific Volume m^3/kg
0.01	6.98	129.20
0.02	17.51	67.01
0.03	23.24	50.09
0.04	28.98	34.80
0.05	32.58	29.27
0.06	36.18	23.73

TABLE VII Typical Expander Duties

	Methane	Propane
Flow Kg/hr	306,935	328,779
Mol weight of gas	17.30	44.10
Inlet Pressure bar a	79.50	13.30
Inlet Temperature °C	7.5	41.0
Outlet Pressure bar a	38.00	2.5
Outlet Temperature °C	-36.2	-17.3
Rpm	16,500	5,500
Efficiency %	85.0	83.5
Gas Power KW	5162	5827
Wheel Diameter mm	305	914
Tip Speed m/s	263	263

TABLE VIII EFFECT OF INLET COOLING & EXHAUST COOLING

	Units	Case 1	Case 2	Case 3
Inlet Cooling	MW	None	2,211	2,211
GT Inlet (30°C 36% RH)	°C	30	14	14
Exhaust Flow	t/h	469.7	499.9	499.9
Exhaust Temperature	°C	552	543	543
ST Exhaust	°C	38	38	22
Moisture	%	11.3	11.3	13.3
Cooling Water Inlet	°C	25	25	9
Temperature Rise	°C	10	10	10
Condenser Pressure	bar	0.066	0.066	0.0266
Condenser Duty	MW	40.89	42.41	41.14
Stack Temperature	°C	120	120	120
Steam Turbine	MW	19.576	20.173	21.479
Gas Turbine	MW	35.991	39.98	39.98
Combined Cycle	MW	55.567	60.153	61.459

FIGURE 1

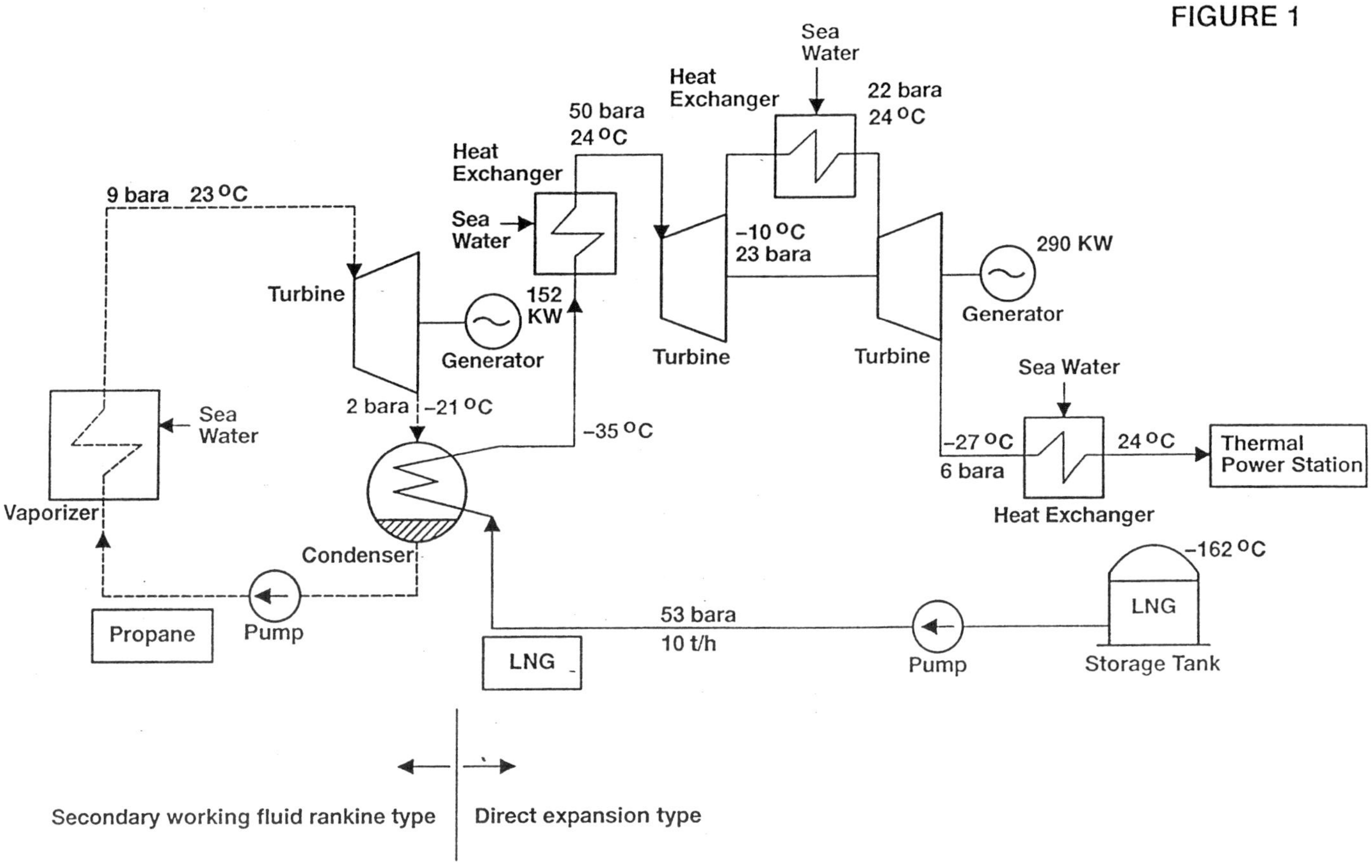

FIGURE 2

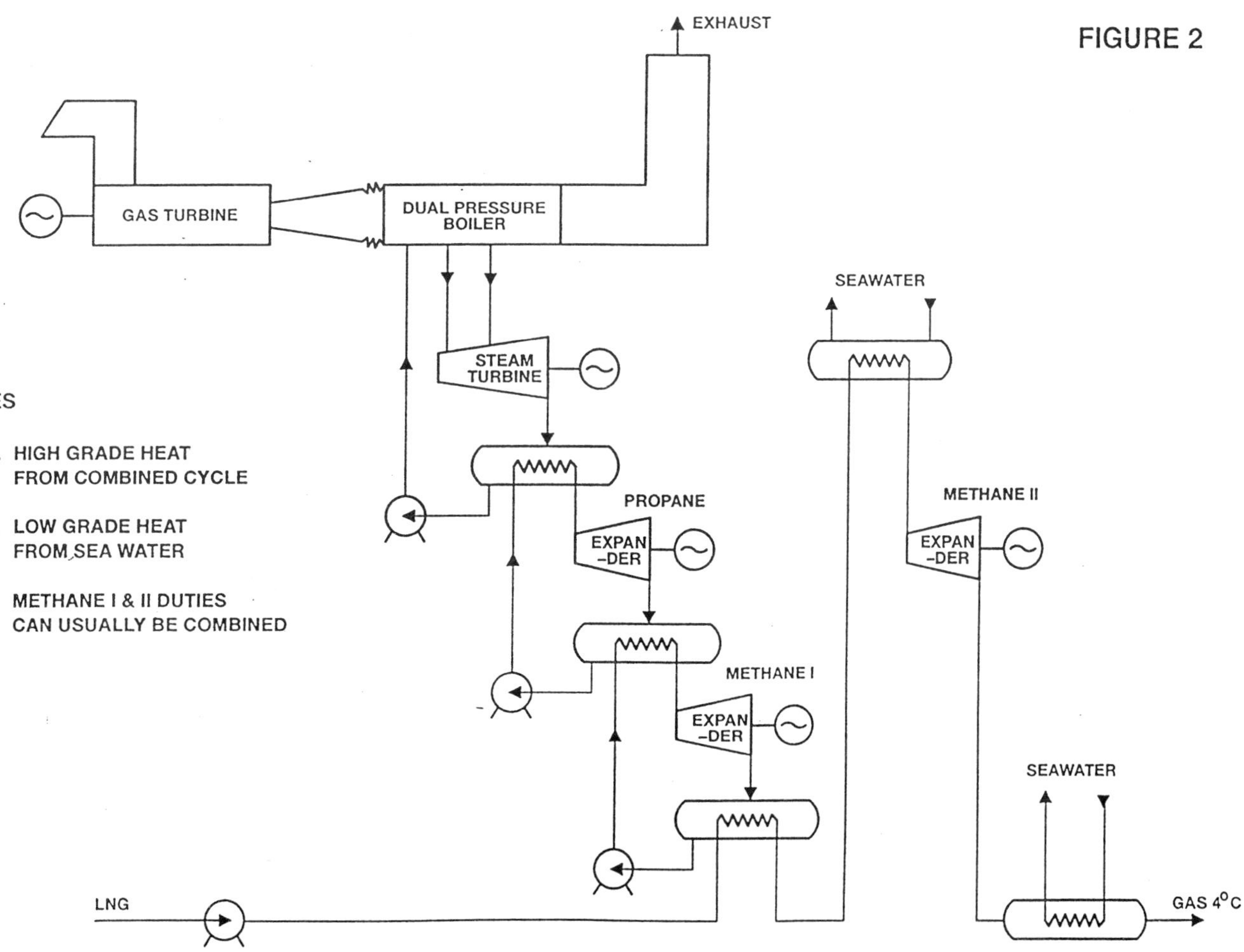

NOTES

1. HIGH GRADE HEAT FROM COMBINED CYCLE
2. LOW GRADE HEAT FROM SEA WATER
3. METHANE I & II DUTIES CAN USUALLY BE COMBINED

11500 V – 50 Hz – 3000 RPM

COOLING BY AIR IN CLOSED CIRCUIT AND AIR/WATER COOLERS

FOR TOTAL TEMPERATURES : STATOR < 124 DEGREES C (BY EMBEDDED DETECTORS)
ROTOR < 120 DEGREES C (BY RESISTANCE VARIATION)

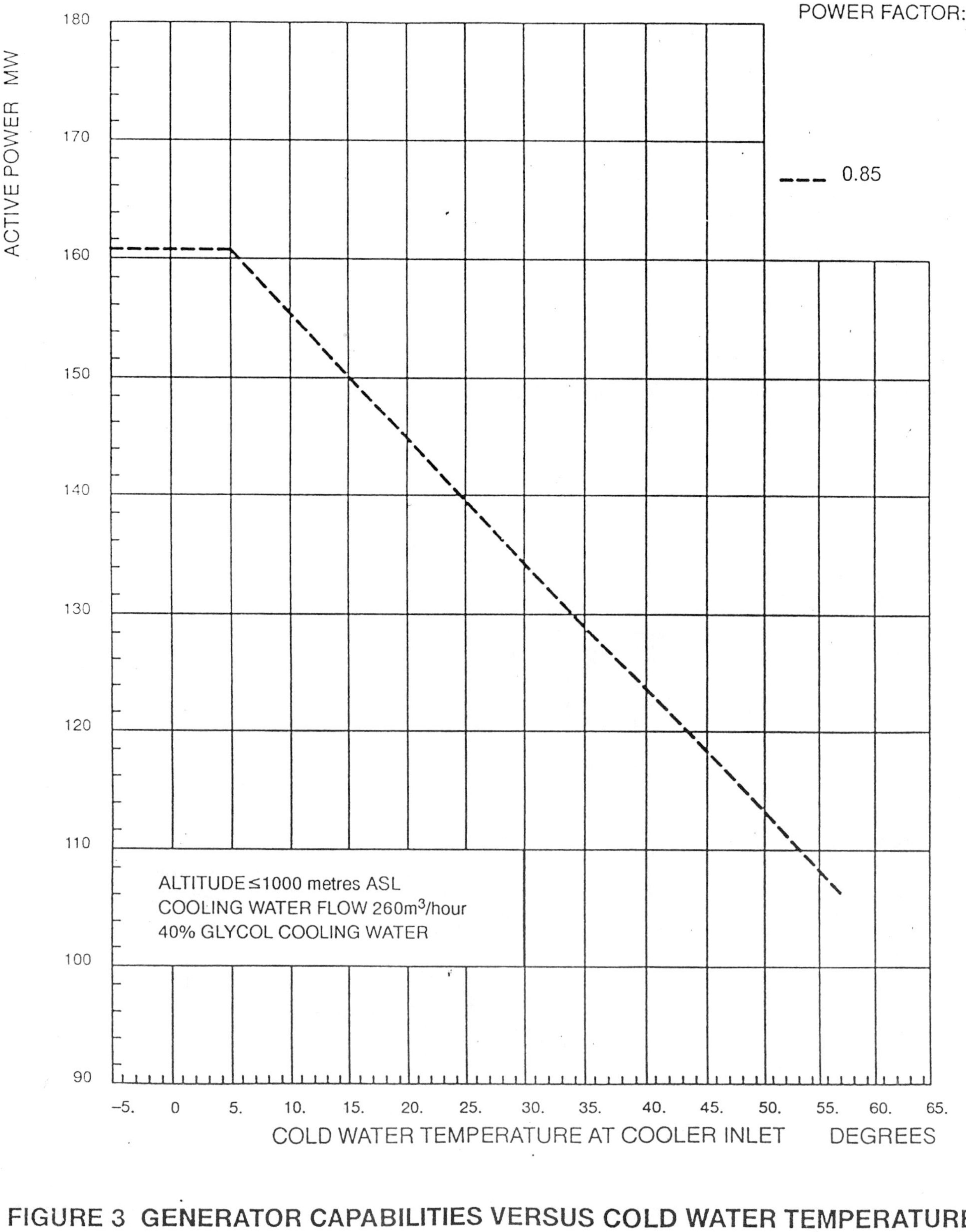

FIGURE 3 GENERATOR CAPABILITIES VERSUS COLD WATER TEMPERATURE

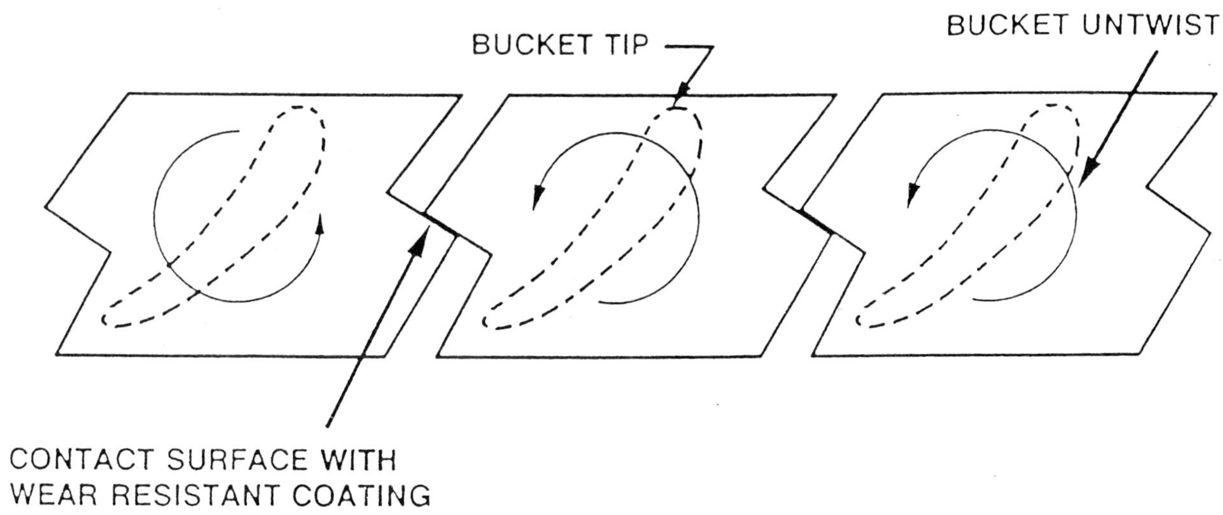

FIGURE 4. Z-LOCK COVER DESIGN

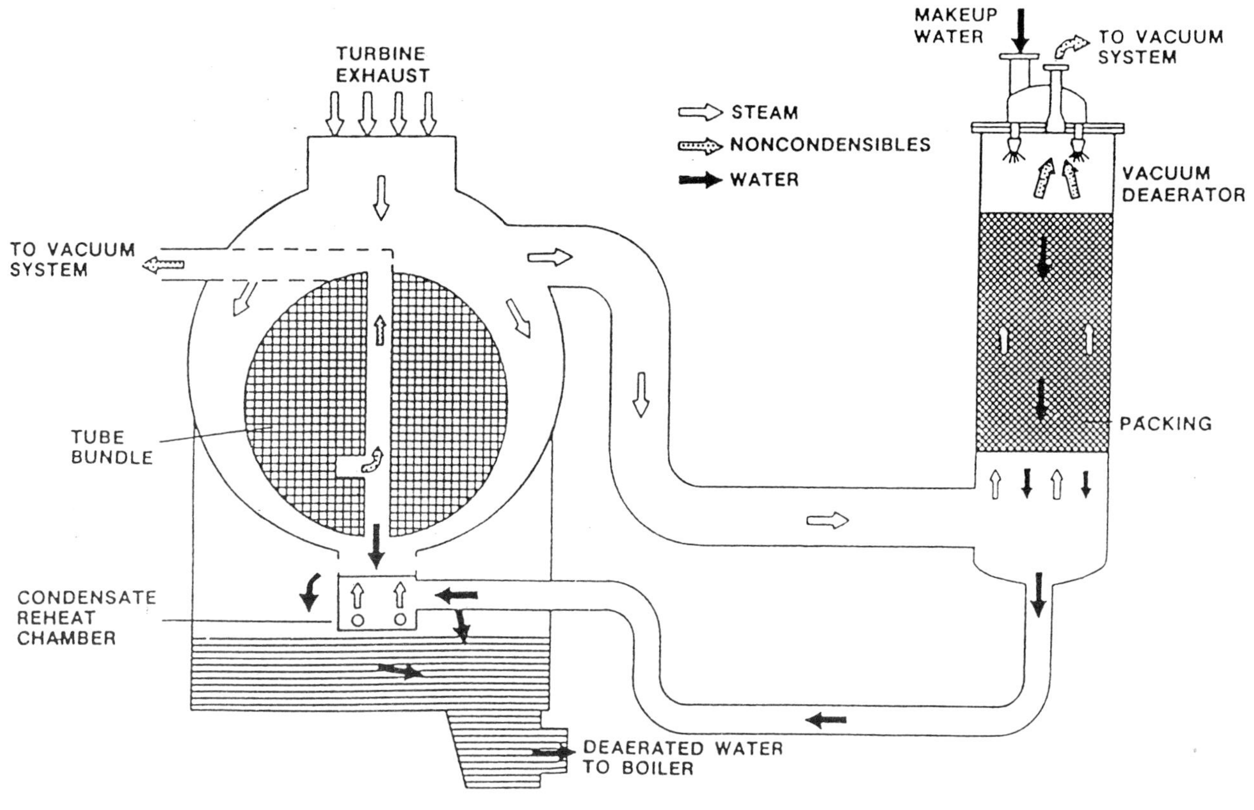

FIGURE 5. CONDENSER/VACUUM DEAERATOR FLOW SCHEMATIC

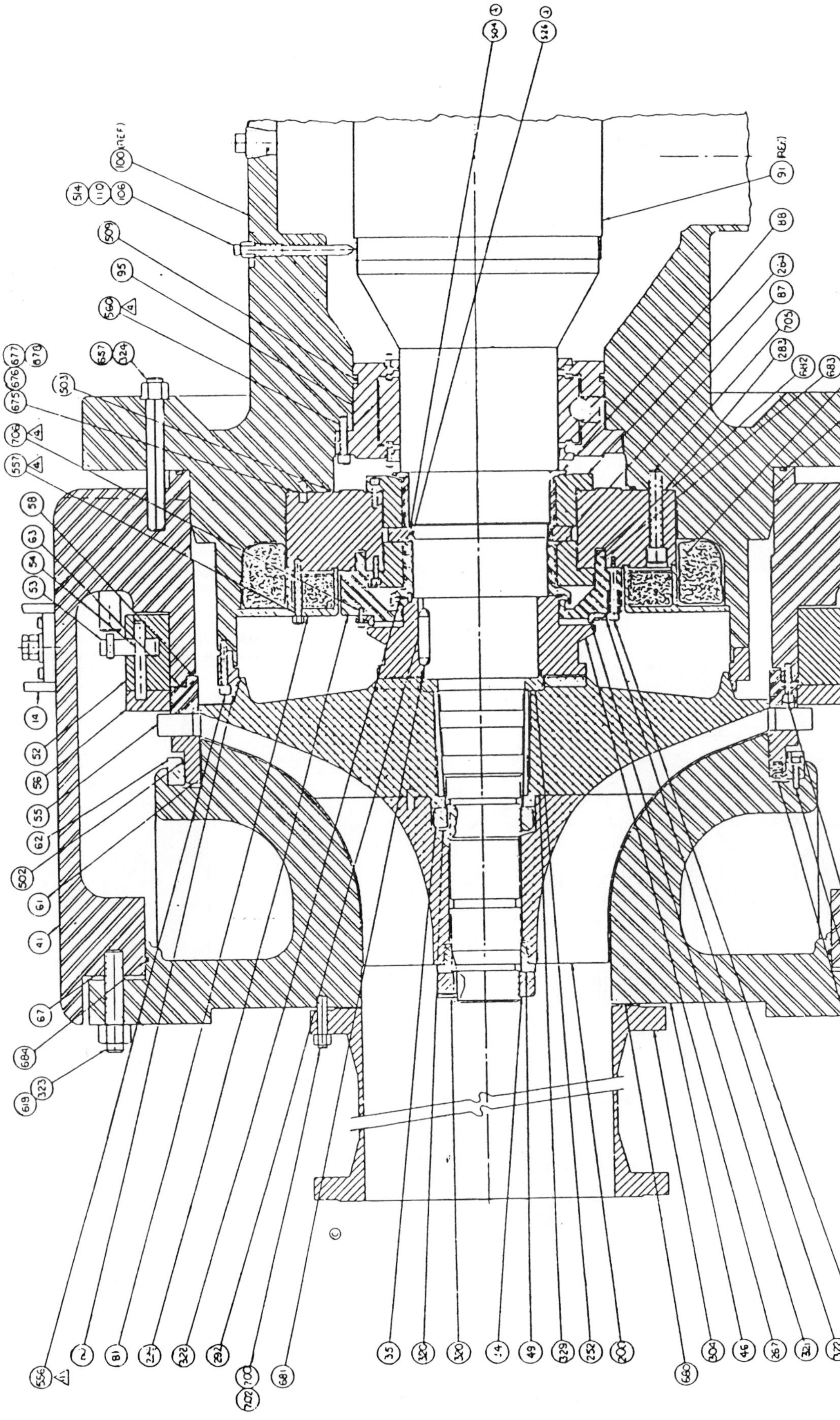

EXP. POWER SHAFT
SERVICE ASSEMBLY

FIGURE 6

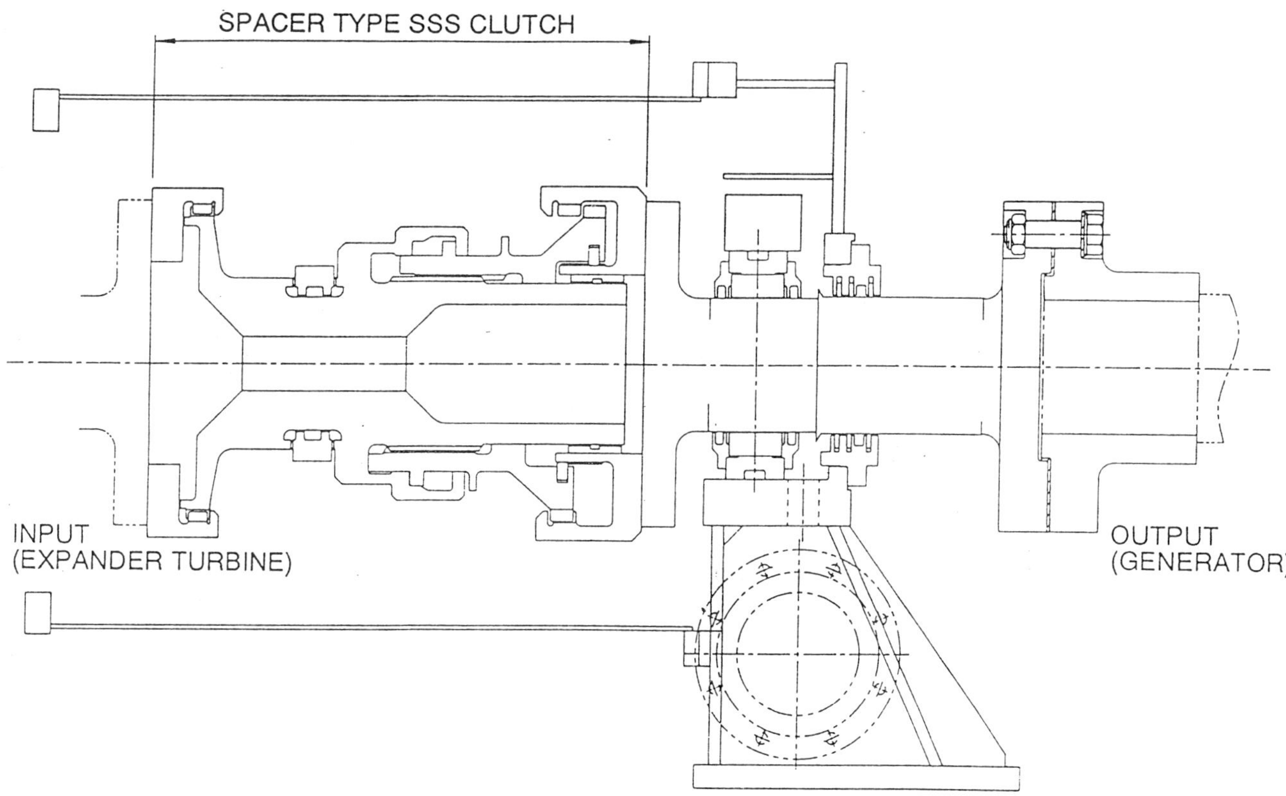

FIGURE 7

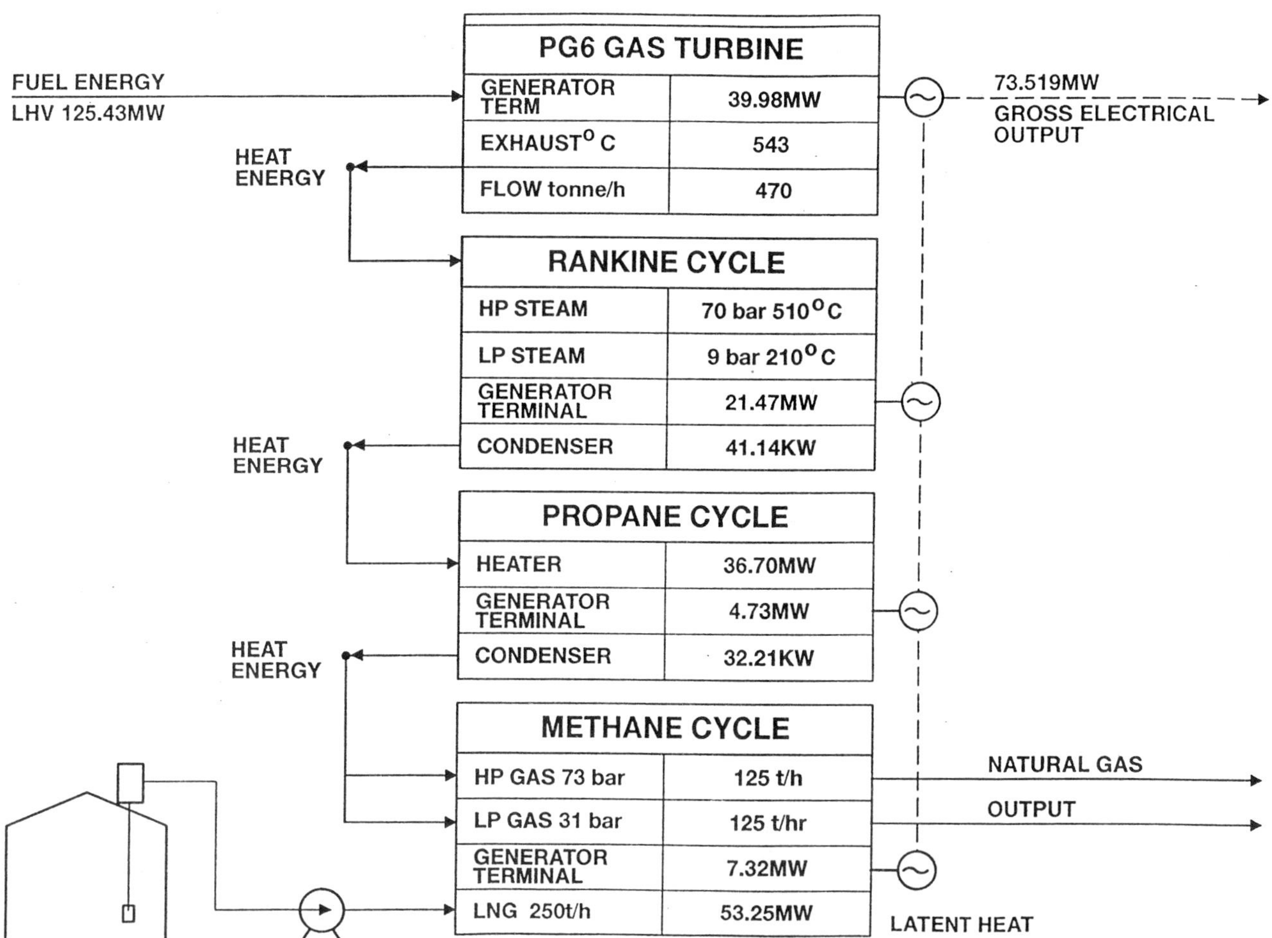

FIGURE 8

C449/029/93

Re-engineering surplus equipment for application to a 12 MW (16000 h.p.) butane compression train: a case history

J BIELK, BS, MS, PhD, S DAPKIEWICZ, BSME, J CRICKENBERGER, BSES
Conmec Inc, Bethlehem, Pennsylvania, USA

SYNOPSIS

The petro-chemical industry is increasing its reliance on re-engineered surplus rotating equipment for application in new plant construction critical applications. This practice requires that surplus equipment be carefully selected, inspected, evaluated, and re-engineered for any new application. Even if a unit's nameplate performance data closely matches the demands of the new application, a complete range of turbomachine engineering analysis and design specialties must be applied to the surplus equipment to insure unit operation within contractual specifications. A case study is presented in this paper that details the history, from quotation to train mechanical testing, of a 12 MW (16000 HP) re-engineered surplus equipment train used to compress butane in a Texas (U.S.) chemical plant and refinery.

1. INTRODUCTION

Frugality may be defined as the conscious effort to avoid or minimize waste in pursuing an objective. A frugal approach to satisfying the rotating equipment needs of the petrochemical and gas pipeline industries is increasingly employed in North America: reclamation of surplus new/unused and used equipment for applications where only specially constructed new equipment previously had been considered suitable. Application of surplus equipment in new plant construction has increased over the past few years for a variety of reasons:

(a) Decreased cost of equipment

In recent years, U.S. oil refinery capital expenditure budgets have been substantially decreased to offset declining refined product revenues. In some applications where existing equipment exactly matches customer specified process conditions the cost of the surplus equipment may be less than 60 per cent the cost of new equipment because the subject unit only requires refurbishment and re-certification.

(b) Technology transfer to non-traditional vendors

The distinction between specially contracted 'new' and re-engineered surplus equipment has blurred with the increased engineering capabilities of non-traditional turbomachine vendors. Some surplus equipment applications require that equipment be completely re-engineered in order to meet new process performance requirements. A re-engineered unit may possess completely different internals from the original unit: the authors' company routinely rerates existing rotating equipment by replacing complete aerodynamic bundles (rotating assembly and diaphragms) with custom-engineered designs that maximize efficiency and mechanical reliability. Turbomachinery can, however, only be properly re-engineered if the vendor possesses sufficient engineering capability to effectively function as an original equipment manufacturer (OEM). The technology to engineer high performance turbomachinery, once the sole domain of the large OEM, has spread into the general marketplace with the advent of inexpensive computing and turbomachinery engineering research and development performed at public universities.

(c) Improved delivery schedules

Surplus equipment may be procured and re-engineered in substantially less time than an OEM takes to build new equipment. Reduced lead times offer the petroleum industry increased flexibility, revenues, and decreased payback time.

The advantages listed above for re-engineered surplus equipment are enjoyed by a customer only if the equipment delivered is at least equal in quality to specially constructed new equipment. Quality of a train is evaluated by contractual aero-thermodynamic guarantees, by customer specified mechanical standards (e.g., current applicable API standards), and by the coverage of the equipment warranty. The chance that re-engineered surplus equipment will not meet contractual specifications is a function of the engineering competence of the vendor. Vendor engineering is tested in two ways: first, machinery capable of performing the required service must be selected from available surplus equipment and second, the selected unit(s) must be properly refurbished, re-engineered, and re-certified.

Despite the guarantees and warranties offered by some non-OEM turbomachine vendors, a significant prejudice remains in the petrochemical industry against using re-engineered surplus equipment in new plant construction. In some cases such concern is warranted. The authors can recount cases where non-OEM vendors have sold modified equipment without performing necessary mechanical and/or thermodynamic analyses. The resulting problems were remedied only after great expense to the customer. In one notable case, a steam turbine was re-designed without the benefit of detailed engineering analysis. The modified steam turbine operated on its second critical speed with an overloaded thrust bearing and did not generate required power. The capability of any aftermarket turbomachinery vendor to solve unexpected problems or address customer concerns that inevitably occur during equipment re-design should be carefully examined by potential customers prior to equipment purchase.

One result of the perceived risk that customers assume when purchasing re-engineered surplus equipment is the price penalty the non-traditional turbomachine vendor must pay in order to be considered competitive with specially constructed new equipment. It is not uncommon for potential customers to insist that re-engineered surplus equipment be discounted more than 10 per cent the cost of comparable new equipment even if the re-engineered unit's guaranteed performance is superior and the mechanical warranty is equivalent. In general, custom-engineered surplus equipment receives greater engineering attention than a competing new unit because the non-OEM vendor does not have the benefit of in-depth experience with nearly identical equipment. For this reason, non-OEM vendors should re-design equipment using optimal design methods to insure that the equipment possesses maximum possible design margins. Given this practice, it is discouraging to non-OEM vendors that some potential customers insist that re-engineered surplus equipment be highly discounted (e.g., 40 per cent or greater) relative to specially constructed new equipment. A non-OEM vendor generally can not provide such large discounts if the equipment requires significant re-engineering for the new application.

Customers are not alone in assuming economic risk during re-engineered surplus equipment jobs. The non-OEM vendor assumes risk because the precise amount of refurbishment, re-design, and scope of hardware supply is never known until after the contract is secured and a complete evaluation of the equipment is performed. The exposure of the aftermarket vendor is minimized by obtaining as much information about the unit as possible prior to developing the quotation for work. Ideally, a complete history and engineering evaluation of all train elements would be available prior to quotation so that a precise scope of design modifications and hardware refurbishment could be developed. In reality, however, information on a given unit may be limited to nameplate data, a preliminary engineering assessment of the equipment and experience with similar equipment. As this case study suggests, even if the performance data of a unit closely matches the customer's specifications for an equipment train, extensive re-engineering of the unit may have to be performed to produce a properly re-engineered unit that is capable of efficient and reliable operation.

This paper describes the experience of the authors' company in producing a 12 MW (16000 HP) re-engineered surplus equipment compressor train for application in a Texas (U.S.) oil refinery.

2. EQUIPMENT SELECTION AND QUOTATION

Bids were solicited by a Texas (U.S.) based oil refinery for a steam turbine driven butane upgrade process (BUP) compressor train in September 1990. Marketing personnel searched equipment lists from storage yards for compression equipment that was properly sized to meet the process specifications published by the refinery. Two surplus new/unused compressors, one with a slit-case and one with a barrel case, were located that would generally meet published process requirements for tandem operation according to the data sheets and performance curves that accompanied the units. Traditional OEM competition for this job initially quoted a single body re-entry barrel compressor.

The lube/seal/control oil system proposed for the train was equivalent to systems quoted by the competing OEM vendors: the customer required completely new auxiliary systems be supplied with the train and that they conform to all specifications listed in the 2nd Edition of API 614 (1). The steam turbine governor controls quoted with the train included state-of-the-art digital electronic panels mated to a new hydraulic actuator.

Suitability of the compressors to the new process was initially determined by comparing 'nameplate' aero-thermodynamic and structural data to the specified BUP process conditions. Minimum acceptance criteria for the compressors were defined by the casings:

(a) the casings had to be capable of withstanding the maximum specified discharge pressures and
(b) the specified inlet volumetric flows had to produce an inlet nozzle Mach number less than 0.1.

Once the casings were judged suitable for application to the process, other compressor features were examined for compliance to customer bid specifications including the process gas sealing systems, casing materials, and nozzle arrangements. Available aerodynamic performance and mechanical specifications for each unit were examined to estimate the amount of modification to the compressor's aero-assembly (rotating assembly including impellers, inlet guide vanes, diaphragms) required to adapt the units to the new process. Initial unit aero-thermodynamic assessments revealed that substantial flow-path modifications would be required to re-engineer the split-case compressor. Few flow-path requirements appeared necessary to re-engineer the barrel compressor. Initial unit mechanical assessments indicated, however, that the rotating assemblies for both compressors (shaft, impellers, and sleeves) would have to be replaced because of impeller stress requirements and material limitations for H_2S service.

It is axiomatic that supplying custom-engineered surplus rotating equipment is generally more expensive than constructing a unit with a set of stock components because the full development engineering cost of the custom components is amortized over a single production run and the production run for each component is extremely limited (usually only one or two units). Competitive pricing and profitability of re-engineered surplus equipment is, therefore, generally realized only if the number of modifications to the surplus equipment is limited. Barring special requirements, if a compressor's volumetric flow capacity and polytropic head are both greater than or equal to the demands of a new process then the unit is probably well-suited to the desired application and can be adapted in a cost-effective manner.

Tables 1 and 3 display the aero-thermodynamic design information for the two compressors selected for the quotation. Tables 2 and 4 display casing construction data.

Table 1 First body barrel compressor original and new process design conditions

	Original Process	New Process
P_{inlet}	.279 MPa 40.5 lbf/in^2	Proprietary
T_{inlet}	-14.4 °C 6.0 °F	51.1 °C 124.0 °F
$P_{discharge}$	1.289 MPa 187.0 lbf/in^2	.353 MPa 51.2 lbf/in^2
$T_{discharge}$	74.4 °C 166.0 °F	122.8 °C 253.0 °F
Inlet Flowrate	34.53 m^3/s 73171.0 ft^3/min	34.66 m^3/s 73441.0 ft^3/min
Shaft Power	22.924 MW 30742.0 HP	6.010 MW 8059.0 HP
Shaft Speed	4670.0 r/min	4600.0 r/min

Table 2 First body barrel compressor construction data

Casing $P_{maximum}$	2.758 MPa 400 lbf/in^2
Casing $T_{maximum}$	193.3 °C 380.0 °F
Inlet Nozzle Diameter	1.22 m 48 in
Discharge Nozzle Diameter	.76 m 30 in

Information accompanying the barrel-case compressor included the instruction book data sheets and the OEM predicted performance curves while an instruction manual and design data sheets were located for the split-case compressor. The close match between the barrel compressor's 'nameplate' data (molecular weight, volumetric flow, and speed) suggested that all the existing stationary elements in the unit (guide vanes, shaft seals and impeller packings) could be reused in the re-engineered unit. As previously noted, H_2S material requirements to resist stress corrosion cracking necessitated replacing the rotating assemblies (shaft, impellers and sleeves) of both units. It was understood from the start of the project that the barrel compressor's impellers would be particularly difficult to design because of the following:

(a) unique customer impeller stress criteria (60 per cent yield stress at the normal operating speed) and
(b) ambitious train efficiency goals (quoted efficiency greater than 81 per cent).

Table 3 Second body horizontal split compressor original and new process conditions

	Original Process	New Process
P_{inlet}	.101 MPa 14.7 lbf/in^2	Proprietary
T_{inlet}	33.3 °C 92.0 °F	51.7 °C 125.0 °F
$P_{discharge}$	.800 MPa 116.0 lbf/in^2	.922 MPa 133.7 lbf/in^2
$T_{discharge}$	135.6 °C 276.0 °F	116.1 °C 241.0 °F
Inlet Flowrate	9.99 m^3/s 21178 ft^3/min	11.49 m^3/s 24341 ft^3/min
Shaft Power	3.436 MW 4608 HP	5.313 MW 7126 HP
Shaft Speed	4781.0 r/min	4600.0 r/min

Table 4 Second body horizontal split compressor construction data

Casing $P_{maximum}$	3.103 MPa 450 lbf/in^2
Casing $T_{maximum}$	33.3 °C 92.0 °F
Inlet Nozzle Diameter	.76 m 30 in
Discharge Nozzle Diameter	.51 m 20 in

Modifications to the second body split case compressor were better defined because the complete instruction manual was available and because major modifications to the unit were planned in order to meet performance goals. The instruction manual contained a complete cross sectional view of the unit internals, included critical impeller dimensions, and displayed calculated unit performance. As noted above, H_2S content of the process gas compelled designers to replace the split case compressor's rotating assembly. The shaft had to be replaced for other reasons as well: the rotor had to be converted into a 'drive-thru' unit with the ability to drive the barrel compressor. The split case compressor was selected as the 'drive-thru' unit because it could share a baseplate with the steam turbine. Schematics of the original and proposed re-engineered compressors are presented in Figures 1 and 2. Plots that specify the thermodynamic performance of the barrel and split-case compressors are displayed in Figures 4 and 5, respectively.

As required by API specifications, a 10 per cent margin was added to the calculated compressor power requirements in order to size the steam turbine driver. Given the required power, customer-specified steam conditions, and the design speed of the compressors, a surplus 8 stage marine steam turbine was located to drive the train. A schematic of the turbine is presented in Figure 3. Original and new process turbine design conditions and construction data are presented in Tables 5 and 6.

Table 5 8 stage marine steam turbine driver original and new process conditions

	Original Process	New Process
Shaft Power	16.554 MW 22200 HP	12.453 MW 16700 HP
Shaft Speed	4957 r/min	4600 r/min
Steam Conditions	5.722 MPa 507.2 °C 830 lbf/in^2 945 °F	5.722 MPa 507.2 °C 610 lbf/in^2 750 °F

Table 6 8 stage steam turbine driver construction data

Inlet Flange $P_{maximum}$	6.205 MPa 900 lbf/in^2
Inlet Flange $T_{maximum}$	507.2 °C 945.0 °F
Inlet Flange Diameter	.30 m 12 in
Discharge Flange Diameter	.66 m 26 in

The minimum acceptance criteria used to evaluate prospective steam turbines was based on the turbine's steam chest structural limits. As noted above, the marine unit's steam chest permits inlet pressures much greater than the demands of the customer's steam supply. A second test was also applied to determine if the turbine possessed acceptable corrosion life: bulk flow steam velocity through the inlet flange at design conditions had to be less than 45.72 m/s (150 ft/s). Other factors important in turbine selection included exhaust flange orientation and velocity and a general speed match with the compressors.

Note that while the turbine's rated power was substantially greater than the demand of the process and possessed an operating speed close to the speed dictated by the compressors, the new steam supply has much lower enthalpy than the turbine's design steam supply. No additional information on the turbine was located prior to quotation. Bulk flow calculations indicated that the mass flow through the turbine would have to increase by replacing or re-machining first stage components (nozzle block and blades) if the unit were to provide required power.

A scope of work and supply for necessary rerate modifications to all units was formulated and a quotation issued to the customer in April 1990. This quotation was accepted on a fixed cost basis one month later. Scheduled delivery of the complete train was set to occur less than one year later. Published customer acceptance criteria, based on API standards, insured that the completed train would function safely and reliably. These acceptance criteria were detailed in a customer-approved quality plan, complete with witness tests and documentation.

3. EQUIPMENT ASSESSMENT

The proposed scope of work for the subject train was based on available information. As noted above, in the case of the first body barrel compressor and the steam turbine, available information was limited to 'nameplate' thermodynamic data. As soon as the quotation was accepted by the customer, the units were shipped to the shop for disassembly and dimensional inspection so that detailed engineering assessments could be accomplished. After performing unit evaluations, complete scopes of work and supply for the re-engineered units could be determined.

3.1 Inspection

During disassembly, all parts were inventoried and photographic and written logs of the procedure were generated to assist in developing instruction manuals for the completed units. Once the cases and bundle were split, all packing clearances, axial floats, and casing fits were measured to determine the condition of the existing parts. Draftsmen and design engineers then worked together to create accurate layouts of the received units. The layouts of the existing units served as important production tools by graphically presenting information needed to accomplish the detailed unit engineering assessments.

3.2 Detailed Engineering Assessments

Once disassembly and inspection of the three units was completed, compressor and turbine design engineers performed aero-thermodynamic analyses of each unit to evaluate existing flow-path components (labyrinth seals and packings, nozzle blocks and inlet guide vanes, etc.). The design engineers also identified improper clearances, damaged internal parts, and gross differences between the actual flow-path and the flow-path used in the performance analysis used to formulate the scope of work proposed in the quotation. Engineering specialists performed detailed analysis of critical mechanical components to determine potential for re-use. These analyses included:

(a) rotor dynamics design audits,
(b) shaft end stress evaluations,
(c) centrificated blade natural frequency calculations,
(d) impeller stress analyses, and
(e) stage performance predictions.

The detailed engineering assessments indicated that the proposed scopes of supply would have to be substantially increased. Comparison of the proposed and actual scopes of supply for each unit are given in Tables 7 - 9.

Table 7 Proposed and actual scopes of supply for first body barrel compressor

Proposed Scope of Supply	Actual Scope of Supply
New Parts - 4 impellers - 1 shaft - 4 impeller eye packings - 1 balance piston - 1 balance piston packing - 4 interstage shaft sleeves	New Parts All new parts in Proposed Scope of Supply with the following additions: - 3 interstage seal packings - 1 inlet diffuser - 2 eye packing adapters - 1 balance piston packing adapter - 1 IGV cover plate - 1 thrust end cover labyrinth seal
Re-machined Parts (none)	Re-machined Parts - 3 inlet guide vanes - 2 bearing housings - 1 compressor end cover

Table 8 Proposed and actual scopes of supply for second body split-case compressor

Proposed Scope of Supply	Actual Scope of Supply
New Parts - 5 impellers - 1 shaft - 5 impeller eye packings - 1 balance piston - 1 balance piston packing - 6 shaft sleeves - 2 diaphragms - 4 inlet guide vanes - 1 guide tube - 6 shaft seal packings	New Parts All new parts in Proposed Scope of Supply with the following changes: - 1 diaphragm only - 3 inlet guide vanes only - 1 inner seal labyrinth - 2 radial tilting pad bearing assemblies - 2 bearing wiper plates - 1 thrust end cover labyrinth seal
Re-machined Parts - 1 discharge diffuser	Re-machined Parts - 1 discharge diffuser - 2 bearing housings - 3 diaphragms - 1 diffuser plate - 2 thrust bearing pads

Some expansion of the proposed work scope was accounted for in the quotation to the customer. Unfortunately, three major problems were uncovered during the engineering assessments that could not be anticipated and that proved costly to remedy:

(a) non-radial inlet guide vanes (IGVs) were discovered in the last 3 stages of the barrel compressor,
(b) the split-case compressor possessed a highly amplified first critical speed, and
(c) the calculated turbine first critical speed was found to interfere with the operating speed range of the unit.

Each of these problems is discussed in detail below.

Table 9 Comparison between proposed and actual work scopes for steam turbine

Proposed Scope of Supply	Actual Scope of Supply
New Parts - 1st stage nozzle segment - 1st stage blades	New Parts - 2 radial tilting pad bearing assemblies - 1 interstage labyrinth packings
Re-machined Parts (none)	Re-machined Parts - 1st stage nozzle segment - shaft

3.3 Non-Radial IGVs

Non-radial IGVs produce flows into the adjacent impeller with a circumferential velocity component; radial IGVs are designed to generate flows with a purely axial flow velocity. The non-radial vanes are used by original equipment manufacturers to change the flow and head characteristics of a given compressor stage so that economical 'stock' impellers can be used in a given application. The magnitude of the change produced by the non-radial vanes is a function of the flow's angle to the axial direction as it exits the vanes. This type of IGV, regardless of orientation, always degrades stage efficiency.

As previously mentioned, non-radial IGVs were discovered in the final 3 stages of the first body barrel compressor of the subject train. These IGVs could not be replaced because the resulting manufactured cost of the unit would increase by about 10 per cent. The preliminary performance assessment and redesign of the barrel compressor had been accomplished assuming radial IGVs directed the flow from the return channels into the impellers. The presence of the non-radial vanes dictated that the 2nd, 3rd, and 4th stage impellers be redesigned in order to match vane exit flow velocities with blade absolute inlet flow velocities so that stage efficiencies would not be degraded. The flow and head characteristics of the redesigned stages also had to be increased to make up for the decreased mean axial flow velocity caused by the non-radial vanes.

3.4 Original Split Case Compressor First Mode Amplification

The rotor dynamics analysis performed for the split case compressor was accomplished using a lumped parameter beam model of the rotor. Shear deformation was included in the analysis. Linearized bearing dynamic coefficients used to represent the tilting pad bearings were calculated using a finite element model of the oil film. Rotor dynamics analysis was accomplished using software based on the Transfer Matrix (Myklestad-Prohl) method. The existing unit was analyzed as well as the proposed design in order to determine the relative change in the unit's calculated rotor dynamics characteristics. Identification of the original customer indicated that the units had likely been subjected to a mechanical acceptance test and/or a full performance test. Thus, the original unit was thought to have met rotor dynamics acceptance criteria outlined in API 617 (2). The original split case compressor's midspan response to a general unbalance distribution is plotted in Figure 6. Table 10 summarizes the calculated rotor dynamics characteristics of the original unit.

Table 10 Lateral analysis summary original second body split case compressor

1st Mode Response		1st Mode Stability	
Amplification Factor	Critical Speed (r/min)	Log Dec	Damped Frequency (r/min)
17.5	2880	0.170	2810

The high amplification factor associated with the fundamental critical was caused by the following:

(a) excessive midspan shaft flexibility and
(b) moderately high bearing unit loadings (.827 MPa/120 lbf/in^2 and .972 MPa/141 lbf/in^2 for the thrust and coupling end bearings, respectively).

Although the high amplification factor did not automatically imply problematic operation, it did indicate that the unit would display response sensitivity to unbalance and might be prone to labyrinth seal damage during process upsets. Unlike previous versions, the latest revision of the API Standard Paragraphs (3) does not limit the amplification of response when a critical speed is traversed. If the unit is subjected to frequent shut-downs, however, the high rotor vibrations encountered during operation through the highly amplified critical might induce a permanent rotor bow that would degrade unit balance during normal operation.

One customer specification required that the units be continuously operated on nitrogen during train start-up. Compressor discharge temperature limits mandated that the train operate below 3600 RPM during nitrogen operation so all units must be capable of operating in close proximity to their calculated fundamental critical speeds.

Given the unfavorable numerical assessment of the existing unit, the conversion of the unit to a drive-thru configuration, and the requirement for reduced speed operation with nitrogen the rotor dynamics characteristics of the re-engineered split case compressor had to be improved.

3.5 Original Steam Turbine Rotor Dynamics Analysis

Rotor dynamics analysis of the steam turbine was accomplished for the steam turbine in a manner similar to that for the split case compressor. The original turbine's midspan response to a general unbalance distribution is plotted in Figure 7. Table 11 summarizes the calculated rotor dynamics characteristics of the original unit.

Table 11 Lateral analysis summary original marine steam turbine

1st Mode Response		1st Mode Stability	
Amplification Factor	Critical Speed (r/min)	Log Dec	Damped Frequency (r/min)
6.60	3950	0.221	1940

These calculations indicated three distinct problems:

(a) Critical Speed Interference

Customer specifications required an operating speed range of 70 to 105 per cent of the rated speed. If the unit operates above its fundamental critical and the amplification factor (AF) associated with the critical is greater than 3.55 then API 611 (4) specifies the following minimum critical speed separation margin:

$$SM - 100 - \left(84 - \frac{6}{AF - 3}\right)$$

Table 12 codifies the turbine operating speeds and the restrictions on critical speed placement if the amplification factor associated with the fundamental critical is equal to 9.0.

Table 12 Steam turbine operating speed and critical speed definition

Maximum frequency of an amplified critical (85% of Minimum Operating Speed)	2737 r/min
Minimum operating speed (70% of rated speed)	3220 r/min
Rated Speed	4600 r/min
Maximum Continuous Operating Speed (MCOS) (105% Rated Speed)	4830 r/min
Minimum Frequency of an Amplified Critical Speed (125% MCOS)	6038 r/min

Results presented in Table 11 indicate that the existing turbine fundamental critical speed is 3950 r/min. Sister units to the marine steam turbine currently operate on board crude oil supertankers. The approximate location of the turbine's fundamental critical speed was confirmed when

an engineer at the shipping company that operates the tankers was queried.

(b) Location of Rigid Bearing Critical

The new operating speed was nearly coincident with the rotor's first rigid bearing critical speed. Although this concern is not addressed by API 611 (4) the coincidence of a rotor's rigid bearing critical speed with rated speed is potentially catastrophic if a process upset or a blade failure occurs.

(c) Unconventional Bearing Design

The fluid film bearings supporting the rotor were of unconventional design with the following negative design features:

(1) high clearances (greater than .051 mm/mm/.002 in/in of journal diameter),
(2) high unit loadings (.806 MPa/117 lbf/in^2 and .972 MPa/141 lbf/in^2 for the thrust and coupling end bearings, respectively), and
(3) narrow oil supply grooves.

Schematics of the bearings are contained in Figure 7. These bearing generated extremely asymmetric linearized dynamic coefficients at unit operating speed. For example, the ratio of the vertical and the horizontal bearing stiffnesses was approximately 4.5:1 at 4600 r/min. This stiffness asymmetry implies that the response characteristics of the unit will be non-optimal.

4. REMEDIAL AND RERATE ENGINEERING DESIGN

The problems discovered during unit engineering assessments may be divided into two groups:

(a) problems that might have affected the mechanical reliability of the existing equipment and
(b) problems that arose as a consequence of redesigning the original equipment for application to the new process.

Remedial engineering design efforts focused on solving those problems that affected the unit's mechanical reliability. Remedial engineering efforts were principally aimed at improving the rotor dynamic characteristics of train elements. This was accomplished by minimizing, within casing envelope and economic constraints, the amplification factors associated with the fundamental critical speeds of the three units.

4.1 <u>Split Case Compressor Remedial Engineering</u>

The existing rotor-bearing assembly in the split case compressor required significant modification to lower the amplification factor associated with the fundamental critical. Fortunately, process requirements mandated that the original shaft be replaced so rotor dynamics engineers were permitted to alter shaft geometry to improve adverse rotor response characteristics. Economic constraints, however, did not permit shaft re-design inboard of the floating ring oil seals.

Split case compressor rotor remedial design modifications included:

(a) increasing journal diameters from 139.7 mm (5.5 in) to 152.4 mm (6.0 in),
(b) replacing original load-on-pad tilting pad journal bearing assemblies with increased length load-between-pads tilting pad bearing assemblies, and
(c) installing a reduced moment coupling to join the second body split case compressor with the first body barrel case compressor.

The re-engineered compressor's response to a general unbalance distribution is plotted in Figure 6. The high amplification of the first critical was reduced by increasing the stiffness of the shaft through the constrained shaft geometry changes and by reducing the moderately high bearing unit loadings (.827 MPa/120 lbf/in^2 and .972 MPa/141 lbf/in^2 for the thrust and coupling end bearings, respectively) by installing the optimized design geometry tilting pad bearings. The original and retro-fit tilting pad bearings are described in Figure 6. A reduced moment coupling was placed on the new drive-thru end of the unit to reduce the overhang's unbalance sensitivity at rated speed. The calculated damped unbalance responses are displayed in Figure 6 for the original and re-engineered units. The net effect of these changes was striking: the calculated amplification factor associated with the fundamental critical was reduced from 17.5 to 8.1 even with the addition of the drive-thru shaft overhang.

The cost of retro-fitting the replacement tilting pads was minimized by reworking the bearing housings. Increasing the journal OD and changing the type of pivot (from line contact to a ball and socket arrangement) resulted in a significant decrease in housing thickness. A finite element stress analysis of the housing was accomplished to insure that the housings did not undergo significant radial deflections even with applied radial loads equivalent to the dynamic loads the rotor would generate under trip level vibrations.

4.2 <u>Turbine Remedial Engineering</u>

Recommendations for resolving the critical speed separation problem uncovered in the turbine included:

(a) decreasing the effective midspan shaft diameter and
(b) replacing existing 4 axial groove sleeve bearings with optimized design tilting pad bearing assemblies.

These recommendations were formulated only after detailed design studies were conducted and with consideration of cost-effectiveness. 'De-tuning' the rigid bearing critical could only be accomplished by changing the mass-elastic characteristics of the rotating element. Note the midspan shaft diameter of the turbine was decreased only between the bladed disks after it was determined that the calculated critical speed was relatively insensitive to shaft geometry diameter outboard of the disks. This action resulted in significant cost savings by salvaging the main labyrinth packings. The shaft diameter selected for the re-engineered unit represented a compromise between maximizing the critical speed margin and minimizing the associated amplification factor.

The re-engineered compressor's response to a general unbalance distribution is plotted in Figure 7. Decreasing the effective midspan shaft diameter lowered both the rigid bearing and the observed critical speeds of the unit. Installing optimized design 5 pad load-between-pads tilting pad bearings also decreased the frequency of the observed first critical and reduced the associated amplification factor.

5.0 RERATE ENGINEERING

While the remedial engineering efforts emphasized maximizing unit mechanical reliability, rerate engineering efforts were concerned with adapting the existing unit to the new process conditions. Rerate engineering efforts focused principally on solving problems related to flow-path design. The bulk of these efforts was spent designing the barrel compressor's impellers to meet customer imposed maximum stress criteria. As previously noted, the presence of non-radial IGVs in the final 3 stages of the barrel compressor nullified all flow-path analysis that had been accomplished prior to unit disassembly and inspection. Barrel compressor impellers were redesigned instead of the non-radial IGVs because new impellers were budgeted for in the proposed scope of supply while the unit cost would have increased by nearly 10 per cent if new IGVs were provided.

5.1 Impeller Design Procedure

Each of the 4 barrel compressor impellers were designed to deliver the performance specified by compressor designers and displayed in Figures 4 and 5. Impeller designs were constrained to fit envelopes defined by diaphragms and the shaft. The impeller blade inlet flow angles were fixed according to the inlet flow angles fixed by the inlet guide vanes and the operating speed of the rotating element. Impeller discharge geometry including blade angle and exit area was fixed according to the impeller's inlet geometry, maximum OD, and the specified performance. Once the entrance and exit geometry of the impeller was determined the design was evaluated and re-designed as necessary using three different computer flow codes. The extensive computer analysis insured that algorithmic or modeling limitations in the software were not manifested in faulty designs.

Once impeller flow-path geometries were designed the precise geometries were communicated to a structural analyst so that stress analysis of the impellers could be accomplished. 2D impellers were initially designed to produce the required performance. This type of wheel offers simpler and more cost effective construction than 3D designs because the blades can be fabricated from flat plates.

5.2 Impeller Stress Analysis

The impeller stress analysis was accomplished using one of many well-validated finite element analysis packages. Key features of the impeller model included:

(a) Pressure loads and thermal stresses were considered negligible relative to centrifugal body forces.
(b) Load sharing between the disk and cover through the impeller blades, shear forces, and stress concentrations were all considered.
(c) All impeller elements were modeled as 2D axi-symmetric finite elements. Blades do not generate hoop stiffness, however, and were represented as plane stress elements, incapable of supporting circumferential loads.
(d) The contact interface between the shaft and the impeller hub was modeled with non-linear gap elements.
(e) A conservative value for the coefficient of friction, $\mu = 0.12$, was used at the gap element shaft to impeller hub interface.
(f) An average impeller temperature, 27.8 °C (50 °F) higher than the shaft temperature, was used to evaluate shrink fit pressures at operating speed.
(g) Discontinuous features such as balance rings, lifting eye holes, and keyways were ignored.

The 2D finite element model generated through this process has been demonstrated to adequately represent the actual 3D impeller.

5.3 Maximum Impeller Stress Acceptance Criteria

Results of an impeller stress analysis include the following:

(a) maximum stresses and strains,
(b) burst speed, and
(c) maximum torque transmission from shaft to impeller for a given fit

Given these calculated results, conservative acceptance criteria have been empirically formulated. These criteria apply to covered impellers that are not subject to low cycle fatigue and are manufactured using joining techniques (usually welding) that do not significantly degrade material properties. The stress acceptance criteria traditionally used by the authors' company requires the maximum calculated disk bore hoop stress and the maximum element Von Mises stress of both the disk and cover to be less than the material 0.2 per cent yield strength when the impellers operate at the overspeed defined in API 617 (2). The customer, however, required the maximum impeller stress at 100 per cent rated speed be less than 60 per cent the yield strength of the impeller. If the impeller material is stressed in the elastic range of the material then the 100 per cent yield stress criteria at overspeed translates into about 68 per cent yield stress at rated speed. If the nominal minimum yield strength of the impeller material is 552 MPa (80 000 lbf/in^2) then the customer specifications limited the maximum impeller stress at rated speed is 331 MPa (48 000 lbf/in^2).

5.4 Impeller Design Optimization

The finite element stress analysis of the 2D impeller initially designed by the aerodynamicist for the barrel compressor's 1st stage predicted a maximum impeller stress at rated speed nearly 90 per cent of the minimum yield strength of the material. The calculated impeller stresses in the disk and cover at rated speed are displayed in the contour plot presented in Figure 8. The difference between the calculated maximum stress for the 2D impeller and the maximum acceptable stress was so great that producing an acceptable impeller design appeared difficult to accomplish.

The maximum stress in the 2D impeller was located on the inner side of the cover near the impeller eye. This stress was greater than 90 per cent of the material yield strength. The maximum stress in the disk was located at the bore on the side of the impeller opposite the eye. All efforts to reduce the maximum impeller stress focused on reducing the stress in the eye region of the cover. Aerodynamic and rotor dynamic considerations effectively eliminated any possibility of substantially altering the shaft diameter and the size of the flow passage. In an attempt to minimize changes to the flow passage, all initial attempts to decrease the cover stress retained the original 2D blade geometry. The eye labyrinth seal land length was changed to reduce the stress on the inside of the cover and the thickness of the cover outside the seal area was decreased to reduce the cover's 'pull' on the areas near the impeller eye. Calculations indicated that while these changes reduced the maximum cover stress the revised 2D impeller design was not close to meeting the customer's impeller stress criteria.

Given the failure of intermediate measures to solve the impeller stress problem, a more comprehensive redesign of the impeller was undertaken. The large stress on the inside of the cover near the sealing land was a combination of the high tangential stress plus bending in the area of the sealing land. These stresses were reduced by moving the leading edge of the blades closer to the front of the impeller disk. Such action reduced the cover stresses by sharing the radial cover loads with the impeller disk through the blades. Such designs had to be carefully accomplished, however, as load sharing decreased the interference pressures between the impeller hub and the shaft and decreased the maximum transmitted torque. Additionally, lengthing the blades changed the impeller aerodynamics so the flow passages and blades had to be designed with cooperation between the aerodynamicist and the structural analyst. Over one dozen design iterations were required to produce a first stage impeller design with minimum maximum (minmax) stress and acceptable aerodynamic characteristics.

The effort expended to minimize the maximum first stage impeller stress generated an impeller with a significantly reduced maximum stress. The maximum calculated stress, however, remained marginally above the allowable stress based on a minimum material yield strength of 552 MPa (80 000 lbf/in^2). After consulting with impeller forging vendors, the guaranteed impeller minimum material yield strength was specified at 552 MPa (80 000 lbf/in^2). Raising the minimum yield strength to this level required extremely tight control of the heat treating process because the corrosive H_2S environment restricts the maximum material yield strength 621 MPa (100 000 lbf/in^2) and the material hardness (22 HRC). The increase in the minimum yield strength of the impeller material permitted a larger maximum impeller stress at rated speed. The finite element impeller stress analysis confirmed that the 3D impeller designed for minimum stress possessed a maximum stress at rated speed less than the 372 MPa (54 000 lbf/in^2).

The final first stage impeller design for the barrel compressor is displayed in Figure 9. Stress contours are also displayed in Figure 9.

5.5 Split Case Compressor Rerate Engineering

As previously discussed, the scope of work originally proposed for the split case compressor proved rather accurate, partly because the unit's internal geometry was well-understood prior to quotation and partly because the quoted scope of supply was quite extensive. Rerating the unit proved easy relative to the barrel case unit because the compressor design engineers and aerodynamicists were given a 'blank sheet of paper' and permitted to replace all impellers and most stationary flow-path components including diaphragms.

The scope of supply of 'new' parts for the split case compressor was actually reduced by salvaging a diaphragm and the inlet guide tube. The 2nd stage diaphragm was re-machined to increase the size of the diffuser paths and alter the return channels and vanes to match the stage 3 impeller. The inlet guide tube originally had a non-radial bend at the vane exits. These vanes were trimmed back to eliminate the bend. The trimmed blade tips were also dressed to assure an efficient inlet flow design.

6. TRAIN TESTING

After constructing and assembling the re-engineered units they were bolted to the job baseplates and mechanically tested. As the authors' company does not possess permanent train testing facilities, significant effort was expended in developing temporary test facilities:

(a) a boiler capable of generating sufficient steam to drive the train was rented and required steam piping was fabricated and rigged,
(b) a vacuum pump was purchased along with associated piping so the units could be tested without the experiencing windage losses,
(c) the lube console was placed in a pit several feet deep to permit lube oil drainage, and
(d) instrumentation was rented to monitor and record all vibration data.

Train mechanical testing proceeded without major incident. Measured vibrations were well within acceptable limits defined by applicable customer specifications. Figures 10 and 11 compare calculated and measured rotor vibrations for the one shaft displacement probe on the split case compressor and the steam turbine. These figures display excellent qualitative correlation between the calculated and measured rotor vibrations. Full performance testing of train units is scheduled to proceed on-site immediately after unit commissioning in May 1993.

7. CONCLUSIONS

The following specific conclusions are offered from the experiences related in this paper:

(a) Methods used to estimate the scope of work required to re-engineer surplus rotating equipment may significantly understate actual work scope if precise information describing unit internals and performance is not available.

(b) A complete engineering audit of each rerated unit must be accomplished to remedy any problems that might exist in the original equipment.

(c) The non-OEM rotating equipment vendor must possess the full range of engineering capabilities to insure that the delivered equipment meets all customer specifications and is capable of efficient, reliable operation.

(d) The effort required by a non-OEM vendor to re-engineer surplus equipment is generally equal to, if not greater than, the OEM's original design effort spent because the OEM relies on extensive experience with nearly identical equipment. OEM turbomachine vendors often design rotating equipment with stock impellers that necessitate flowpath modifications. The benefit of custom engineering is the degree that the final design may be optimized for efficiency and reliability.

REFERENCES

(1) AMERICAN PETROLEUM INSTITUTE Lubrication, Shaft-Sealing, and Control Oil Systems for Special Purpose Applications, API Standard 614, Second Edition, Washington, D.C., January 1984.

(2) AMERICAN PETROLEUM INSTITUTE Centrifugal Compressors for General Refinery Service, API Standard 617, Fifth Edition, Washington, D.C., April 1988.

(3) AMERICAN PETROLEUM INSTITUTE Standard Paragraphs, API Standard Subcommittee on Mechanical Equipment, Revision 19, Washington D.C., September 1991.

(4) AMERICAN PETROLEUM INSTITUTE General-Purpose Steam Turbines for Refinery Service, API Standard 611, Third Edition, Washington D.C., August 1988.

Figure 1 Original and Re-engineered First Body Barrel Compressor

Original Compressor

Re-engineered Compressor

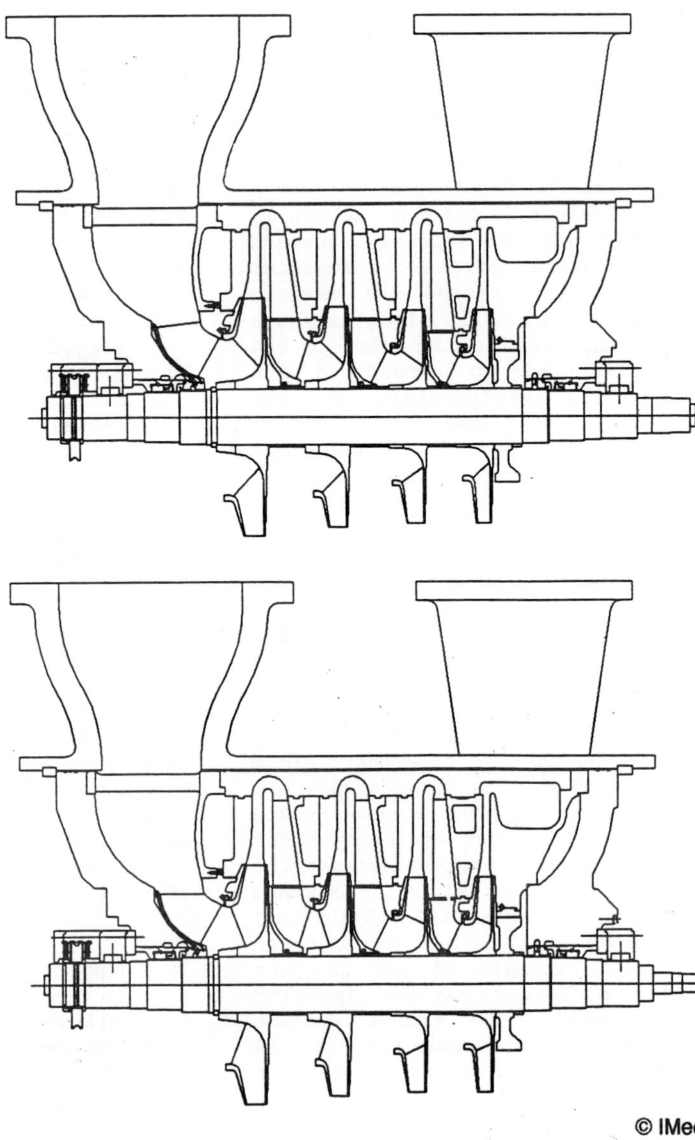

Figure 2 Original and Re-engineered Second Body Split Case Compressor

Original Compressor

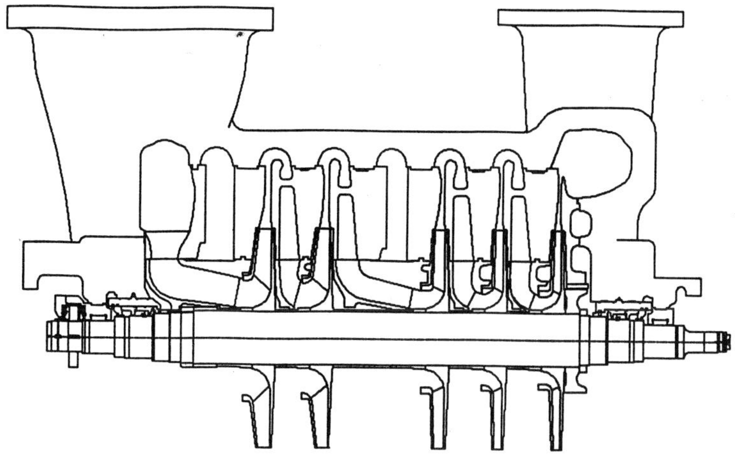

Re-engineered Compressor

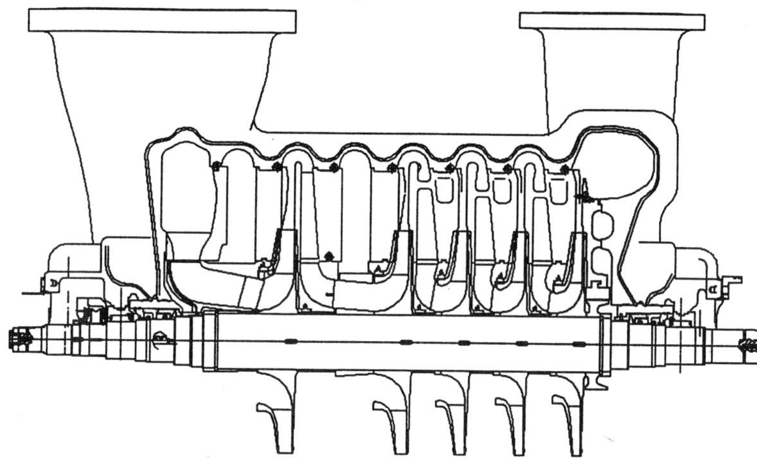

Figure 3 Marine Steam Turbine

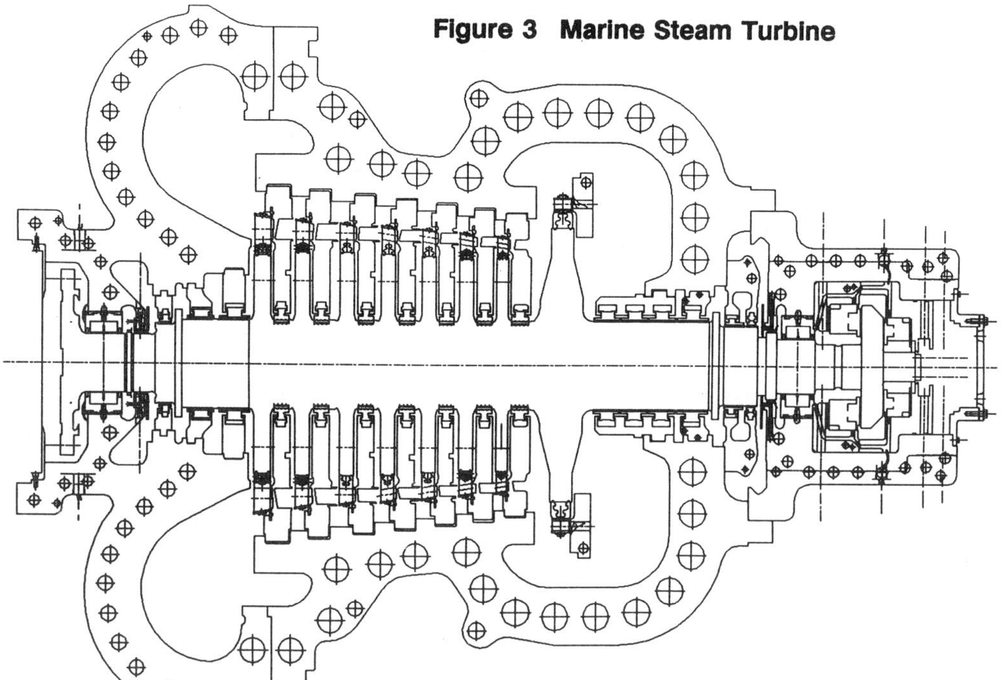

Figure 4

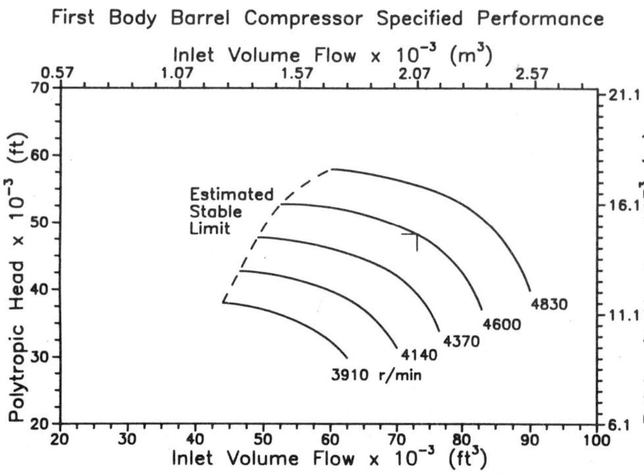

Figure 5

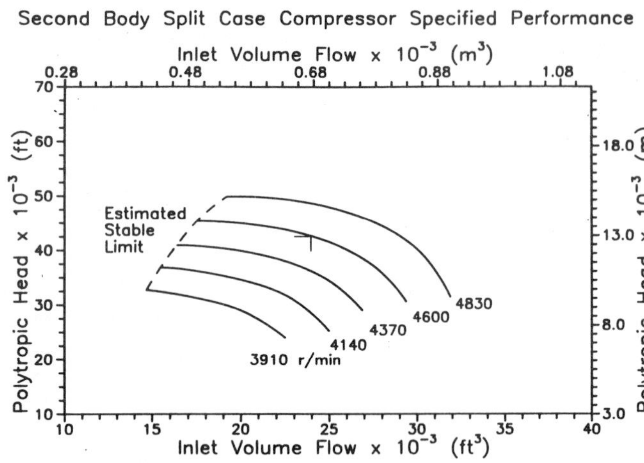

Figure 6 Comparison of Original and Re-Engineered Split Case Compressor Rotor Midspan Responses

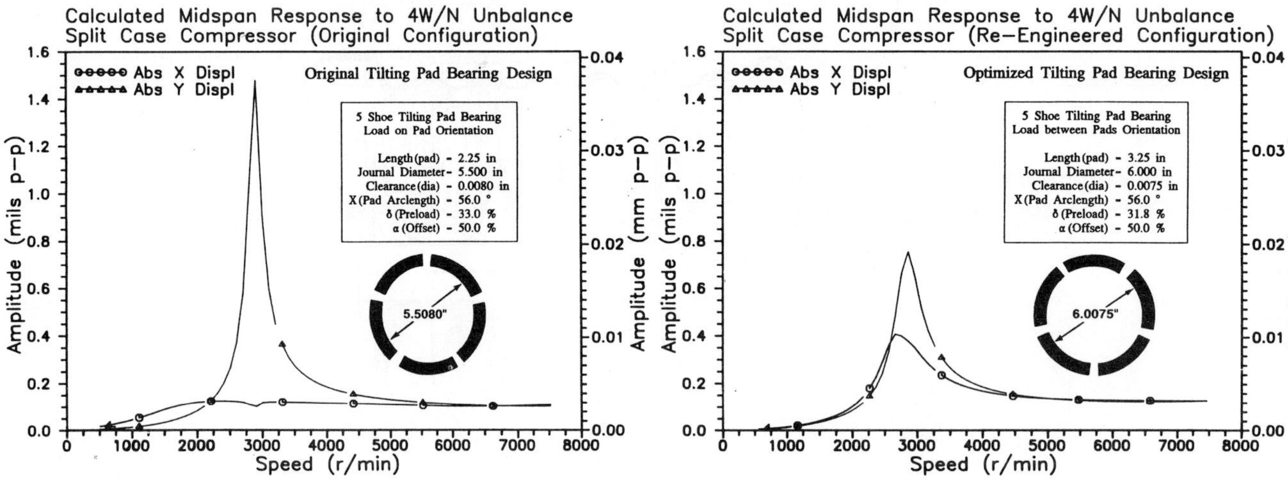

Figure 7 Comparison of Original and Re-Engineered Marine Steam Turbine Rotor Midspan Responses

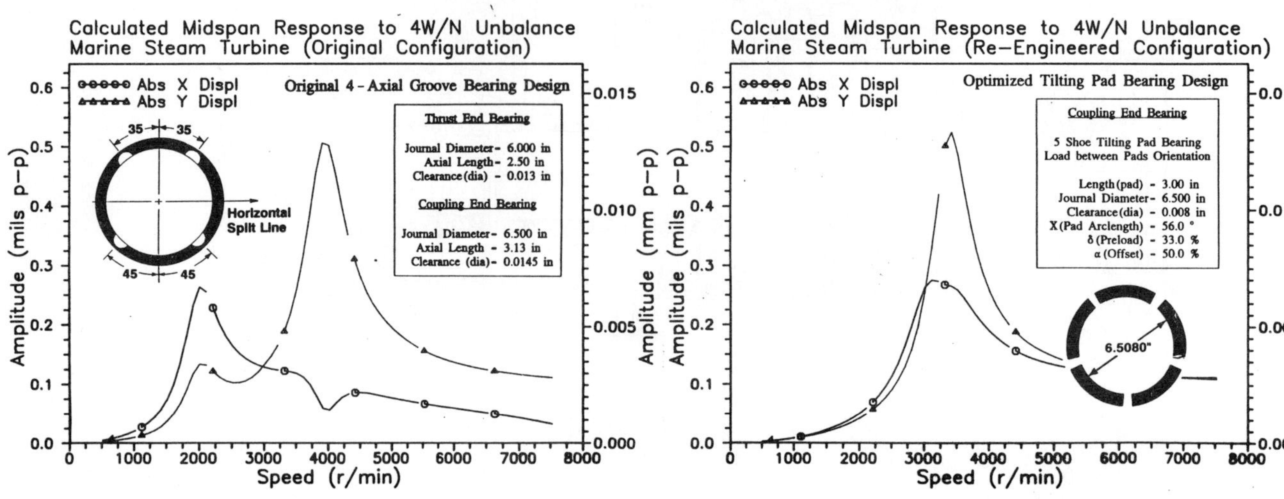

Figure 8 2D Impeller Deflection and Stress Analysis

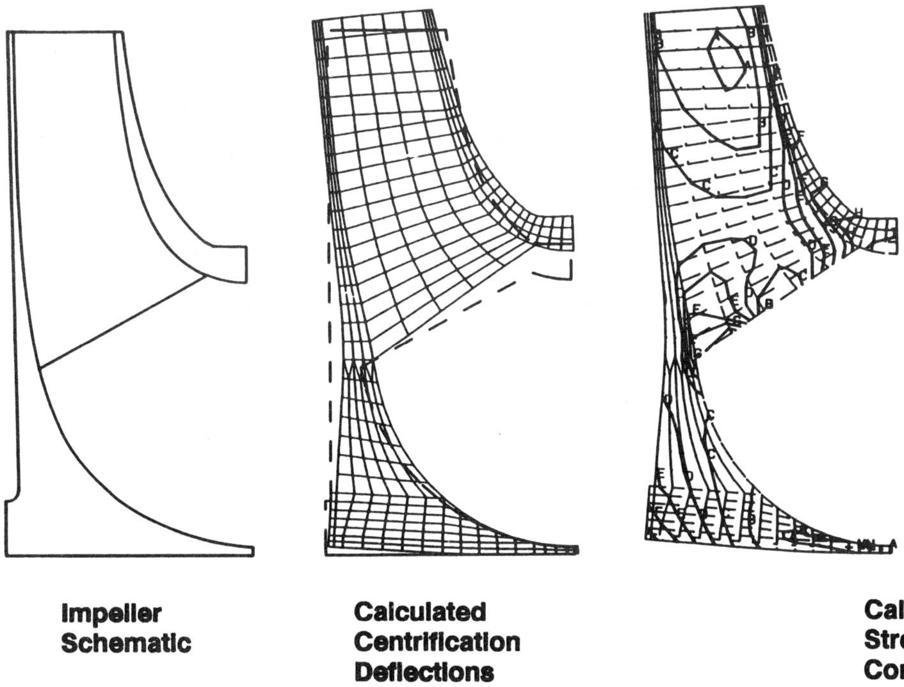

Contour	Calculated Stress
A	34.6 MPa 5021 lbf/in^2
B	69.0 MPa 12910 lbf/in^2
C	143.4 MPa 20799 lbf/in^2
D	197.8 MPa 28687 lbf/in^2
E	252.2 MPa 36576 lbf/in^2
F	306.6 MPa 44465 lbf/in^2
G	360.9 MPa 52353 lbf/in^2
H	415.4 MPa 60242 lbf/in^2
I	469.7 MPa 68131 lbf/in^2

Figure 9 3D Impeller Deflection and Stress Analysis

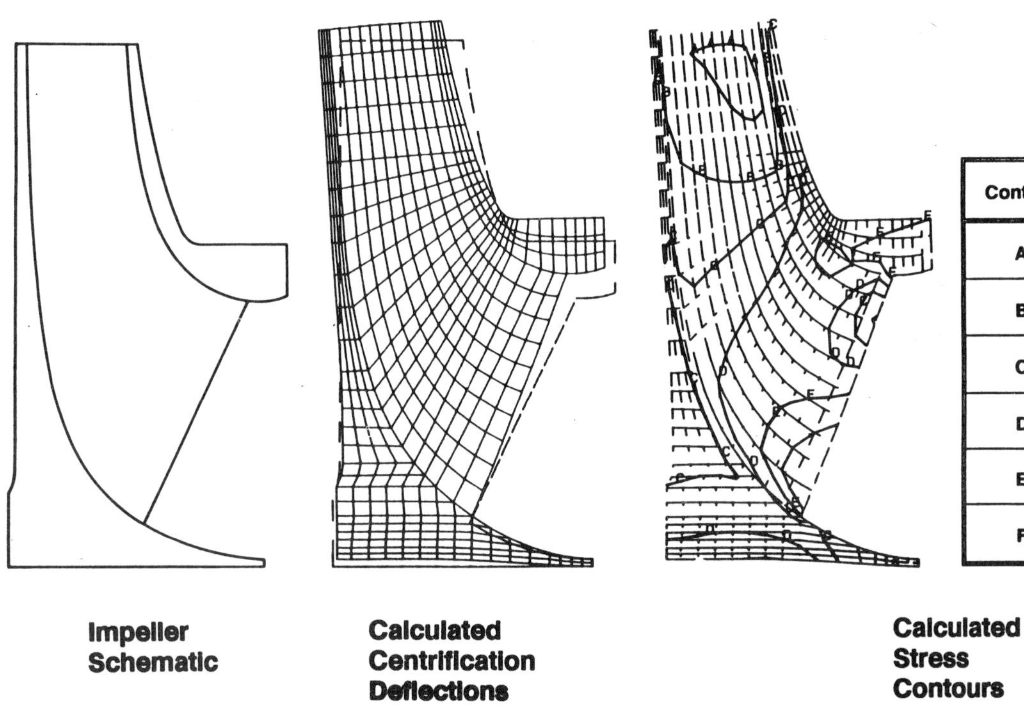

Contour	Calculated Stress
A	38.6 MPa 5604 lbf/in^2
B	108.2 MPa 15696 lbf/in^2
C	177.8 MPa 25787 lbf/in^2
D	247.4 MPa 35878 lbf/in^2
E	317.0 MPa 45970 lbf/in^2
F	386.5 MPa 56061 lbf/in^2

Figure 10 Comparison of Calculated and Measured Split Case Compressor Discharge End Rotor Responses

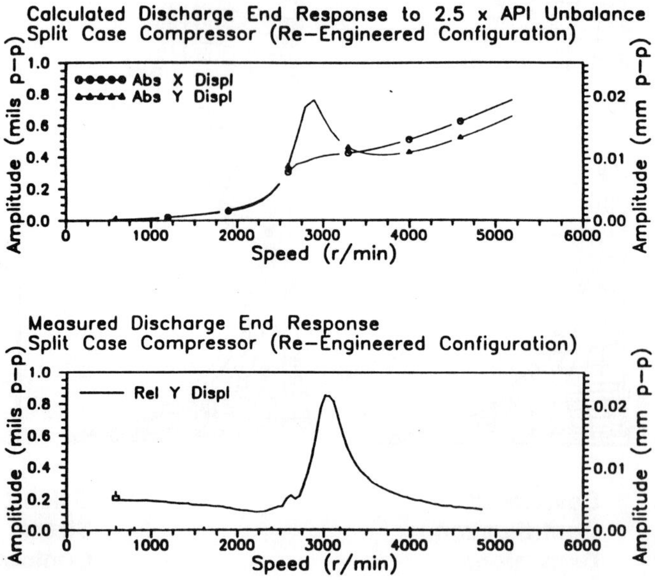

Figure 11 Comparison of Calculated and Measured Marine Steam Turbine Plain End Rotor Responses

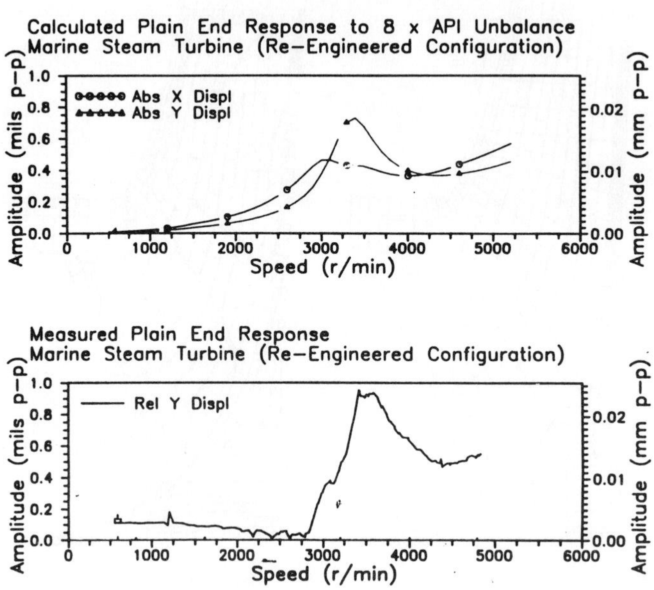

C449/026/93

The integral gear compressor for oxygen service

N J HARYETT, BSc, CEng, MIMechE
Air Products plc, Walton-on-Thames, Surrey
U FINGERHUT, Dipl Ing
Mannesmann Demag, Duisburg, Germany

1.0 INTRODUCTION

The use of integral gear compressors for compression of air and nitrogen is well established within the air separation industry. With the introduction of six stage integral gear compressors, the typical duty for a product oxygen compressor, 8,000 Nm^3/hr to 40,000 Nm^3/hr at a delivery pressure of between 20 barg and 45 barg, came within the operating range of the integral gear compressor.
Aspects of the design of standard integral gear compressors made them unsuitable for oxygen service and dictated the continued selection of single shaft designs. However, the performance, packaging and installation advantages of the integral gear compressor were substantial enough that Mannesmann Demag started to develop an integral gear compressor for oxygen. Additionally, some of the design features required for oxygen service would enable this type of compressor to be used over a range of process gas compression duties.

In 1988, Air Products identified a requirement at Ghent in Belgium for an oxygen compressor in a new air separation plant which would be suitable for a six stage, integral gear design. Reliability of the oxygen supply would be critical, and as is usual with major plant compressor's, the compressor would be unspared.

The possible selection of an integral gear oxygen compressor for this duty was subject to an extensive risk assessment by Air Products. All areas of the design which were perceived to carry an increased risk over a conventional single shaft compressor were fully and openly discussed between Owner and Manufacturer. Such aspects included material selection, shaft seal design, rotor dynamics and the consequences of a compressor fire.

The Owner concluded that the risks associated with the integral gear oxygen compressor were not significantly greater than those associated with a single shaft compressor. Additionally, the Ghent plant is owned and operated by Air Products, giving full control over the compressor operation, and the existing plant could provide reserve oxygen compression capacity in the event of problems with the integral gear compressor.

The Manufacturer received their first order for an integral gear oxygen compressor in July 1988. The compressor discussed here was ordered in February 1989. These compressors have now been in operation for over two years, establishing valuable references for both operators and manufacturer.

2.0 MECHANICAL DESIGN

2.1 General

The mechanical design of the integral gear oxygen compressor follows the fundamental principles used for all other gases. However, all aspects of the mechanical design were subject to a thorough design review involving the Owner, who brought extensive operating experience to add to the Manufacturer's design and manufacturing standard.

The areas of material selection and seal design received particular attention.

2.2 Materials Selection

During the last few years, a great deal of research has been carried out into the principles of material selection. The fundamental aim of this research was to investigate the conditions under which ignition or fire occurs on metal after the creation of an unoxidized, fresh and active surface. Such active surfaces can be created by rubbing, fracture, or by the impact of a foreign particle and this was simulated experimentally using a test rod in heated high-velocity, high-pressure oxygen [1].

The result of this investigation is the Manufacturer's philosophy [2] to classify the materials used in centrifugal compressors for oxygen service, excluding aluminium and bearing metal alloys, into four safety-groups:

- I cast iron, unalloyed and low-alloy steels
- II martensitic and austenitic stainless steel
- III copper and nickel alloys
- IV silver

For these various material groups, the development of a fire depends on the metal temperature. The temperature at which fire is possible decreases as the pressure increases.

From this it may be surmised that centrifugal compressors should be built exclusively from nickel alloys such as Monel K-500, in order to obtain the highest possible degree of safety. However, this would lead to an expensive solution. Therefore, the material selection for the Manufacturer's compressors is based on the criterion of a rubbing risk. From this point of view, three risk groups as shown in Fig. 1 can be defined. Where two material groups are shown, at least one of the rubbing partners must be of the higher grade.

Materials with the lowest ignition temperatures are used in areas where there is no risk of rubbing with rotating parts. The most expensive metal and alloys, such as copper, nickel alloys and silver, are used in those areas where the possibility of rubbing cannot be excluded during normal operation of an oxygen compressor facility. The combination of nickel alloy and silver has proved to be safe under the service conditions of existing single shaft compressors, and was used for shaft and impeller labyrinth seals of the integral gear oxygen compressor.

2.3 Seal Design

The integral gear compressor design requires each stage to be sealed against the atmosphere. Fig. 2 [3] shows the impeller and the shaft seal of the third pinion of an integral gear oxygen compressor. All labyrinth seals in contact with oxygen or nitrogen are of stepped type to reduce leakage. The extensive use of pressure balancing across all the labyrinth type shaft seals optimises efficiency and reduces the product losses to atmosphere.

The high pressure leakage, which occurs in the fifth and sixth stage only, is returned to the fifth stage inlet. Medium pressure leakage is returned to the third stage suction and the low pressure leakage of all six stages is fed back to the first stage suction. A vent port is arranged after the intermediate sealing ports, followed by a nitrogen buffer port (Fig. 3.).

The positive differential pressure between the buffer port and the vent port is one of the most significant safety features of an oxygen compressor.

Additional nitrogen buffering is provided for all shaft seals of the gear case. This is an additional safety precaution which ensures an inert atmosphere in the gear case.

3.0 PERFORMANCE

Fig. 4 shows a typical comparison between the operating range of a single casing, six stage integral gear compressor, and that of a two casing, single shaft compressor train with six impellers arranged in three stage groups per casing. The single shaft compressor is equipped with a main and an intermediate gear unit. Each compressor installation, i.e. the integral gear compressor and the single shaft compressor, is equipped with five intercoolers.

The operating range of the integral gear compressor is of light shade and that of the single shaft compressor is dark in the illustration. Due to the inlet guide vane control, the integral gear compressor features a much larger turndown range compared to the variation range of the single-shaft compressor, when equipped with suction throttle control, as is typical. Moreover, the counter rotational flow setting of the guide vanes of the integral gear compressor offers an overload range of some 110% V/V_0. The aerodynamic advantages of the integral gear compressor lead to higher efficiency at the design duty point, and therefore a power advantage. A power difference of approximately 30% may be seen in the extreme part-load range.

The integral gear oxygen compressor for this application was fitted with inlet guide vanes on the first and third stages, which allowed all specified operating points to be met without recycle. For additional increase of stage efficiencies, all six stages were fitted with fixed vaned diffusers.

4.0 PACKAGE CONCEPT

4.1 Scope

The objective of the package concept specified by the Owner was to minimise site erection, cleaning and testing, by maximising the scope of work carried out at the factory.
Equipment included in Manufacturer's scope was as follows:-

Compressor
Coupling and guard
Five gas intercoolers and one aftercooler
Interstage piping
Suction piping, including strainer
Discharge piping, including non-return valve
Recycle valve and piping
System dump valves
Discharge relief valve
Cooling water manifold piping
Skid for the above items.

A free standing lube oil system and interconnecting piping

A free standing seal gas control console.
Process, seal gas and lube oil system instrumentation.

As usual, the main drive motor was supplied by the Owner.

Control of the compressor is by a plant distributed control system (DCS).

4.2 Spares Philosophy

In addition to purchasing the usual capital spares, including complete rotor assemblies, bull gear, bearings and gas seals, the Owner also purchased rough-machined 5th and 6th stage volute castings. The highest pressure stages were perceived to be at the greater risk of an oxygen fire. Holding spare castings for these volutes would allow the compressor to be rebuilt in about 6 weeks, even allowing for considerable damage to the compressor gearcase.

4.3 Cleaning and Maintaining Cleanliness

In addition to correct design and a suitable choice of materials, freedom from contaminants is critical to the safety of an oxygen compressor.

Particles which are driven against metal surfaces in a high velocity oxygen stream may initiate a fire. Flammable materials, particularly oil or grease, may ignite. Additionally, accumulation of dust with its large surface area can represent a hazard, especially if dust collects in cavities.

The compressor package was fully cleaned and assembled before the works performance and mechanical test. In order to maintain the cleanliness of components before and during assembly, the assembly of the various compressor components was carried out in a clean work area set aside for this purpose (Fig. 5).

The use of gas in shell intercoolers with externally finned tubes is essential for low interstage pressure drops and close approach temperatures within a compact design. This type of cooler would require a major effort to reclean if contaminated after assembly. Hence, these coolers were subject to stage inspections for cleanliness during bundle assembly and bundle installation.

In order to reduce the extent of cleaning required on interstage piping, stage suction piping, (some of which uses mitred bends with internal turning vanes), was fabricated in stainless steel rather than carbon steel.

After testing and mechanical inspection, all termination flanges were fitted with special blanks which remained undisturbed during erection of plant piping, and were replaced with spacer rings just prior to commissioning.

During transportation and erection the most likely cause of contamination was from rust on carbon steel components, due to the ingress of damp air. Throughout this period a dry nitrogen purge was provided.

Moisture indicators were fitted at various tapping points to show any ingress of water. The Owner's Ghent plant site is only 200km from the Duisburg works of the Manufacturer, and hence the short shipping time facilitated the maintenance of a dry gas purge.

4.4 Testing

Works testing of a single shaft oxygen compressor is usually carried out on a stage by stage basis. Single shaft oxygen compressors are supplied as field erected units with elevated compressor casings and floor mounted intercoolers. In many cases, this does not allow the full prefabrication of interstage piping. Hence, this piping is not available for shop testing.

Testing of oxygen compressors using a mixture of nitrogen and carbon dioxide, rather than either nitrogen or air to simulate the molecular weight of oxygen is an established practise. This allows the compressor to be run at its design pressure ratio and gas density, which gives more accurate performance measurement and verification of rotor stability. This test is inevitably difficult to set up on site. Site testing would be followed by a full mechanical inspection of bearings and seals, involving opening of the compressor casing. The single shaft oxygen compressor therefore requires an extensive programme of site testing and mechanical inspections, which is both time consuming and expensive.

The testing programme for the integral gear oxygen compressor was developed jointly between the Owner and Manufacturer. The aim of the testing programme was to give the maximum level of confidence in the thermodynamic and mechanical performance, whilst minimising testing at site and eliminating mechanical inspections at site.

Testing and inspection at the factory was as follows:-

Mechanical test of compressor.
Nitrogen test run.
Nitrogen/carbon dioxide test run.
Shutdown simulation.
Post test inspection of shaft seals, bearings and gears.
Nitrogen test run after rebuild.

Testing at the Ghent site was as follows:-

Shakedown nitrogen test run.

No mechanical inspections were performed.

The mechanical test was carried out on the compressor only, prior to oxygen cleaning. All subsequent tests were carried out using the full compressor package in an oxygen clean condition.

The first works nitrogen test run established base line data for performance and mechanical running, including shaft vibration.

The nitrogen/carbon dioxide test run simulates the molecular weight of oxygen and was used to measure the full performance envelope and surge characteristic. In order to test rotor stability, the carbon dioxide concentration was increased to operate the compressor at a gas density equivalent to a discharge pressure 25% higher than design. Increasing the test loop suction pressure to increase gas density would have caused interstage pressures to exceed equivalent design pressures.

To reduce the possibility or duration of an oxygen fire, the system pressure must be reduced to less than 1 barg within 10 seconds of a shutdown. The inclusion of recycle line and recycle valve, and the system shutdown vent valves, allowed the simulation of a compressor trip and the measurement of the time taken to depressurise the system.

The post test stripdown included inspections of all gas seals, bearings seals, and a cleanliness check. The seal inspection was additional to those checks normally undertaken for an integral gear compressor on air or nitrogen duty. This inspection was considered necessary due to the vital function of the shaft seals and the possible consequences of a seal rub.

Results from the post-rebuild nitrogen test run were compared with the first nitrogen test run, effectively proving the compressor was rebuilt correctly.

The site shakedown nitrogen test run was compared again with previous nitrogen test runs, and proved that transportation had not disturbed the compressor and that erection had not introduced piping or skid frame stresses in the compressor.

No mechanical inspections were made at site, and following a satisfactory nitrogen test run the compressor was released for oxygen service.

5.0 SITE INSTALLATION

5.1 Layout

Any oxygen compressor requires an enclosure to restrict personnel access and retain material or jets of molten metal in the event of an oxygen fire. Ideally, the enclosure should contain all oxygen equipment, but leave the main drive motor and lube oil system outside the enclosed area, to allow access to those items and to prevent consequential fire damage.

These considerations led to mounting the main drive motor outside the compressor enclosure, on a low concrete table top foundation alongside a floor mounted lube oil system. The compressor package was mounted on a single frame located on a flat concrete foundation.

The low level installation of the integral gear compressor compared with an equivalent single shaft compressor, gave significant cost reductions for foundations and associated access platforms.

5.2 Enclosure

The enclosure for the compressor serves as both a safety enclosure and an acoustic building. The Owner's usual construction for such a building is a steel frame, with steel sheet and fire proof boarding, lined with acoustically absorbent material.

For this project a cast concrete building was found to give significantly increased safety and acoustic performance at a lower cost than a steel frame building. This was due in part to the low level layout of the integral gear compressor. Keeping compressor components oxygen clean during maintenance is always of concern. Consequently, floor space was allowed within the enclosure for cooler bundle withdrawal and for laying down parts during maintenance.

For acoustic reasons, the enclosure was fitted with a roof. Ventilation air was drawn into the enclosure and allowed to vent via open acoustic louvres installed in the roof ridge. This open area was also designed to limit the over-pressure within the enclosure in the event of a compressor fire.

5.3 Erection

The compact design of the package allowed delivery to site by road. Construction of the compressor enclosure was completed except for the roof, and the compressor skid was lifted directly from the transport into place through the roof. The enclosure roof construction was then completed.

The total time taken between arrival at site and the compressor being ready for test run and commissioning was three weeks. This compares with a minimum duration of twelve weeks for the equivalent site-erected oxygen compressor, which would then require a more extensive programme of testing and inspection.

6.0 OPERATING EXPERIENCE

The oxygen compressor was commissioned according to plan in August 1990 and delivered oxygen to the pipeline in October 1990. Since then the compressor has effectively run continuously except for 8 hour shutdowns, once or twice a month, when there is no product oxygen demand. As usual, all starts were carried out using nitrogen.

One significant incident occurred in February 1991. During a nitrogen restart at extremely low ambient temperatures (between -10°C and -15°C) a heavy seal rub occurred on the 5th/6th stage pinion. Although lube oil supply temperature had reached the required starting minimum of 20°C, it is suspected that the oil temperature at the bearings themselves may have been below 20°C, as prior to starting the auxiliary oil pump had been running for only a short time. Lateral rotor system analyses showed that rotor stability would be affected adversely at minimum bearing clearances and 10°C lube oil temperature. This is considered the most likely cause of the failure. As a result, minimum lube oil temperatures for starting have been revised.

7.0 SUMMARY

To date, the Manufacturer has sold three six stage integral gear oxygen compressors, each with discharge pressures of more than 30 barg.
Based on the excellent operating experience, and subject to the total installed and operating costs showing an overall saving against the single shaft compressor alternative, the Owner certainly will consider installing further integral gear oxygen compressors.

References

[1] Pears, C.D., Monroe, R., Wren III, J. E.,and Bates, C.E., "Final Report to the Department of Energy on Structural Materials.
Evaluation for Oxygen Centrifugal Compressors".
Project 3528, Report XXXVI,
Southern Research Institute, Birmingham,
Alabama, 1978

[2] Dr. Klaus Boddenberg, Dr. Jürgen Waldmann
Mannesmann Demag,
"Special Construction Features of Centrifugal Compressors for Oxygen Service",
Standard Technical Publications 1111,
American Society for Testing and Materials, 1991

[3] Udo Fingerhut, Dr. Ernst Rothstein, Gerard Sterz "Standardized Integrally Geared Turbo Machines - Tailor-Made for the Process Industry",
20th Turbomachinery Symposium, Dallas, 1991

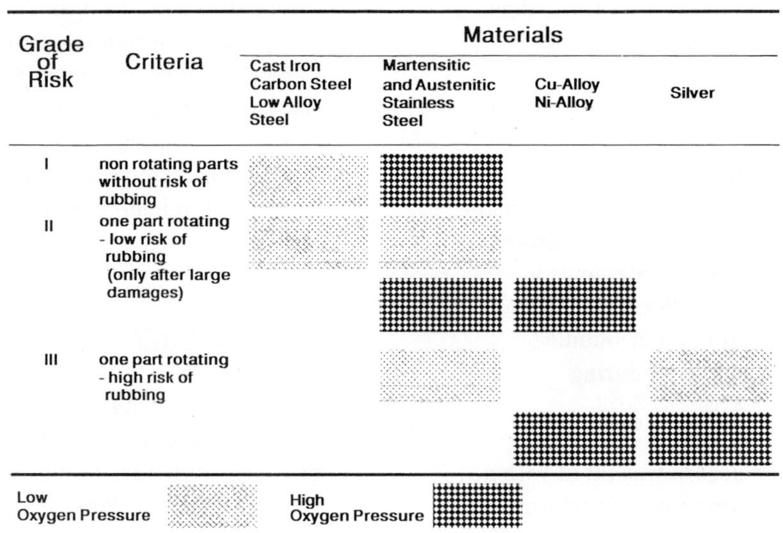

FIG. 1. - Materials selection matrix for oxygen compressors.

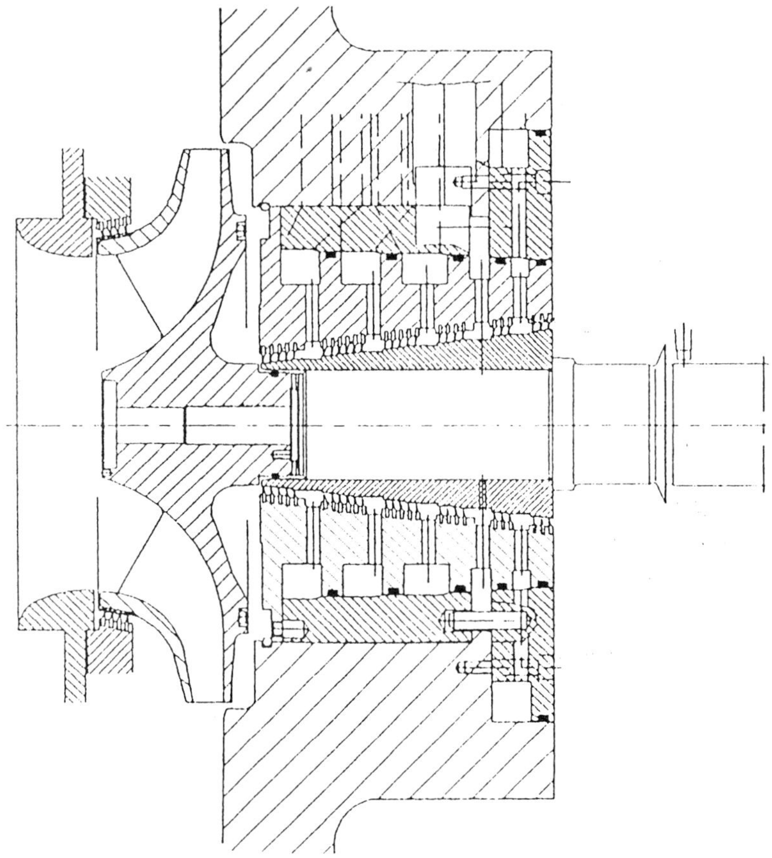

FIG. 2. - Shaft seal, 6th stage.

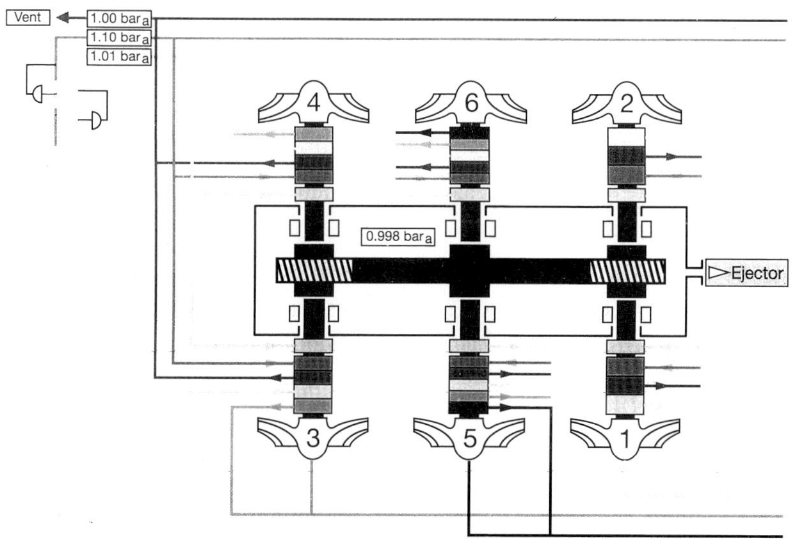

FIG. 3. - Seal gas schematic.

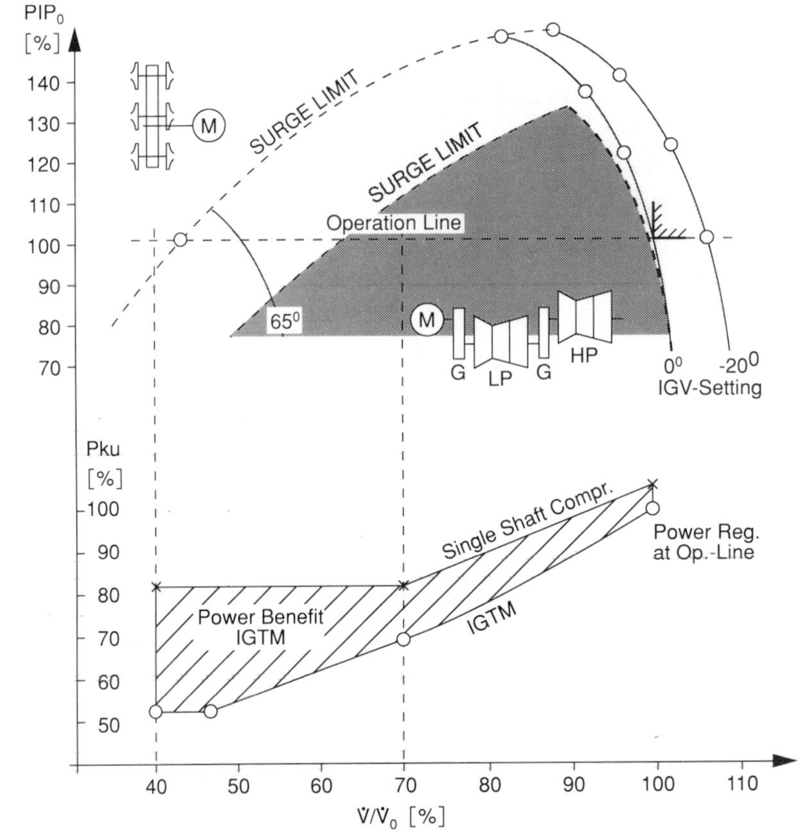

FIG. 4. - Performance comparison between integral gear and single shaft compressors.

FIG. 5. - Compressor clean room assembly.

Engineering of turboexpander-compressor units for offshore installation

R R AGAHI and B ERSHAGHI
Rotoflow Corporation, Gardena, California, USA

SYNOPSIS Turboexpander design for off-shore applications requires special engineering and design features. Past experience on off-shore installations has extremely improved the turboexpander reliability and productivity.

1. Introduction

Turboexpander application in natural gas plants began in the early 1960's [1]. It took almost another twenty years before off-shore platform process designers used turboexpanders in their designs.

The first application, to our knowledge, was in 1976 for a dense phase application [2]. On the Piper-Alpha platform, a turboexpander was used to cool down natural gas to -30°C (-22°F) and dropout 46% of the total mass flow into liquid at the discharge. Since then, there have been several successful off-shore turboexpander installations, such as in Qatar and Norway. Additionally, others such as the Valhall platform in The Netherlands.

Structural design of any off-shore platforms impose certain contraints on the turboexpander package. The turboexpander package requires special pipe nozzle loading and structural integrity to handle platform movements. Off-shore turboexpanders are usually handling untreated corrosive process gas at a very high pressure. Suitable material selection guideline and sealing methodologies have been developed to cope with the latter off-shore demand.
Over the last thirty years, turboexpander manufacturers have improved the core unit design to be more efficient and reliable. Auxiliary system design has also improved to be more being more compact, lighter, self contained, and reliable.

2. Structural Support

Turbo machinery installed on off-shore platforms are exposed to much higher piping forces due to space limitations. Design and fabrication of structural steels needs special attention to eliminate sharp-edged areas and welded plates. Turboexpander package is a compact design to minimize the weight and fit into a small a space. Design of turboexpander support structure is analyzed under different loading conditions such as: gravity loading, lateral loading and maximum allowable forces and movements from the piping. Normally, piping loads are the dominant load. Table 1 shows a typical piping load requirement for an off-shore turboexpander installation. Vibration analysis of the system is also required in order to assure that the system frequency does not match the operating speed of the turboexpander. Table 2 shows frequencies versus running speed calculated for the same installation.

3. Material Selection

Untreated natural gas usually contains a high concentration of hydrogen sulfide. If the partial pressure of H_2S is greater than 3.4×10^{-3} bar (0.05 psi), then the material selection falls within the National Association of Corrosion Engineering (NACE) specification [10]. In this case, materials which resist sulfide stress cracking must be used. In the cases where natural gas contains CO_2, corrosion is even more severe than H_2S corrosion [3]. CO_2 corrosion should be considered when the partial pressure of CO_2 to H_2S is greater than 100 times.

Furthermore, platform environments are also exposed to sulfur, salt, condensation, dust, and pollutants. Sea salt carried by the wind increases the corrosion rate as a function of temperature and solar radiation. Condensation in the presence of oxygen and bacteria will result in pitting corrosion. Cyclic stresses from environmental factors such as waves, tides and operating loads causes fatigue and effects the corrosion rate of welded joints. Corrosion rates on structural steel are normally in the range of 0.05 to 0.64 mm (.2-25 mills)/year. Atmospheric corrosion is minimized by using special coatings (e.g. cadmium coated bolts). Use of dissimilar metallic material should be avoided. Dissimilar metal can create galvanic effect - corrosion problems.

Turboexpander casing and labyrinth seal housing normally cast from stainless stèel. Internal part materials are also chosen to be compatible with the process gas and the functional usage. Auxiliary system piping, fittings and components are usually made from 316L stainless steel. If a given component is not available in 316L material suitable coating, plating or in some cases purge environment is used.

4. Turboexpander Design

Natural gas extracted from deep sea reservoirs is a rich gas. Rich gas is normally processed off-shore before being exported through a submerged pipe line to the on-shore facilities. Processing of the gas involves dehydration and removal of heavy hydrocarbon condensates by Turboexpander.

The turboexpander duty is to recover NGL from the gas and an integral compressor recovers turboexpander energy before the residue gas compressor supplies lean gas with the required dewpoint at normal pipeline pressure of 75 BAR (1100 psi) is about 5°C (40°F). Turboexpander is used in dual function; primarily the most efficient method of supplying refrigeration, and also producing mechanical work to drive the intergral compressor. A typical expander compressor process P & I Diagram is shown in Figure 1. Using expander is proven to be an optimum means of reducing process power consumption, process cooling duties, and additional equipment. A typical turbo expander-compressor cross section is shown in Figure 2.

4.1 STAGNANT DRAINER OPERATIONS

Application of turboexpanders in off-shore installations usually encounters the problem of seal gas availability. The dry-face seal system may seem a promising alternative in place of the conventional labyrinth seals. The former does not require warm, compatible and high pressure seal gas as it is the case with the latter.

Present dry-face seal installation lists do not show any off-shore installations. High differential pressure, sour gas, and hydrocarbon liquid are the three major problems with dry-face seal. The existing dry-face seal design has operational experience of up to 75 Bar (1000 psi). Sour gas (corrosive fluid) damages the sealing materials. This may be prevented by injection of nitrogen that is either not available or not economical on off-shore platform. Hydrocarbon liquid is worse than corrosive fluid because it damages the sealing carbon surfaces and causes permanent damage to the sealing materials [5].

The conventional labyrinth seal design does not have the shortcomings of the dry-face seal. The only problem, however, is that it requires warm compatible seal gas that is not readily available. Although it is feasible to overcome the problem by taking the well head process gas and treat it, warm it, regulate it, and use it for seal gas, but the cost is prohibitive.

Stagnant drainer operations concept has been developed to solve the sealing problem in the off-shore installation of turboexpander. In this design, oil drainer pot, bearing cavity, labyrinth seal cavity, expander wheel back pressure, and the interconnecting piping are all equalized to the same pressure level. Therefore, theoretically, diluting process gas flow is eliminated and lubricating oil is not diluted beyond the acceptable limit.

Process gas containing heavy hydrocarbons, when mixed with lubricating oil reduces viscosity. In any turbomachinery, bearing stiffness is sensitive to lubricating oil viscosity. If the oil viscosity drops beyond the allowable limit, turbomachinery may experience numerous problems including unacceptable and harmful vibration levels.

Using flash calculation method [5], lubricating oil dilution level up to saturation with hydrocarbon is predicted. Initial charge of lubricating oil will have a higher viscosity. The auxiliary oil system also operates with heavier oil viscosity a few hours at start-up. It is not easy to predict the transition period from original viscosity to normal operating viscosity, because it is a function of gas and oil temperature, process gas pressure, process gas pulsation in the piping, and the turboexpander load. Recently, a continuous viscosity monitoring system has been designed [8] to provide viscosity information while the unit is in operation. This system may be equipped with an alarm or shutdown annunciation.

4.2 EMERGENCY OIL SUPPLY SYSTEM

In turbomachinery, "coastdown" is defined as the time interval between shutting-off the source of power to the time when the rotor comes to almost a stop, slow roll, i.e., a few hundred rpm. Coastdown time varies with the size and characteristics of the machine.

Usually it is 10-25 seconds.

In oil lubricated bearings, particularly with high speed machinery, bearing protection during coastdown time is a major concern. Bladder type accumulator has traditionally been used as the reserve oil volume during coastdown.

Off-shore turboexpanders, as was discussed earlier, are handling very high pressure process gas of 150 bar (2,200 psi) or more. Using the conventional labyrinth seal design, high process gas pressure and start-up thrust forces require high pressure lubricating oil supply, usually 100 bar (1500 psi) or more. Supplying high oil pressure against the high bearing cavity pressure of 80 bar (1200 psi) and more, requires in-line accumulator bladder(s) to be precharged to high pressure of 65 bar (1000 psi). High bladder pressure and high lubricating oil supply pressure results in a very low usable volume in the accumulator(s). Approximately 10-15% of the nominal capacity [8]. To compensate for the low usable volume per accumulator, several accumulators are needed. Installation of several accumulators, if feasible, increases the cost and adds to turboexpander package weight and neither is appreciated. Additionally, routine check of accumulator(s) to ensure that charge pressure has not dropped is not needed any longer. Figure 4 shows emergency oil supply system diagram.

Emergency oil supply system [9] resolves the above dilemma.

The conventional accumulator is used in this design. The bladder, however, is not precharged. Instead it is vented to atmosphere or any other low pressure source. Therefore, in normal operation, almost the full volume of the accumulator is filled with oil and, as will be explained, total volume will be used to lubricate the bearings during coastdown.

A high pressure source, turboexpander inlet process gas, high pressure nitrogen bottles, etc. is connected to the bladder via an on-off valve. The on-off valve is recommended to be a tight-shut-off type, ANSI class IV leakage rate. Upon the shutdown due to oil pressure loss (lubricating oil pump failure, lubricating oil pump driver failure, power failure, etc.), the on-off valve opens and high pressure gas flows into the bladder. Inflated bladder pushes lubricating oil out of the accumulator and thus supplying oil to the bearing during coastdown.

Control sophistication of the emergency oil supply system depends on the overall off-shore platform control philosophy. It may be equipped with pressure sensors, diaphragm valves with solenoid valve operator, etc.

4.3 AUTOMATIC THRUST BALANCING SYSTEM

The process gas fluctuation in day to day operations and variation over time have always concerned the end users and the turbomachinery manufacturers. The former are of concern because of the direct impact on hydrocarbon recovery and overall plant operations. The latter are of concern because of the design parameters, optimization in general, and axial forces in particular. Integral turboexpander-compressor units are sensitive to the process gas pressure fluctuation and changes in the expander discharge and compressor suction pressures. Such variation changes the axial forces and may exceed the thrust bearing capabilities.

The axial position detector

probes and subsequent annunciation of the excess axial forces has been the conventional approach. Provision of such detection method is extremely difficult. The conventional method lacks any immediate corrective action and usually either shuts down the plant and/or bypasses the turbomachinery. Neither action is desirable. The former involves loss of liquid recovery and stopping pipeline gas flow to on shore. The latte, on the other hand, results in a reduction of liquid recovery and may feed the pipeline with natural gas containing heavy hydrocarbons.

In off-shore platform operation, due to the limited length of process gas piping (from well head to the processing unit) and compact design, process gas pressure fluctuation is more pronounced and the fluctuation is readily sensed by the equipment. turboexpander therefore, is more exposed to the axial force fluctuations. In most off-shore platform operations, when the turboexpander unit shuts down, heavy natural gas is routed to the flare to burn. This characteristic of off-shore operation makes axial force control and prevention of turboexpander shutdown extremely critical.

Turboexpanders equipped with an automatic thrust balancing system have successfully been installed in off-shore operations. The use of an automatic thrust balancing system eliminates the need for an axial probe detector. Furthermore, this system continuously monitors and balances the axial forces.

The automatic thrust balancing system concept is simple and effective. Static pressure of a cavity that considerably effects the axial forces balance is connected to a lower pressure source through a linear characteristic control valve. Control valve opening is controlled by the signal sensing the difference in hydrodynamic pressure of the thrust bearings. Figure 5 shows the schematic diagram of the automatic thrust balance control.

4.4 Dust-Free Design

Process gas flowing through the turboexpander should be relatively clean. Normally it is recommended that solid particle size not exceed 5-10 (.2-.4 mills) microns, with a maximum concentration of 40 mg/m^3 (2.5x10^{-6} lb/ft^3). This is required, not only to prevent erosion, but also because higher concentrations have a measurable effect on expander performance. Large particles cause disturbances in flow which may result in a 2-3% drop in thermal efficiency. High gas velocities carrying solid particles through the inlet guide vanes will cause erosion and damage to seals and impeller blade tips. Figure 6 shows a design to harmlessly discharge dust particles to the expander discharge and prevents accumulation of dust particles behind the impeller. Also, erosion of labyrinth seals may result in an efficiency loss of 2-3% and cause thrust unbalance which may result in serious damage to the turboexpander if it goes undetected.

5. Conclusion

Past installations have shown that the turboexpander is a reliable piece of equipment. Initial commissioning may require some trouble shooting, mostly process related. Turboexpander design for off-shore applications requires additional steps due to the nature of the gas and process conditions. Several design features have been implemented into the existing conventional design for safety reasons and for ease of operation required in the off-shore environment.

References:

1. Beverly K. Stokes Expander Plants, Exxon Production Research Company, July 1981.

2. J. Burnwell and W. Wong 5th Off-Shore Southeast Asia Conference. Singapore, February 1984.

3. "Metal Handbook", ASM 9th Edition, Volume 13.

4. Ulrich Grundmann and Myron Sidersky, Experience with Gas Lubricated (Dry face) seal, Romag 91, Washington, D.C. July 1991

5. "Engineering Data Book", Natural Gas Processors Tulsa, Oklahoma 1987

6. J.S. Swearingen U.S. Patent No. 4,442,704

7. Zeke Zahid, Accumulators and Their Applications, Greer Hydraulics 1976

8. J.S. Swearingen U.S. Patent No. 4,390,082

9. J.S. Swearingen U.S. Patent No. 4,152,092

10. NACE Standard MR-01-7S, National Association of Corrosion Engineers, 1984

Table 1
Maximum Allowable Forces and Moments
at Inlet-Outlet Connections

Nozzle Size	Flange* Rating	Forces (N)	Moments (Nm)
8"	900	F_x = 12,100 F_y = 30,150 F_z = 24,120	M_x = 18,384 M_y = 10,455 M_z = 10,455
10"	900	F_x = 17,795 F_y = 31,687 F_z = 25,350	M_x = 25,755 M_y = 18,210 M_z = 18, 210

* For flange rating refer to Table 3.

Table 2
Natural Frequencies

Mode (Cycle/Sec.)	Free Condition Frequency (Cycle/Sec.)	Fixed Condition Frequency
1	21.5	21.8
2	21.5	21.9
3	40.1	43.6
4	43.6	43.6
5	43.6	75.2
6	58.1	75.6
7	60.4	174.4
8	75.3	176.0
9	76.6	179.0
10	153.7	196.9
11	176.0	197.2
12	187.3	237.5
13	196.8	239.0
14	198.0	239.5
15	230.3	392.2
16	236.5	334.4
17	239.1	401.7
18	242.8	436.0
19	314.3	449.2
20	336.2	454.0

Turboexpander manufacturer does not normally have any knowledge of structural properties of the piping system on the off-shore platform. As a result, two sets of nodal analysis are usually studied. Column one represents free condition for nodal points and column two represents fixed condition for nodal points.

The actual frequencies of the structure under operating conditions will be somewhere between the above two limits.

TABLE 3 CLASS 900 PRESSURE - TEMPERATURE RATINGS

Pressures are in pounds per square inch, gage (psig)

Mat'l Group	1.1	1.2	1.3	1.4	1.5	1.7	1.9	1.10	1.13	1.14
Materials / Temp. °F	Carbon				C ½Mo	½Cr-½Mo Ni-Cr-Mo	1 1¼Cr-½Mo	2¼Cr-1Mo	5Cr-½Mo	9Cr-1Mo
-20 to 100	2220	2250	2085	1850	2085	2250	2250	2250	2250	
200	2025	2250	1970	1685	2035	2250	2135	2150	2250	
300	1970	2185	1915	1640	1955	2185	2020	2030	2185	
400	1900	2115	1850	1585	1920	2115	1975	1945	2115	
500	1795	1995	1745	1495	1865	1995	1925	1920	1995	
600	1640	1815	1600	1370			1815			
650	1610	1765	1570	1345			1765			
700	1600	1705	1555	1345			1705			
750	1510	1510	1420	1325			1595			
800	1235	1235	1175	1110			1525		1490	1525
850		805					1460		1315	1460
900		515					1350		1060	1350
950		310			845	1030	1130		780	1110
1000		155			495	640	670	805	575	875
1050						565	410	595	420	565
1100							290	340	310	340
1150							155	310	205	225
1200							105	165	135	155
1250										
1300										
1350										
1400										
1450										
1500										

2.1	2.2	2.3	2.4	2.5	2.6	2.7	3.1	3.2	3.4	3.5	3.6	3.7	3.8	
Type 304	Type 316	Type 304L / Type 316L	Type 321	Types 347 348	Type 309	Type 310	Cr Fe Mo Cu Cb 20Cb	Nickel Alloy 200	Ni Cu Alloys 400 405	Ni Cr Fe Alloy 600	Ni Fe Cr Alloy 800	Ni Mo Alloy B2	Nickel Alloys	Temperature °F
2160	2160	1800	2160	2160	2015		1800	1080	1800	2160	2160	2250		100
1800	1860	1520	1830	1910	1815		1670	1080	1585	2015	1990	2250		200
1585	1680	1360	1635	1765	1705		1570	1080	1485	1915	1870	2185		300
1410	1540	1240	1485	1665	1600			1080	1435	1845	1800	2115		400
1310	1435	1145	1375	1555	1510			1080	1435	1780	1735	1995		500
1245	1355	1080	1310	1475	1435			1080	1435	1720	1720	1815		600
1225	1330	1050	1280	1440	1395				1435	1690	1705	1765		650
1210	1295	1030	1260	1405	1370				1435	1670	1690	1705		700
1195	1270	1010	1245	1385	1340				1405		1595			750
1180	1245	985	1240	1370	1305				1375		1520			800
1165	1215	965	1225	1330	1275									850
1150	1180		1215	1295	1245									900
1125	1160		1160	1160	1160									950
965	1090		1070	1090	1010	1050								1000
925	1080		1040	1080	875	1000								1050
770	965		905	965	670	875								1100
585	825		710	825	515	740								1150
465	620		545	515	390	620								1200
330	545		420	370	300	485								1250
245	410		320	280	235	360								1300
185	310		245	205	175	235								1350
145	225		185	155	135	165								1400
105	175		145	125	95	115								1450
70	125		115	105	70	70								1500

Reference ANSI Standard B16.5, 1981

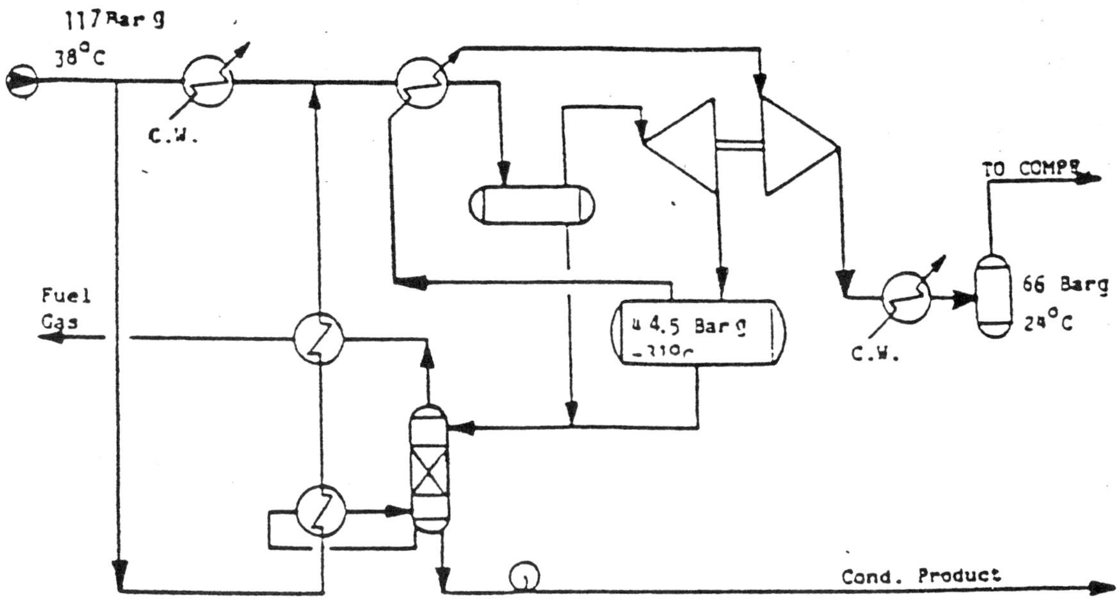

Fig. 1 - Typical expander compressor process diagram

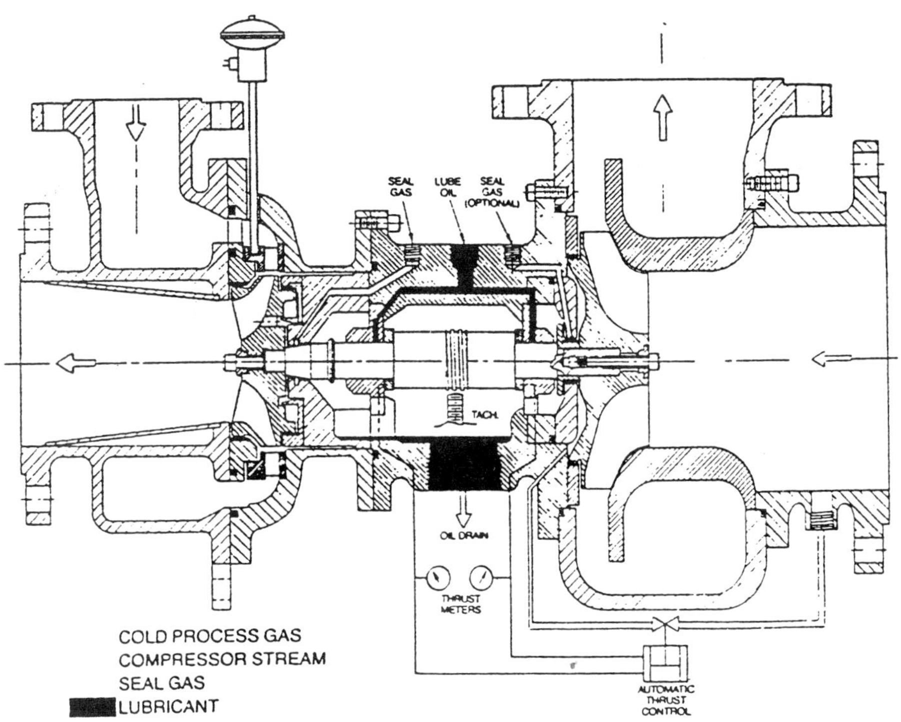

Fig. 2 - Expander-compressor cross section

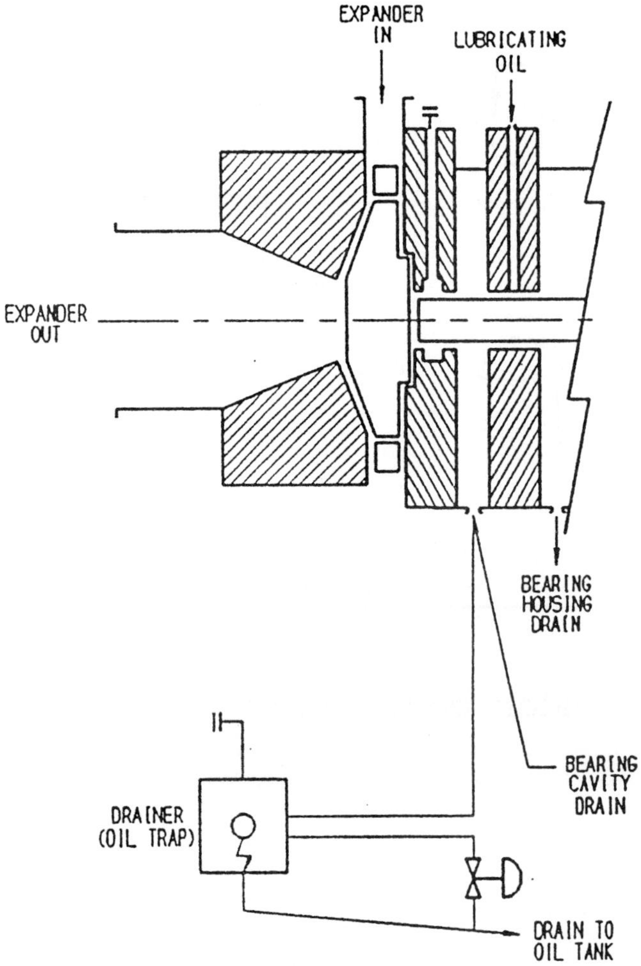

Fig. 3 - Stagnant drainer system

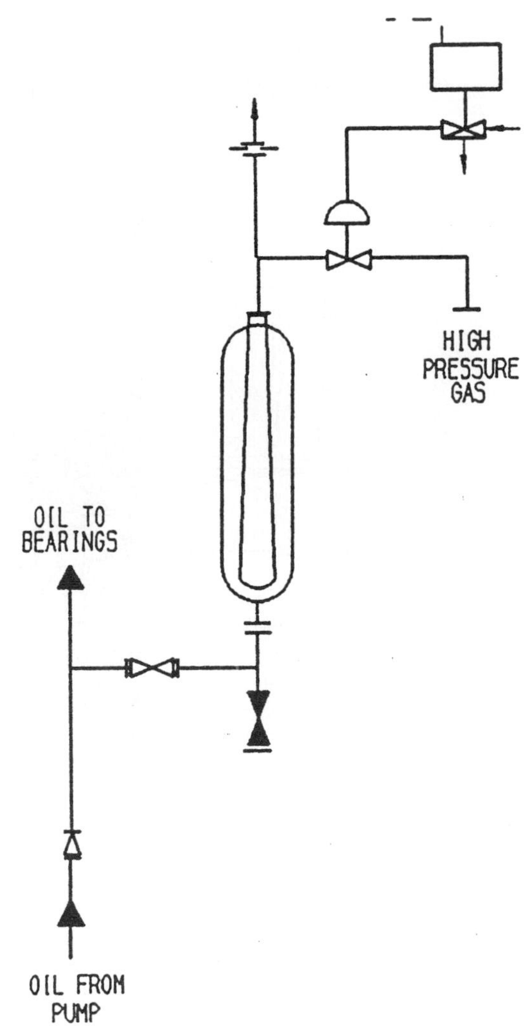

Fig. 4 - Emergency oil supply system

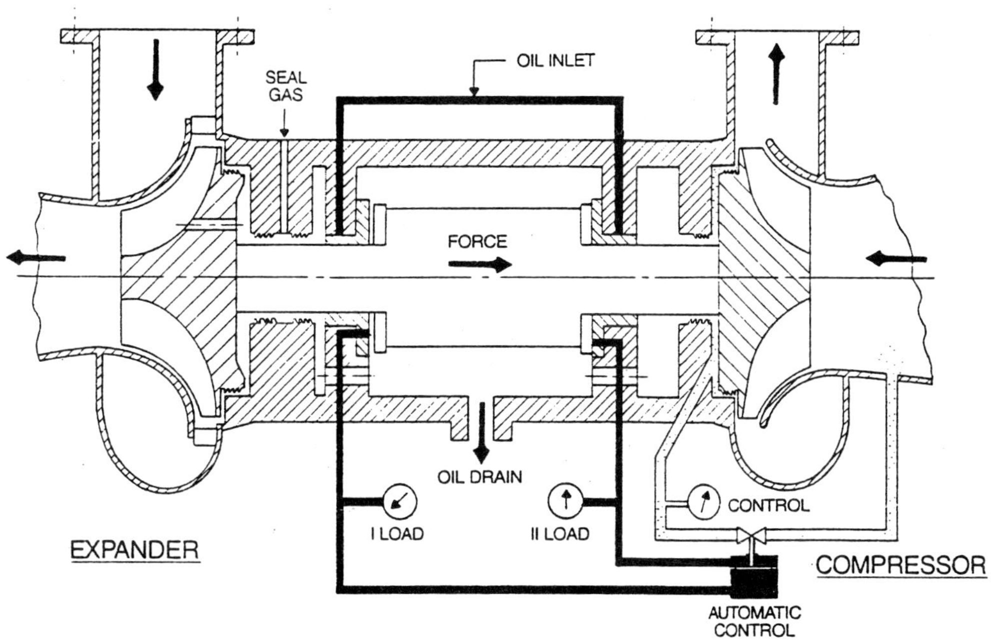

Fig. 5 - Automatic thrust control

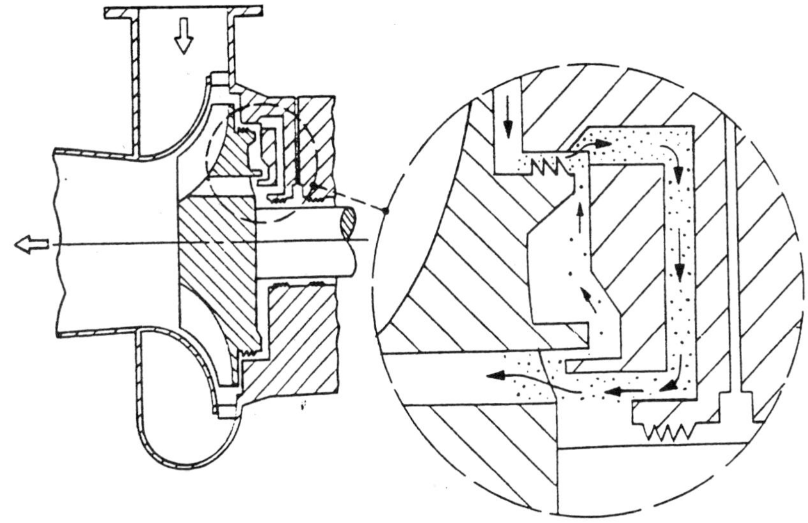

Fig. 6 - Dust free seal

C449/008/93

The centrifugal pump as a source of pulsations in pipe systems – a comparison of measurements and simulation results

E Van BOKHORST and H KORST, Ing
TNO Institute of applied Physics, Delft, The Netherlands

SYNOPSIS A variety of fluid machinery generate pulsations, which can cause vibrational problems in pipe systems. High pulsation and vibration levels can have a negative effect on the performance of the machine, cause malfunctioning of instrumentation, may result in fatigue failure and will therefore on the long run influence the lifetime of the installation as a whole.
This paper describes the results of measurements and simulation of the pulsating flow caused by centrifugal boiler feed pumps as a case study. The measurements have been carried out in a test loop with a one-stage and a seven-stage centrifugal pump.
The results on the one-stage booster pump show high amplification factors for the pulsations in the test-loop at the vane-passing frequency. An acoustical model of both pumps is build up with the digital simulation program PULSIM. The simulation results confirm the model used to predict the pulsations generated by an impeller at the design condition.
The study proves that a simulation of the centrifugal pump and the connected piping is an effective means to prevent resonances in the piping and to locate pipe sections on which considerable pulsation induced vibration forces are acting.

1. INTRODUCTION

Pulsations or pressure-waves are flow-dynamic phenomena, which may cause vibration-related problems in fluid machinery and connected pipe systems. Pulsations and pulsation induced vibration forces can cause fatigue failure in piping, may have a negative effect on the efficiency of running equipment, disturb flow-meter readings and can be a source of noise.
Pulsations are well-known in positive displacement compressors and pumps. Nowadays a pulsation study is an important step in the design of each pipe system using reciprocating machinery [ref.1]. Rotary-type machinery, such as centrifugal compressors and pumps are not well known source of pulsations. Though it has been demonstrated in practice that in pipe systems with large rotary-type machinery also serious pulsation problems may occur. With this project the possibilities of the PULSIM program to tackle pulsation problems in centrifugal machinery and connected pipe systems has been demonstrated.

Pulsations in pipe systems and fluid machinery has been a research topic at the TNO Institute of Applied Physics (TPD) since the late sixties [ref.2].
A powerful tool, the digital simulation program **PULSIM (PULsation SIMulation)** is used on a day to day basis for the studies of pulsations in pipe systems.
The PULSIM program has been extended with a model for the pulsations generated by centrifugal pumps, so that it is possible to predict pulsations in systems similar to the boiler feed system discussed in this paper.
The program is based on the method of characteristics [ref.3], which is a very suitable method for the solution of one-dimensional flow-equations on a digital computer. This method results in a minimum number of computations and is very efficient in the simulation of complicated and extended pipe systems with a number of excitation sources.

In this paper the results of pressure pulsation measurements of a cold-loop test with a booster pump and a hot-loop test with a high speed main boiler feed pump are described. The pumps are part of a series of three sets of pumps to be located at the EPZ AMER 9 power station at Geertruidenberg (The Netherlands). A simulation of the test-loops has been performed to compare the results of simulation and actual measurements.
An overview of the results of the pulsation study for the boiler feed system of the AMER 9 power station is given. The objective of this study, carried out in the design stage, is to calculate pulsation levels and pulsation induced vibration forces

caused by the centrifugal pumps in the system. As there are no formal criteria for allowable pulsation levels caused by centrifugal pumps, like in API 618 for reciprocating compressors, pulsation levels should be judged on the vibration forces which are acting on the pipe system. It is recommended to prevent acoustical resonances, which can result in high amplification factors. Though it should be noticed that even low forces can cause unallowable vibration levels and cyclic stresses if acoustical and mechanical resonances coincide.
The dynamic behaviour of the piping under high-frequency forces should be examined further.

2. DESCRIPTION OF MEASUREMENTS IN THE TEST LOOP

The investigated BW/IP pumps have the following design data:

Booster-pump 12 x 6 x 23L DSJH				
Speed [RPM]	Shaftpower [kW]	Head [m]	Stages	Vanes
1560	438	100	1	4
Boiler feed pump 12 x 12 x 14 HDB				
Speed [RPM]	Shaftpower [kW]	Head [m]	Stages	Vanes
5760	18000	4230	7	3 - 5

Cold-loop test

The measurements are intended to determine the acoustic characteristics of the pumps, such as source strength of pulsations and acoustic impedance of the pumps.
The optimum condition for the determination of the source strength of the pump is a reflection free pipe system, which could not be obtained as possibilities for modification of the test-loop were very limited. The isometric of the cold-loop in fig. 1 shows the position of the pump and the pressure-control valve (PCV) in the loop. The double-volute one-stage booster-pump has four blades on the impeller.

The measurements show high pulsation levels for the 1-stage booster-pump up to 25% peak-to-peak of the static pressure, due to a standing wave to the PCV-valve in the system. An almost purely sinusoidal pulsation occurs at the vane passing frequency of approximately 100 Hz, which is the 4th harmonic at a pump test speed of 1500 RPM.
The frequency spectra for a location at suction side P1 and at discharge side P3 are shown in fig. 3. The 2nd and 3rd harmonic of the vane-frequency are clearly visible, though much lower than the 1st harmonic. The contribution at the impeller frequency (25 Hz) is negligible. The measurements, carried out at five different operating points, are plotted in fig. 4 and show a maximum level at location P5 at the highest flow in the range measured (approximately 1400 m^3/hr). The point of maximum efficiency during the test is at approx. 1050 m^3/hr, which shows a considerable lower pulsation level than at the maximum flow.
The pulsation levels at lower flows are not investigated, though it is observed that pulsation levels increase gradually at the lower flows, during the run-out of the pump pulsation levels up to 4 bar pp are noticed.

Hot-loop test

The main boiler feed pump has seven stages with three blades on the 1st and five on the 2nd up to the 7th stage.
An isometric lay-out of the test loop for the boiler-feed pump is shown in fig. 5, indicating the position of pump, PCV-valve and measuring locations for pressure (P1 up to P5) on suction- and discharge side of the pump. Measuring locations on the hot-loop were in this case limited because of the high-pressure piping involved. Only three suitable connections are available for the connection with a pressure-transducer, two on suction and one on discharge side.
The measurements performed on the 7-stage high pressure boiler feed pump at elevated temperature show pulsations up to 2% pp in a frequency range of 0-1000 Hz. The contribution at the vane frequencies of 210 and 350 Hz is low, as shown in the spectra in fig. 6.

The measurements carried out on three locations at eight different operating points on the Q-H curve, plotted in fig. 7 show an increasing pulsation level at a decreasing flow.

3. SIMULATION OF TEST LOOPS

The test loop and pumps are simulated with the PULSIM program, the internal geometry of the pump is modelled from the detailed drawings. The program uses a one-dimensional model of wave propagation, which implies that the simulation is only valid for sources which generate waves at a wave-length, which is large compared to the inner diameter of the piping. Or in other words the limiting frequency for the one-dimensional model is:

$$f < 0.586\ c\ /D \text{ [ref 4].}$$

We assume that at normal operating conditions cavitation does not occur in the pump. The main sources of hydraulic noise or pressure pulsations are hydraulic unbalance, at shaft frequency and its harmonics, and vane frequency pulsations due to interaction between impeller blades and the volute.

Unbalance of the rotor can result in shaft vibrations and thus eccentric rotation of the impeller in the volute channel or whirl. The dynamic forces acting on the impeller vanes cause pressure pulsations in the medium. In literature we have not found a formula for the source strength at the impeller frequency. If the pump is well designed unbalance forces will be minimized, and therefore the pressure pulsation at the impeller frequency will be small.

If this is the case fluctuating hydrodynamic forces on impeller blades and the pump volute will dominate. The interaction between the wake of the vanes and the volute results in pressure pulsations at the vane frequency.

This excitation occurring in a volute pump can be described by an empirical correlation formula, which can be used for the determination of the amplitude of the fluctuating pressure at the vane passing frequency at the design point [ref.5,6]:

$$Lp = 3.5 + 35lg\,(dp) + 10lg(Q) - 40lg(N) - 20lg(a^2t)$$

in which:

Lp = excitation amplitude at vane frequency [dB RMS]

dp = pressure difference across the impeller [Pa]

Q = flow [m3/s]

N = rotational speed [Hz]

a = outer diameter of the impeller [m]

t = width of the vane at outer diameter [m]

This formula is used in our simulation for the determination of the source strength of the pump at the vane frequency.

The interaction between acoustical and mechanical behaviour of the liquid/piping systems is not included in this study. It is assumed that the magnitude of the pulsation forces is such that the vibrations of the piping have no considerable effect on the dynamic behaviour of the fluid. Previous measurements have confirmed this assumption as the vibration levels are below 10 mm/s pp.

A correction in the speed of sound in the fluid, due to the wall flexibility of the piping is included.

4. COMPARISON BETWEEN SIMULATION AND MEASUREMENTS

Booster pump in cold-loop test

Measured and calculated pressure pulsations caused by the booster pump have shown the occurrence of standing waves between the pump and the PCV-valve. This valve, which reduces the discharge pressure (9-14 bar) to the suction pressure (0.55-1 bar) has a reflection coefficient close to 1. The standing wave resonances are occurring at frequencies corresponding to:

f = 0.5.n.c/l

in which:

n = 1, 2, 3, 4, 5..........

c = velocity of sound

l = the distance between suction and discharge side of the valve

The measurements have shown a dominant vane-frequency component at 100 Hz, with also a contribution at the harmonics at 200 and 300 Hz. The maximum pulsation levels have an amplitude of approximately 1.4 bar pp or 15% of the static pressure. The contribution at the impeller frequency (25 Hz) is negligible.

The simulation results of the test-loop show a max. pulsation level of 1.3 bar pp or 14% at 100 Hz. The simulation results plotted in fig. 7 show a standing wave pattern from the suction side of the PCV valve, through the pump, to the discharge side of the valve. Differences may be caused by inaccuracies in the exact location of the source, which has been modelled at the inlet of each volute channel. The pressure pulsations in the simulation cause considerable vibration forces on the 12 inch piping up to approx. 23 kN pp The model of the booster pump is, in a later stage, also used for the simulation of the actual boiler feed pipe system at the AMER 9 power station.

Boiler feed pump in hot-loop test. The hot loop as simulated is shown schematically in fig. 6 with the node numbers for the simulation and the points P1 up to P3 at which the actual pulsation measurements are performed during the test run. At these locations a comparison between the simulated and measured values can be made.

The 7-stage boiler feed pump consists of a num-

ber of different sources, one for each impeller, with the volute channels between the stages modelled according to detailed drawings of the pump.
The simulation model has been investigated at 8 different conditions, representing the load conditions during the test. The simulation of this model also shows the occurrence of standing wave resonances between the pump and pressure control valve. Though the results show that interaction between suction and discharge side of the pump is less significant than for the booster pump.

The excitation vane-frequencies are 3 times pump speed for the 1st stage and 5 times pump speed for the 2nd up to the 7th stage. During the test with a running speed of approx. 4200 rpm these frequencies are respectively 150 Hz and 350 Hz.

Generally the measured and simulated pressure pulsations at the vane frequencies are low, the maximum overall level in the measurements is below 2% pp of the static discharge pressure (201 - 282 bar). So the absolute pressure pulsation of 4 bar pp is in the same order of magnitude as for the booster pump. In the measurements as well as in the simulation the maximum pulsation is found at the lower flows, the minimum is found close to the design point. The simulation and measurement results are plotted for the three measuring locations as a function of the flow in fig. 8.

The calculated pulsation levels are in the same order of magnitude as the measured pulsation levels.

Near the design point the calculated pulsation levels show a good resemblance with the measurements. For off design conditions the measured pulsations are in general higher than predicted by the simulation. The pulsation induced vibration forces achieve maximum levels of approx. 12 kN pp. So pulsation induced vibration forces on the cold-loop piping are higher than for the hot-loop as a result of the higher pulsation levels.
The model of the boiler feed pump is used for the simulation of the boiler feed system in the AMER 9 power station.

5. SIMULATION OF COMPLETE BOILER FEED SYSTEM

The main objective of the study is to prevent vibration problems in the boiler feed piping in a design stage. Therefore the complete piping layout from feed water tank via booster pump and boiler feed pump to the inlet of the boiler is taken into account in the simulation model (total pipe length approx. 1000 m). Acoustical resonances in the piping are localised and pulsation levels and shaking forces have been calculated on all pipe sections. The boiler feed system contains three sets of a booster pump in series with a main boiler feed pump of which two sets are operated in parallel. The pulsations caused by each pump are recorded separately. As there is no phase relation between the pumps the maximum attainable pulsation level is the sum of the pulsation amplitudes caused by each pump separately.
As a result of the excitation by different pumps a beating pulsation will result in which minima and maxima do occur. The maximum in pulsation levels and forces is obtained by adding the levels caused by each pump individually.

The system is investigated at 4 different system flow points on the Q-H curve.

The pressure pulsations, caused by the booster pump in the suction piping, achieve a maximum level of approx. 9% pp at a suction pressure of 10 bar. In the high pressure discharge system the main source of pressure pulsations are the boiler feed pumps causing pulsations at 3 and 5 times pump speed. The maximum levels occur in closed side branches as a result of local standing waves to closed valves. The maximum levels are approx. 1.7% pp of the local discharge pressure. In such cases it is recommended to relocate the valves as close as possible to the main line to prevent building up of standing waves. It should be noticed that the system is investigated in a range with a deviation of approx. 15% around the nominal velocity of sound. This is done to account for variations in temperature, deviations in the lay-out of the piping and inaccuracy of the simulation method. During the simulation therefore the worst case conditions are analyzed which will not necessarily occur in practice. So there is always a chance that the actual resonance conditions are not met.

The calculated pulsation levels in the piping further downstream of the pre-heaters are below 0.1% of the local static pressure. This implies that the simulation up to the inlet of the boiler is sufficient.
Though the excitation amplitude of each impeller decreases at the lower speeds, according to the formula (1) the highest pulsation levels occur at the minimum flow. This is due to the fact, that as a result of the decreasing flow the frictional damping effecting pulsation levels is decreasing.
At the locations where resonances can build up

pulsation induced vibration forces are significant, the need for modification depends on the mechanical properties of piping and support structure. A further evaluation of the forces in a mechanical study is recommended.

The simulation has shown that amplification factors of maximum 18 times the excitation amplitude can occur at minimum flow conditions, whilst the maximum amplification factor at the design point is maximum approx. 6. It is shown that pulsations at resonance conditions can lead to considerable vibration forces on the piping. A further evaluation is necessary to investigate whether these forces can be restrained effectively. Special attention should be paid to small diameter side-branches and instrument connections.

CONCLUSIONS

This study has shown that simulation of piping systems with centrifugal pumps is an effective means to prevent resonances in the piping and to locate pipe sections in which considerable forces are present. The model used for the source strength of an impeller is valid for the design point. For off design conditions the model should be improved to be able to make predictions with the required accuracy.
Nevertheless the predictions resulting from this model are already sufficient to indicate where potential vibration problem areas in the pipe system may occur. Pulsation and vibration measurements at site are planned to check the results of the studies.
The model of pulsating flow generated by centrifugal pumps can be used in a variety of applications in oil, petrochemical and power industries.

ACKNOWLEDGEMENT

The authors wish to thank BW/IP at Etten-Leur and EPZ at Geertruidenberg for their permission to present the data in this paper, besides they are grateful to the TNO organisation for providing the facilities for the preparation of this paper.

REFERENCES

[1] API Standard 618, 3rd edition February 1986 "Reciprocating Compressors for General Refinery Services"

[2] SMEULERS J.P.M. "Simulation of flow dynamics in pipe systems" Proceedings IMechE Seminar London, 1-2 Dec. 1988 p. 97-106

[3] KIRSHNER J.M and KATZ S. "Design theory of fluidic components" chapter 3. 1975 Academic Press ISBN 0-12-410250-6

[4] SKUDSZYK E. "The foundations of acoustics" Springer 1971

[5] SIMPSON H.C. "A theoretical investigation of hydraulic noise in pumps" Journal of sound and vibration (1967) no. 3, p. 456-488

[6] HUBNER G. "Taschenbuch der Technischen Akustik" Springer Verlag 1971.

[7] MOELKER W.H. and WIJDIEKS J. "Pumpnoise and its control; a review" (in Dutch) April 1985 - TPD report 308.464

[8] BOLLETER U. "Blade Passage Tones of Centrifugal Pumps" Vibrations. Vol. 4, No. 3, September 1988

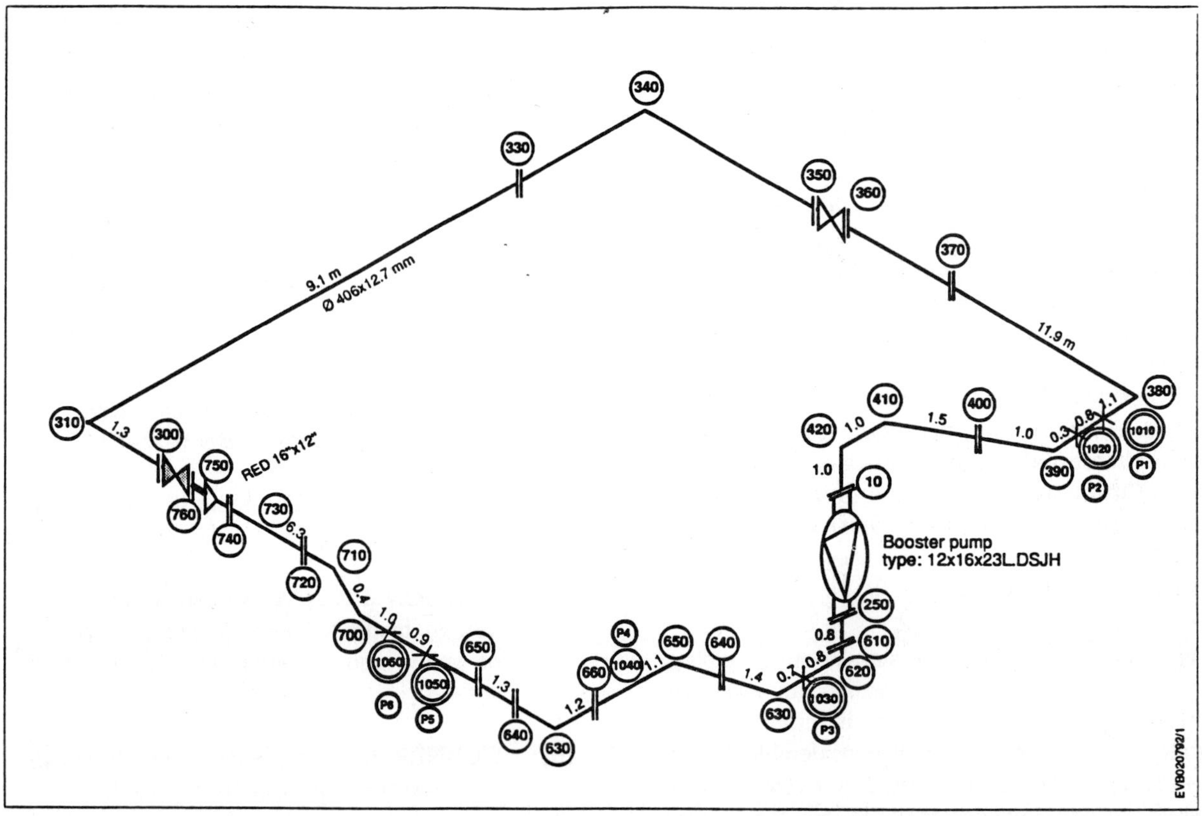

Figure 1 *Isometric sketch of the cold loop test of the booster pump*

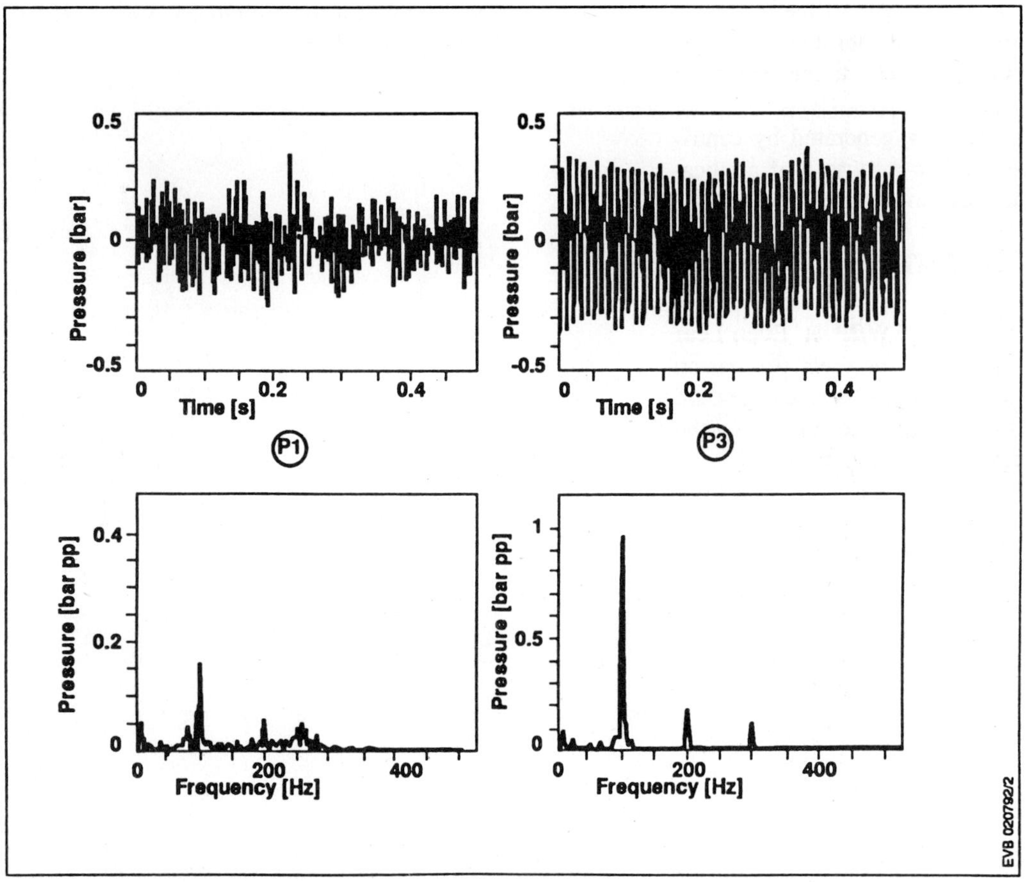

Figure 2 *Measured pressure pulsations versus time and spectra for the booster pump at a flow of 1231 m^3/hr.*

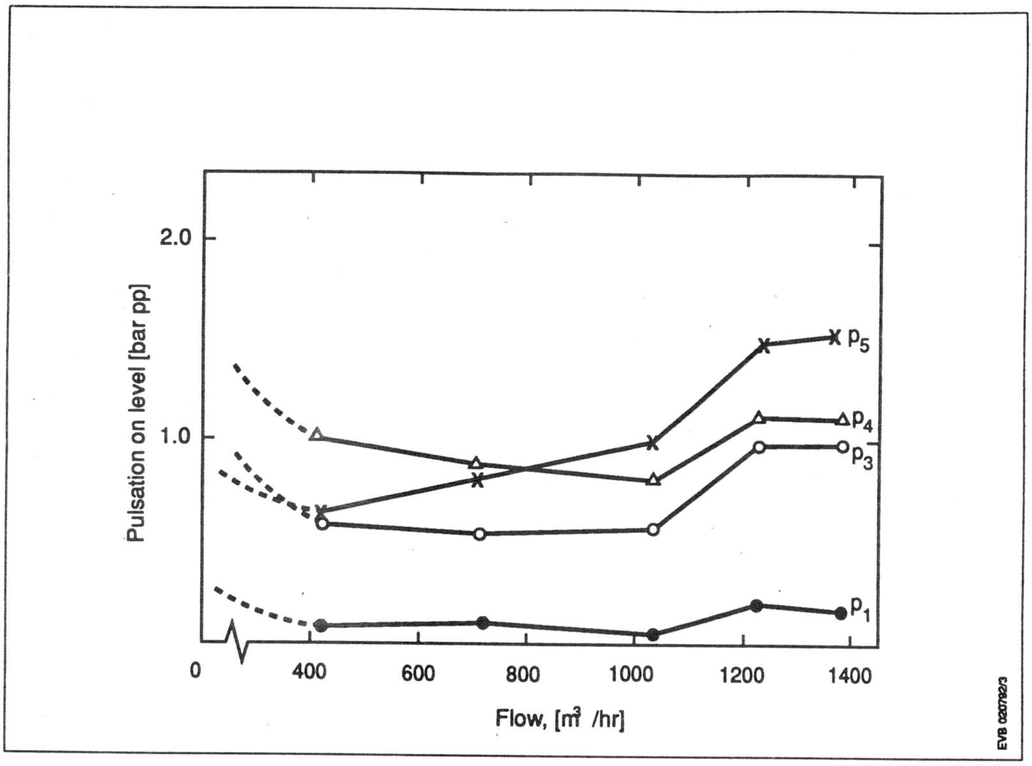

Figure 3 *Measured pressure pulsation levels versus flow of the booster pump (frequency 4N = 100 Hz)*

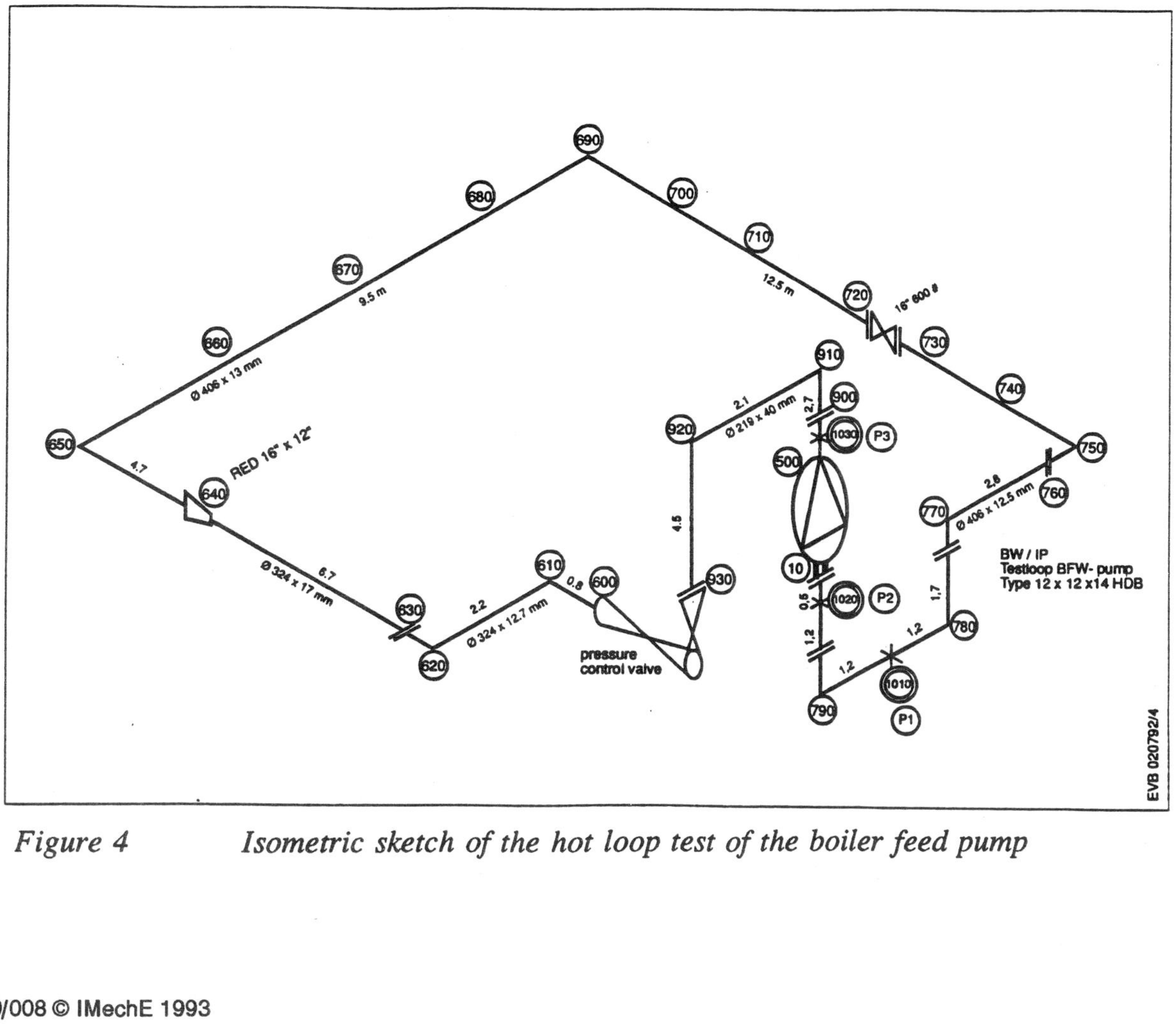

Figure 4 *Isometric sketch of the hot loop test of the boiler feed pump*

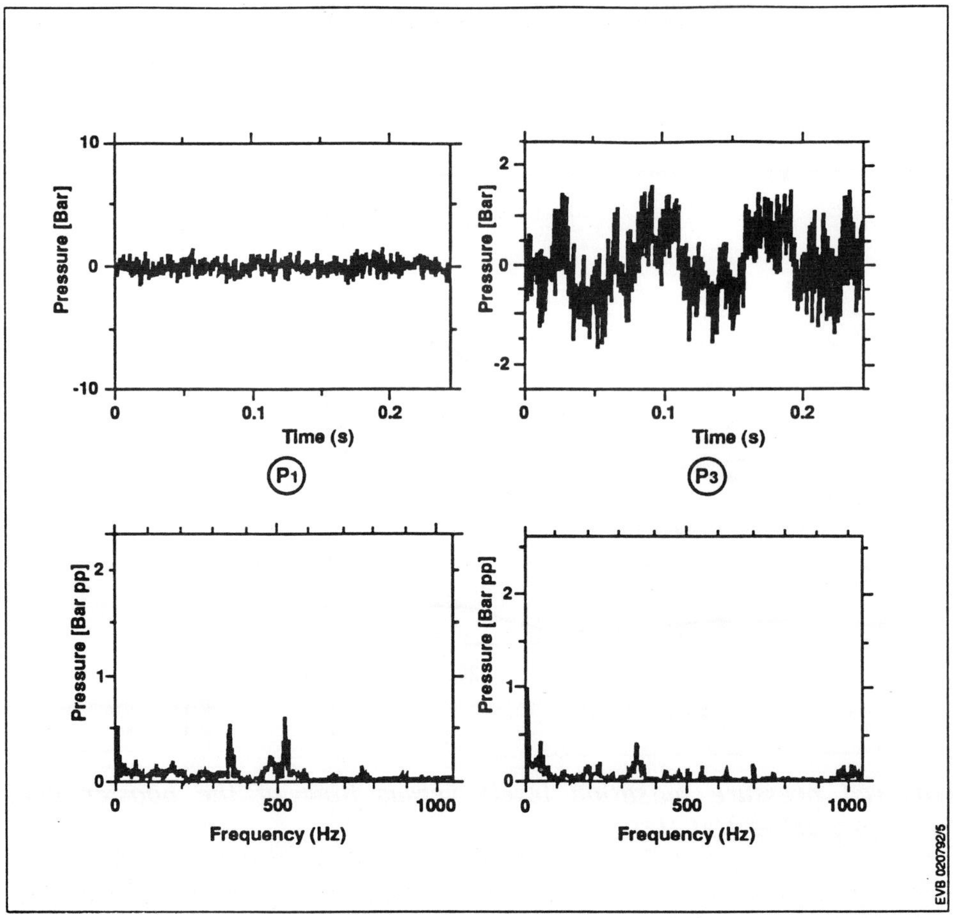

Figure 5 *Measured pressure pulsations versus time and spectra for the BF-pump at a flow of 324 m^3/hr.*

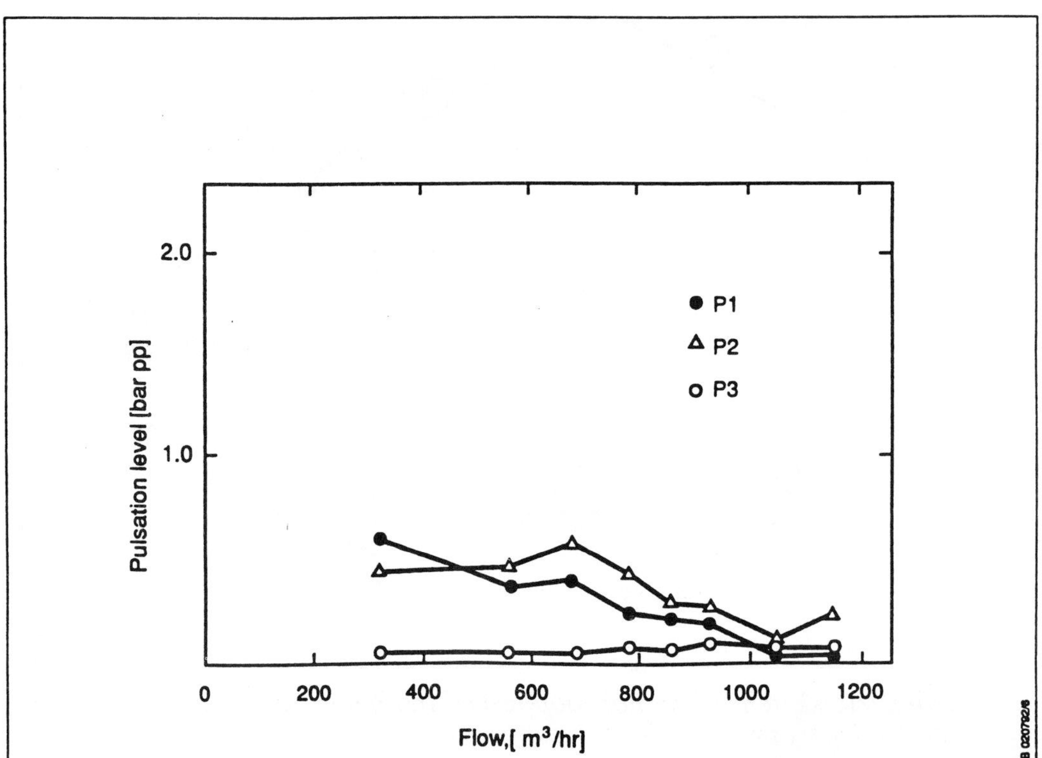

Figure 6 *Measured pressure pulsation levels versus flow of the boiler feed pump*

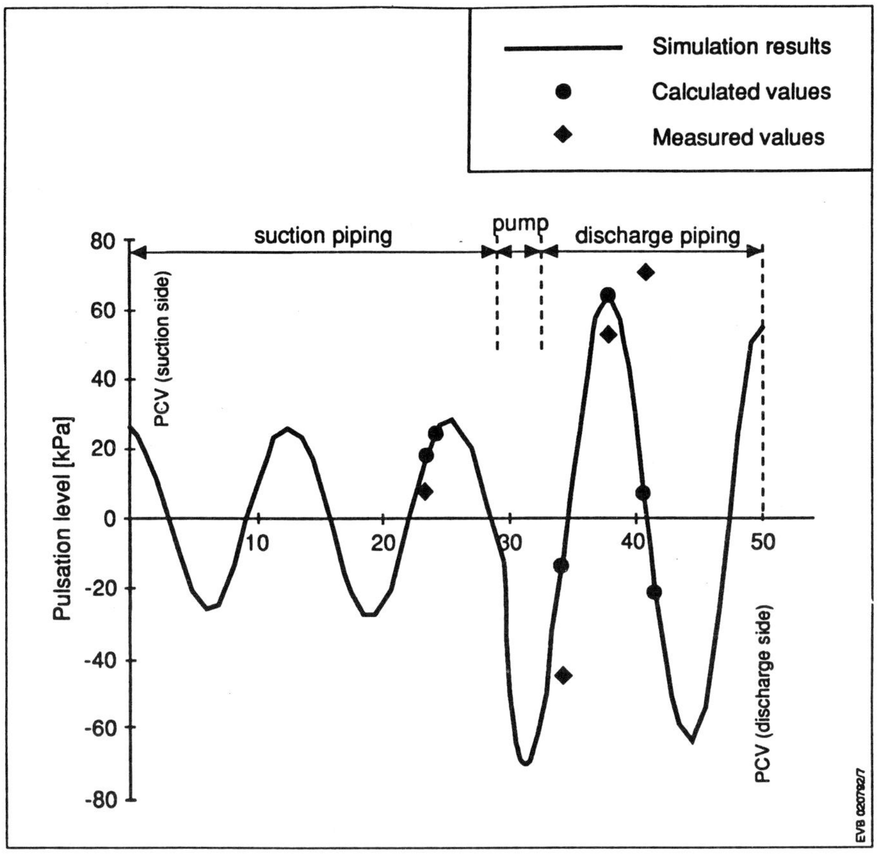

Figure 7 *Simulated standing wave pattern of pressure pulsations in the booster pump test loop and comparison with measurements (flow = 1395 m^3/hr.*

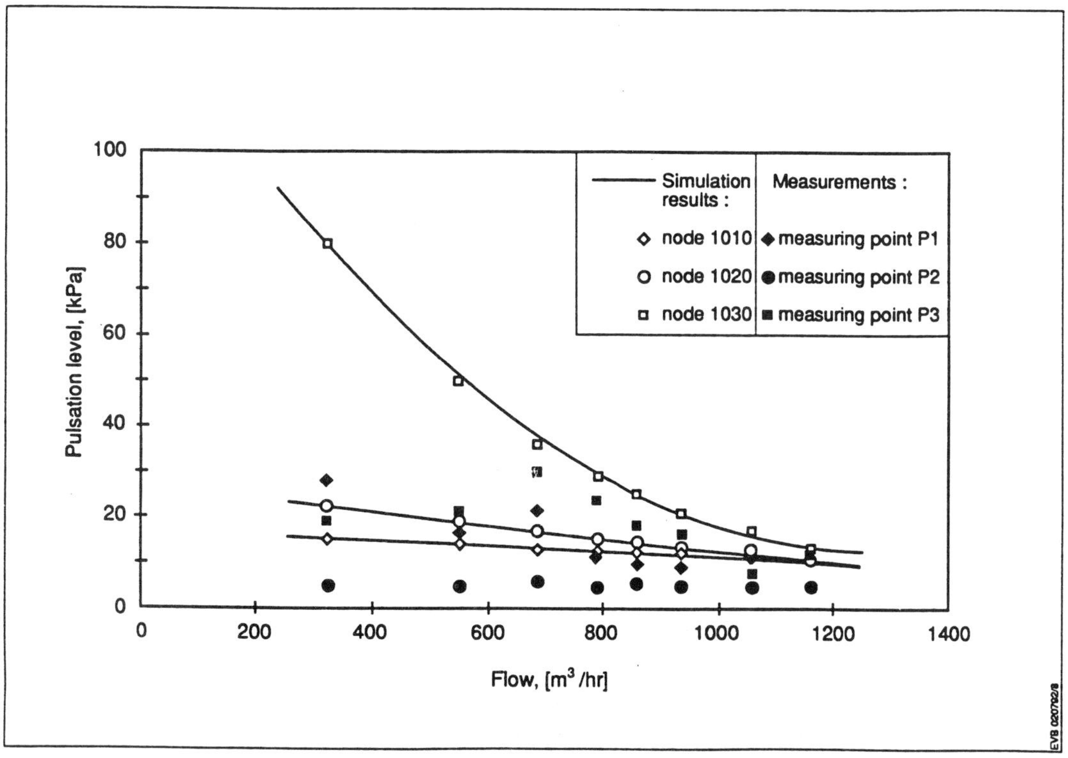

Figure 8 *Comparison between measured and simulated pulsation levels in the boiler feed pump test loop*

C449/024/93

Application of three-dimensional viscous flow calculations in the aerodynamic design of industrial turbocompressor components

P DALBERT, Dr Ing, T KMECL, Dipl Ing and K RUETTI, Dipl Ing
Sulzer Escher Wyss Limited, Zurich, Switzerland

SYNOPSIS The application of the Navier-Stokes code, written by W. N. Dawes, in the daily design and analysis work of radial turbocompressor components for industrial applications is presented. Results of different developments with and without the three-dimensional viscous flow calculations are discussed to demonstrate the benefits and the limitations of such a design procedure.

NOTATION

a_1	speed of sound at inlet
c	absolute velocity
c_P	specific heat at constant pressure
D	diameter
d(S)	entropy change, Equation (8)
m	mass flow
m(rel)	relative meridional length
m_{LS}	mass flow through labyrinth
Mu_2 = Mu(2)	tip speed Mach number = u_2 / a_1^*
n	polytropic exponent
P	input power
p	pressure
P_{RR}	power loss due to disk friction
R	gas constant
T	temperature
u_2	blade tip speed
Vm	= c_m, meridional velocity
α	absolute flow angle
Δh	total enthalpy rise
Φ	flow coefficient
η_R	total-to-static impeller efficiency
μ_0	stage work coefficient (= $P / m / u_2^2$)
μ_R	impeller work coefficient
ρ	density

subscripts

$\ldots_1$	impeller inlet
$\ldots_2$	impeller outlet
$\ldots_u$	circumferential
$\ldots_m$	meridional

Superscripts

$\ldots^*$	referred to total conditions

Note: All angles are indicated with respect to the tangential direction

1 INTRODUCTION

Turbomachinery flows are extremely complex by three-dimensional effects due to the curvature of the flow channels and the high pressure gradients. The flow may be laminar, transitional or turbulent and separations are often encountered. The viscous effects cause regions of high stress. Frequently, regions of incompressible, subsonic, transonic and supersonic flow are simultaneously encountered in a turbomachine. Additionally, the vorticity by secondary, tip leakage and trailing edge flow dominates the event.

In the continuing search for improvements in flow range and efficiency, increasingly sophisticated design procedures and computing methods have to be applied to the development of turbomachines. Apart from one-dimensional velocity triangle evaluation, the simplest application of the discipline of computational fluid dynamics (CFD) on the design of turbomachinery is the solution of the throughflow equations in the meridional plane, applied on a mean stream surface defined by the geometry of the blades. Depending on the kind of turbomachinery to be evaluated, different modifications for the effect of deviation, slip or incidence are applied, and correlations for losses can be introduced. This technique can provide quite reasonable estimates of the flow field in most of a radial impeller, as demonstrated in [1].

This is due to the fact that the velocity gradients and pressure distributions are determined by the curvature of the streamlines which are more or less fixed by the hub and shroud contours.
The increasing demand for more information about the flow structure as well as the enormous progress in the computer capacities led to the amazing popularity of the CFD in turbomachinery design and analysis. For a long time, different streamline curvature or throughflow calculations were the only CFD methods in the radial impeller design with acceptable results at a reasonable expenditure of calculating time. However, since the important deviations between the simple model and the real flow in centrifugal impellers arise as a result of neglecting of the viscous effects, only a full Navier-Stokes analysis promises an acceptable guarantee to capture the flow physics sufficiently accurate.
At the beginning, the development of Navier-Stokes codes and their applications were more or less restricted to universities due to the very high development costs and the computer capacities. An era of three-dimensional viscous flow calculations in the turbocompressor industries was initialized only about 6 to 8 years ago, when the first publications of the full Navier-Stokes solvers for turbomachinery flows came out , e.g. [2], [3].
Soon afterwards, Moore et al. [4], [7], Dawes [5] and Hah et al. [6] applied their own codes to different centrifugal impellers, while Krain and Hoffman [19] used the Dawes code for an application to the impeller which later became known as "Krain" impeller. Extensive L2F flow field measurements were carried out on this impeller, which were compared with the calculated data. All of them were intended primarily to validate the results with the few available measurements of detailed flow fields. Most of these validations were carried out with "inducer type" impeller geometries (usually with the well-known "Eckardt"-impellers "0" and "A") similar to turbocharger compressor wheels. Casey et al. [8] predicted the flow field of a backswept industrial impeller (Eckardt impeller "B") without inducer using Dentons 3D code with distributed body forces to simulate viscous effects [9] and the Dawes code. This validation proved that both codes are able to predict the flow field of an industrial impeller. On this basis, the codes were incorporated in our design system for radial and axial compressor stages.
Since then, several applications to different turbomachinery geometries even with extremely high pressure ratios [10] were reported by turbocharger and gasturbine manufacturers, and thus demonstrate that the Dawes code, in particular, is very widely used in the industry. This may be mainly due to the fact that the code is easy to grid, has a transparent input structure, is reliable, has a relatively low computational time and has no user controllable input (fudge factors) which can cause significant errors in the results.
A lot of experience gathered over the last years proved the utility of the code in the day-to-day design work. Several examples of this work are described below and comparisons of different design calculations shown.

2 DESIGN PROCEDURE

In most industries, the design of turbomachinery components is carried out using a sophisticated design system adapted to the respective requirements, the available data base, the computer codes in use and the possibilities in computer-aided design and engineering. In the AGARD lecture series [11], some of the authors describe comprehensive design procedures for axial compressors (Stow) or for turbines (Hourmouziadis). Examples of state-of-the-art CAD/CAE design procedures, particularly for radial impellers, are described by Krain [12] or Nojima et al. [13].
A typical system used successfully by the authors over the last years in the design of high performance industrial centrifugal compressor stages, updated with the 3D viscous flow calculation, is shown in Figure 1. The design specification includes the desired overall performance and a preliminary one-dimensional design. By using a mixture of experience from similar stage types, taking into account the geometrical and manufacturing restraints imposed by standardization and rotordynamics, a one-dimensional flow model estimates the performance curves and the anticipated stable operating range. This provides an initial estimate of the main impeller and diffuser geometry, such as inlet and outlet width, blade angles, impeller diameter and so on. This information, together with supplementary data to define the shape and the curvature of the channel and the blades, is fed into the geometry definition program "LAYOUT" [14]. This first approximation of the stage geometry, generated in a parametric form, can now be used in succeeding programs for the detailed design. In an iterative fashion, the optimization of the geometry is carried out to achieve maximum efficiency and flow range, using the LAYOUT program as a kind of impeller generator, which creates all necessary input data for the subsequent steps. The design and analysis system consists of the following elements:

- The streamline curvature throughflow program WUFLOW [15] with a blade-to-blade approximation helps for a global flow analysis. It provides accurate data about the several factors directly related to the performance characteristics of the impeller, e.g. incidence, diffusion ratio, blade loading and Mach number on blade etc. The main design loop is carried out with the programs LAYOUT and WUFLOW.

- The full 3D viscous flow calculations, carried out with Dawes' Navier-Stokes solver, are used in the final stage of the design for a detailed flow analysis.
- The stress and blade vibration calculations are performed with the fully three-dimensional finite element model ANSYS [16].
- The specification of the geometry for the CAD for manufacture.

If an optimum design has been realized, one or two prototype impellers are built and thoroughly tested in one of the two available single-stage closed loop test stands (see Dalbert et al. [17]). If the performance objectives are achieved, the new stage can be standardized and the measured performance characteristics incorporated in a stage-stacking procedure to predict the performance of multi-stage compressors.

3 Code for 3D Viscous Flow Calculations, Pre- and Postprocessing

The applied computational code in our design system is the full three-dimensional Navier-Stokes code BTOB3D, written by Dawes [5]. For centrifugal impellers with splitter blades, Dawes has modified his original code to include splitters (BTOB3DSP). Otherwise, there is no difference to the original version. Since the basic formation of this code has been outlined several times (e.g. in [2], [3], [5], [8]) a repetition in this context seems superfluous.

A vital condition to use a 3D code in the daily design work is the user friendliness to handle the input data and the results. The pre- and postprocessing had to be developed to this end. Thanks to the flexibility of the geometry definition programs, they could easily be adapted for an optimum grid generation and the production of a formally correct input for the CFD code. The postprocessor had to be created since there was no appropriate program available. The original version was based on the representation of the different flow properties in two dimensional blade-to-blade planes, cross sections or profiles as shown in [8]. Since a detailed data analysis of three-dimensional calculations on the basis of 2-D plots is extremely cumbersome, it was decided to develop a postprocessor for three-dimensional colored graphics on a Tektronix net station. At that time, this selection allowed the use of the same station as well for postprocessing the ANSYS results without any additional program development.

Today, there is almost no expenditure of time to create an H-mesh for the Dawes code for a smaller impeller with 19 x 75 x 19 up to 19 x 99 x 19 grid points (in pitch-, stream- and spanwise direction) or 25 x 99 x 25 for a wider impeller. Impellers with splitter vanes usually need a grid with 33 x 99 x 19 to 33 x 99 x 25 nodes. The computational time used with 1500 to 2000 time steps of the code is about 15 to 30 hours on a VAX 4000, as the necessary CPU time is almost proportional to the number of grid points and time steps. Since the postprocessing and evaluation of the results takes usually several days, these still very high computational times are regarded as acceptable, as long as the computer capacity is sufficient to obtain the results of a smaller job within 24 hours.

The code was thoroughly checked by Casey et al. [8] before it was established in the design system. The results of this validation can be summarized as follows: It indicates good agreement of the calculations with the measured flow fields from near surge to choke, it is able to identify regions of separation and provides considerable insight into the development of the secondary flows and the loss mechanisms inside the impeller. This gave us the confidence to apply the code on the other impellers of similar geometries.

4 Applications

Since the 3D calculations were incorporated in the design system of radial stages, many analysis and design calculations were carried out for existing centrifugal impellers and new designs. In spite of the good agreement of the flow fields, stated in the assessment of the code in [8], it was recognized already at that time that the losses are slightly underestimated by the calculation. To some extent, this hinders the prediction of the performance of the new designs. To illustrate the deviations in the performance and the flow properties that may increase the efficiency, numerous examples from previous work with known characteristic curves were analysed. Most of them were from work on radial compressor design. Up to now there was only a limited experience with axial compressor stages on hand and the design of radial diffusers with 3D viscous flow calculations was initiated recently.

The following examples concentrate on the experience with centrifugal impellers and refer briefly to the hitherto experience with the radial diffuser application.

4.1 Centrifugal impellers

The test stands, test procedure and the data analysis are described extensively by Dalbert et al. in [17]. All measurements with prototype impellers - and there is no impeller in a multistage compressor which has not been tested thoroughly as prototype - are carried out in one of the two dedicated single-stage closed loop test stands. The tests are carried out with a vaneless diffuser or with a vaned diffuser and different blade angle settings at several tip speed Mach numbers Mu_2 (refering to rotational speed of the impeller). Inlet and outlet temperatures, several intermediate pressures, the mass flow and the power absorbed by the impeller are measured. In order to optimize the matching between the impeller and the diffuser, a procedure is employed to calculate the impeller and diffuser performance characteristic. The details of the different calculations are reported in [17] and only the equations for these values, which are compared with the calculations, are given below. The arithmetic mean value of the impeller static discharge pressures, measured at hub and casing side of the diffuser inlet, can be used to determine the total-to-static efficiency as the ratio of the polytropic head rise to the input power less the kinetic energy at the impeller outlet:

$$\eta_R = \frac{\int_{1^*}^{2} \frac{dp}{\rho}}{\Delta h^* - \frac{c_2^2}{2}} = \frac{RT_{1^*}\left(\frac{n}{n-1}\right)\left[\left(\frac{p_2}{p_{1^*}}\right)^{\frac{n-1}{n}} - 1\right]}{\Delta h^* - \frac{c_2^2}{2}} \tag{1}$$

The impeller work coefficient is equal to the stage work coefficient μ_0 less the friction losses and the leakage flows on the cover discs, calculated by analytical models:

$$\mu_R = \frac{P - P_{RR}}{\left(\dot{m} + \dot{m}_{LS}\right) u_2^2} \tag{2}$$

The flow angle of the absolute velocity at impeller discharge is given by Equation (3):

$$\alpha = \tan^{-1}\left(\frac{c_{m_2}}{c_{u_2}}\right) \tag{3}$$

where the meridional c_m is determined from the mass flow and the tangential velocity c_u from the mechanical power measurement.
The flow coefficient is defined as:

$$\Phi = \frac{\dot{m}}{\rho\, D_2^2\, u_2} \tag{4}$$

The data reduction process of the calculation is slightly different. Since the flow properties are known in all grid points, mass averaged mean values were found to be appropriate for such comparisons. The respective equations are indicated below:

$$\eta_R = \frac{R}{c_p} \frac{\ln\left(\frac{\bar{p}_2}{\bar{\bar{p}}_{1^*}}\right)}{\ln\left(\frac{\bar{T}_2}{\bar{\bar{T}}_{1^*}}\right)} \tag{5}$$

$$\mu_R = \frac{\overline{\Delta h^*}}{u_2^2} \tag{6}$$

$$\alpha = \tan^{-1}\left(\frac{\bar{c}_{m_2}}{\bar{c}_{u_2}}\right) \tag{7}$$

Two impellers of similar design with different flow capacities. The first example for an analysis of the results of the CFD code are two impellers designed for medium flow coefficients. The impellers were developed before the 3D calculations were available. The design flow coefficient of the first impeller (D9) is $\Phi = 0.09$, whereas the second impeller (D7) was designed at $\Phi = 0.07$, both at the stage Mach number of $Mu_2 = 0.9$. Figure 2 shows the measured performance characteristics of the impellers at the tip speed Mach numbers 0.8 and 1.0 in comparison with the calculated values near surge, design point and choke. The efficiencies are normalized with respect to the measured peak value at $Mu_2 = 0.8$. The comparison demonstrates that the calculated efficiency at design flow rate and particularly near surge is about 4 to 7% too high, the impeller work coefficient is well predicted, whereas the absolute flow angle is up to 5 degrees too high. A small underestimation of the losses is a known shortcoming of the code. The incorrect simulation of the leakage flows over hub and shroud disk could lead to the error in the flow angle, since μ_R and therefore the slip factor is accurately predicted. The choke point is calculated by defining a ratio of downstream static pressure to inlet total pressure of unity, leading to the maximum possible flow rate (emphasized by the arrow in the diagrams). The calculated impeller efficiency is negative, indicating that the discharge static pressure of the impeller (at trailing edge) is lower than the total pressure at the inlet (see equation (1), due to extremely high losses in the flow channel. The local relative Mach number reaches values of about 1.2 which are reduced again by a shock to values of around 0.1. The mass flow at choking condition seems to be accurately predicted.

The difference between the measured values of the peak efficiency of the two stages is about 2 points which was primarily attributed to the fact that the axial length of the D7 impeller is shorter by a factor of about 0.8 (Figure 2). Only the 3D calculations, which didn't show any significant differences with respect to the losses or to the flowfield within the impellers or at the discharge, nor to the performance characteristics, led to a review of the measured data. It appeared that the impeller efficiencies measured with a vaneless diffuser are almost identical. It must be assumed that the most important difference was essentially the smaller diffuser channels of the D7 stage which led to higher losses due to smaller hydraulic diameters and a lower pressure recovery due to different area ratios of the diffuser.

Design of low flow impellers. In the design of impellers for low and very low flow coefficient stages, the needed measures for an optimum aerodynamic design are in contrast to the mechanical and operational constraints. Low flow coefficient stages are mainly used for high pressure applications where rotor dynamic considerations require stages with short axial length and a large shaft diameter. Both requirements lead to increased aerodynamic losses. The problems arising in the design of this kind of impellers are described by Casey et al. in [18]. The low flow coefficient impeller of which the development is described in [18] reaches its limit of use at $\Phi \approx 0.02$, where high incidence losses reduce the peak efficiency significantly. Impellers for stages with flow coefficients of $0.02 \leq \Phi \leq 0.03$ are especially difficult to design, since they may be installed as first stages in high pressure compressors with all the respective mechanical restraints or as last stages for instance in isotherm compressors, where a high peak efficiency is of upmost importance.

Over the last ten years, four different impellers (A, B, C and D) have been developed for this application. Figure 3 shows the shape of the meridional channels of the four impellers. The impeller A, designed many years ago, operates in the flow range of $0.02 \leq \Phi \leq 0.04$. The main defects of this stage are its relatively long axial length, flat pressure characteristic and moderate efficiency, compared with todays state-of-the-art. In the development of impeller B, intended to replace the smaller versions of impeller A, all the design possibilities of that time were exhausted with LAYOUT and WUFLOW. This comprised an optimization of the meridional channel as a compromise between the aerodynamically desirable and mechanically detrimental axial length and a blade design with radial standing elements to reduce the stress level, an improved deceleration and blade loading distribution etc. The prototype tests in the single-stage test stand manifested that the design objectives were not met. Subsequently, it was decided to develop two new impellers for the same flow range, but with different design philosophies adapted to the distinct applications using the possibilities of the Navier-Stokes code as the numerical experiment. This resulted in the two geometries of which the meridional channels are shown in Figure 4. One impeller (C) was laid out for applications where a high efficiency is imperative. The design was aimed at a minimum of compromises to the mechanical limitations. In comparison to the failed impeller B, the new wheel could be a little longer, with an extremely smooth curvature along the shroud and a lower hub (or shaft) diameter than impeller A. The optimization led to a three-dimensional blade design. The second of the new impellers should match the design of high pressure compressors with the necessary shortness and the larger shaft diameter (impeller D in Figure 4). The blade design turned out to be a quasi-three-dimensional design where the inlet portion was adapted to the flow angle resulting from the high flow rate and the high curvature just upstream of the leading edge. About 80% of the downstream portion of the blade is "purely two-dimensional"(not twisted) as in the low flow coefficient impeller [18]. The 3D viscous flow calculations were used extensively during the design of these two impellers, and an analysis calculation of A and B exposed the explanations for the different performance of the wheels.

Figure 5 shows the meridional velocity vectors near to the blade suction side of the four impellers. All calculations were carried out with an axial inlet to the impeller, with a slightly reduced diffuser width to increase the stability of convergence and a grid size of 19 x 75(77) x 19. The blade leading edge is located at the streamwise calculating station J = 26 for impeller A and 24 for B, C and D, the trailing edge at 66, 64 and 62 respectively. The calculations were carried out with 1500 time steps for impeller B and 2000 for A, C, D. The results can be summarized as follows:

- Impeller A shows strong secondary flows in the spanwise direction towards the shroud downstream of the bend from the axial to the radial direction. This secondary flow arises from the high curvature of the meridional shroud contour just downstream of the leading edge which is even intensified by a locally very low velocity indicating that the flow is on the verge of separation in this bend. The quite significant gradient survives up to the impeller discharge.
- Within the impeller B, this region of low flow velocity extends downstream over the whole channel and creates an extensive wake zone at the impeller exit. These phenomena could not be recognized with the throughflow calculations. The "circumferentially averaged" values of the throughflow method indeed show a gradient of the meridional velocity (Figure 3), but do not provide the necessary blade-to-blade information further downstream within the impeller where viscous effects dominate the events.
- Impeller C shows extremely smooth distribution of the meridional velocity both in spanwise and pitchwise direction. Figure 6 shows the profiles of the relative entropy production $d(S)/d(S)_{rel}$ in a blade-to-blade plane near to the shroud of the impellers B and C. The entropy d(S) is defined in Equation (8):

$$d(S) = \Delta(s) = c_p \ln\left(\frac{T}{T_1}\right) - R \ln\left(\frac{p}{p_1}\right) \qquad (8)$$

  where p and T are the local values of the static pressure and temperature, while p_1 and T_1 represent the local inlet values at the corresponding pitch and spanwise position. It is obvious that near to the whole suction side of the blade (I=19) and at the leading edge near to the pressure side (J=25, I=1) the impeller B produces significantly more losses than impeller C. Measurements in the test stand demonstrated the clear superiority of impeller C over B.
- The last of the four wheels, impeller D, shows a very limited zone of low meridional velocity near to the suction side of the blade which has to be attributed to the blade design. The spanwise gradients are low and the discharge velocity distribution shows no extended "jet-wake" structure which could have a detrimental effect on the diffuser efficiency. The main problem of the meridional velocity at the leading edge of the blade near to the shroud and the related high incidence could be minimized using the 3D code.

Effect of blade lean. There have been several publications concerning the effect of blade lean in turbines in the last few years. Hourmouziadis, for instance, describes the effects that can be achieved by the selection of the stacking law in the AGARD Lecture Series [11]:

1. Change in the spanwise distribution of flow parameters by the blade force
2. Reduction of the secondary flows and losses

Some years ago, the effects of blade lean in centrifugal wheels were investigated by throughflow calculations in an internal study. The basic equation of streamline curvature throughflow methods is the velocity gradient equation, see Casey and Roth [15]. Among the different terms for the radial equilibrium and the streamline curvature, it contains a blade force term, related to the rate at which the flow is turned in the circumferential direction. It is therefore dependent on the lean angle of the blade. The results of the study were that the leaning of the blades in a radial impeller can substantially change the spanwise pressure and the meridional velocity gradient and consequently the distribution of blade loading. This, in return, is expected to reduce the secondary flows and the losses also in a centrifugal impeller. As far as the secondary flows are concerned, this effect can be investigated by CFD, while the reduction of losses has to be established by measurements.
In order to carry out these investigations three different impellers were developed: one with no lean, a second with a high lean angle and a third one with a reduced lean angle as a compromise to the high bending stresses in the blades of the second impeller. Figure 7 shows the influence of leaning the blades on the spanwise distribution of the meridional velocity. The diagrams show impressively the significant reduction of the spanwise gradient of the velocity due to a reduction of the pressure difference between hub (K=1) and shroud (K=19) as calculated with the Dawes code. This result, which is a confirmation of the results obtained by the throughflow calculations and a clear evidence that the secondary flows are reduced by the blade lean, led to manufacturing the three impellers for testing in the single stage test stand. The results are shown in Figure 8. The measurements were

carried out with the same diffuser at the same circumferential speed. The superiority of the impeller with a high lean angle is obvious. The somewhat distorted characteristics originate in some problems with the stability of the new design of the diffuser which may have an effect on the mean value of the static pressure measurements at the impeller exit.

4.2 Radial Diffuser

One of the crucial elements in a centrifugal compressor stage is the radial, usually bladed diffuser. The many different geometrical designs of these components, for instance, low solidity diffuser, channel diffuser, vane island diffusers etc, with different blade geometries and numbers of blades demonstrate that there is no generally valid geometry to guarantee an optimum design. Additionally, there are very little reliable measured data available on radial diffusers which would allow us to calibrate a simple design system for diffusers. With the exception of the many data on channel diffusers or the data collected by Japikse [20], the possibilities for an analytical diffuser design are very limited. In this environment, it seemed to be a promising idea to study the flow structure and the performance of a radial diffuser with the help of three-dimensional viscous flow calculations. In a industrial sponsored research program at the Federal Institute of Technology in Zurich (Switzerland), a project could be included which investigated the ability of the Dawes Navier-Stokes code to predict the performance of a radial bladed diffuser which was also tested experimentally. The results may be summarized as follows:

1. The code is able to calculate the flow near surge, at the design point and near choke. As an example, several streamlines within the diffuser at the three operating points show in Figure 9 the secondary flows due to the curvature of the channels.
2. The upstream boundary conditions are extremely difficult to define. The measured pressures and flow angles have been used for the present calculations .
3. The accuracy of the pressure recovery prediction is not satisfactory.
4. The study did help us to understand many of the flow phenomena in a radial diffuser.

5 SUMMARY AND CONCLUSIONS

The use of CFD and in particular the application of the Navier-Stokes code of W. N. Dawes in the daily design and development work of radial turbocompressor components was presented. It was demonstrated that in many cases this computational tool is an enormous help to understand the physics in the flow channels and the differences in the overall performance of different stages. In particular, this holds true in the design of radial impellers where the viscous effects dominate the flow. In spite of the simplicity of the turbulence models in the Dawes code, many of the essential features in the flow of radial impellers and radial diffusers can be analysed. The fact that this code can be used in a day-to-day manner is mainly due to its extreme user friendliness which arises partially from the disadvantage that it can only be applied to the axisymmetric elements of turbomachines.

Of course, there are also limits in the application of this single blade row code which have to be realized to assess the possibilities correctly.

- As was discussed in connection with the radial diffuser application, where the interaction of more than one component are important, the use of this code is at least very difficult, if not impossible, as the inlet boundary conditions are not known. This problem can be avoided using the multiblade row code "BTOB3D-MSTAGE-SPLITTER" by W. N. Dawes [21] which has already been applied successfully to axial stages.
- The simplicity of the loss model leads to a moderate underestimation of losses. Experience shows that the relative values show the correct tendency, when similar designs are compared. This makes this limitation to a certain extent to a question of experience.
- There are still very few data available about the problem of the assessment of the flow fields. What flow property results in a high impeller efficiency? There is a prominent need for benchmark-quality data in the area of turbomachinery, including not only the flow fields but also performance data of the components.

This shows that even though neither the computer capacities nor the CFD codes are able to predict the flow field and the performance of the whole turbomachine and that design systems as shown at the beginning of this report will still stay in use for many years, the CFD in the turbomachinery has made enormous progress in the last decade. In some cases, the results from CFD did not meet and may not meet the expectations also in future, but in combination with measurements, 3D viscous flow calculations provide an extremely helpful complimentary tool within a development system for simulation, design, optimization and, most importantly, for the analysis of the extremely complex three-dimensional flows hitherto inaccessible to the designers.

ACKNOWLEDGEMENTS

The authors wish to thank Sulzer-Escher Wyss Ltd. for permission to publish this paper as well as colleagues carrying out all our measurements in the test stands at the Fluid Dynamics Laboratory of Sulzer Brothers Limited, who made this work possible.

REFERENCES

[1] Casey, M. V., and Roth, P., "A streamline curvature throughflow method for radial turbocompressors", in *Computational Methods in Turbomachinery*, IMechE. Publication 1984-3, London, Paper No. C57/84, 1984.

[2] Dawes, W. N., "Application of full Navier-Stokes solvers to turbomachinery flow problems", VKI Lecture Series, No.2, January 20-24, 1986.

[3] Dawes, W. N., "Development of a 3D Navier Stokes solver for application to all types of turbomachinery", ASME paper 88-GT-70, 1988.

[4] Moore, J., Moore, J. G., and Timmis, P. H., "Performance evaluation of centrifugal compressor impellers using three-dimensional viscous flow calculations", ASME Journal of Engineering for Gas Turbines and Power, Vol. 106, pp. 475-481, 1984.

[5] Dawes, W. N., "Application of a three-dimensional viscous compressible flow solver to a high-speed centrifugal compressor rotor - secondary flow and loss generation", IMechE Paper No. C261/87, 1987.

[6] Hah, C., Bryans, A. C., Moussa, Z., Tomsho, M. E., "Application of Viscous Flow Computations for the Aerodynamic Performance of a Back-Swept Impeller at Various Operating Conditions," ASME paper 88-GT-39, 1988.

[7] Moore, J. and Moore, J.G., "Secondary Flow, Separation and Losses in the NACA 48-Inch Centrifugal Impeller at Design and Off-Design Conditions", ASME paper 88-GT-101, 1988.

[8] Casey, M. V., Dalbert, P. and Roth, P., "The use of 3D viscous flow calculations in the design and analysis of industrial centrifugal compressors", ASME Paper 90-GT-2, 1990.

[9] Denton, J. D., "The use of a distributed body force to simulate viscous effects in 3D flow calculations", ASME paper 86-GT-144, 1986.

[10] Masutani, J., Aoki, S., "Dawes code applied to a high pressure ratio centrifugal compressor", Numerical Simulation in Turbomachinery, FED-Vol. 120, ASME 1991.

[11] AGARD Lecture Series No. 167, "Blading Design for Axial Turbomachines", 1989.

[12] Krain, H., "A CAD-Method for centrifugal compressor impellers", ASME Journal of Engineering for Gas Turbines and Power, Vol.106, pp.482-488, 1984.

[13] Nojima et al., "Development of aerodynamic design system for centrifugal compressors", Mitsubishi Heavy Industries, Technical Review, Vol. 25, No. 1, 1988.

[14] Casey, M.V., 1983, "A computational geometry for the blades and internal flow channels of centrifugal compressors", ASME Journal of Engineering for Power, Vol. 105, pp. 288-295.

[15] Casey, M.V., and Roth, P., 1984, "A streamline curvature throughflow method for radial turbocompressors", in *Computational Methods in Turbomachinery*, I. Mech. E. Publication 1984-3, London, Paper No. C57/84.

[16] ANSYS, 1983, "Engineering Analysis System", Swanson Analysis Systems Inc., Houston, USA.

[17] Dalbert, P., Casey, M. V., Schurter, E., 1988, "Development, testing, and performance prediction of radial compressor stages for multistage industrial compressors", ASME Journal of Turbomachinery, Vol.110, pp. 283-292.

[18] Casey, M. V., Dalbert, P., Schurter, E., 1990a, "Radial compressor stages for low flow coefficients", IMechE Publication, London, Paper No. C403/004.

[19] Krain, H., and Hoffman, W., "Verification of an impeller design by Laser measurements and 3D-viscous flow calculations", ASME paper 89-GT-159, 1989.

[20] Japikse, D., 1984, "Turbomachinery diffuser design tehnology", Concepts ETI, Inc., Norwich, Vermont, USA.

[21] Dawes, W., N., "Toward Improved Throughflow Capability: The use of Three-Dimensional Viscous Flow Solvers in a Multistage Environment", Trans. ASME, Vol. 114, pp. 8-17, January 1992.

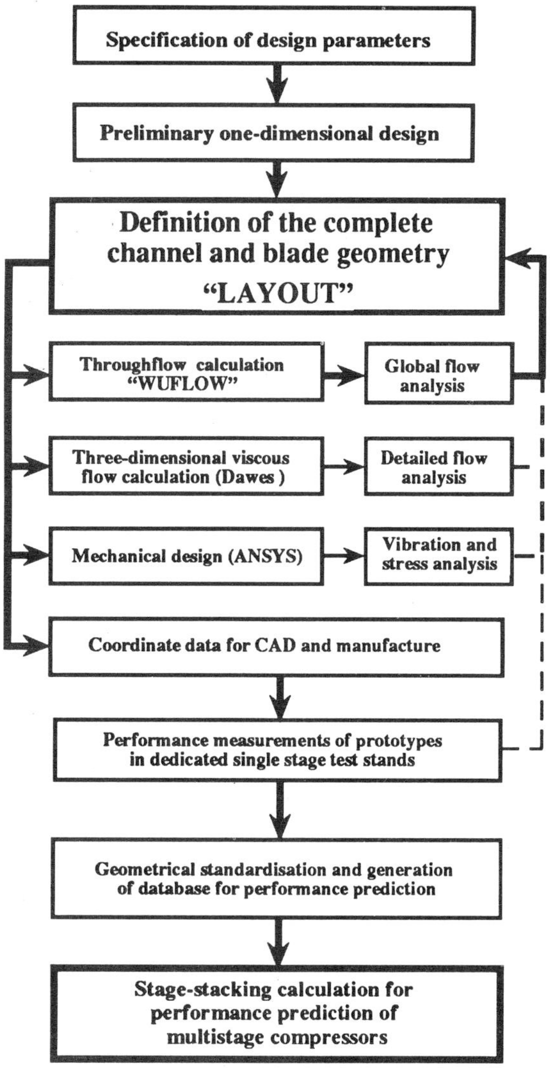

Figure 1. Design system for radial stages

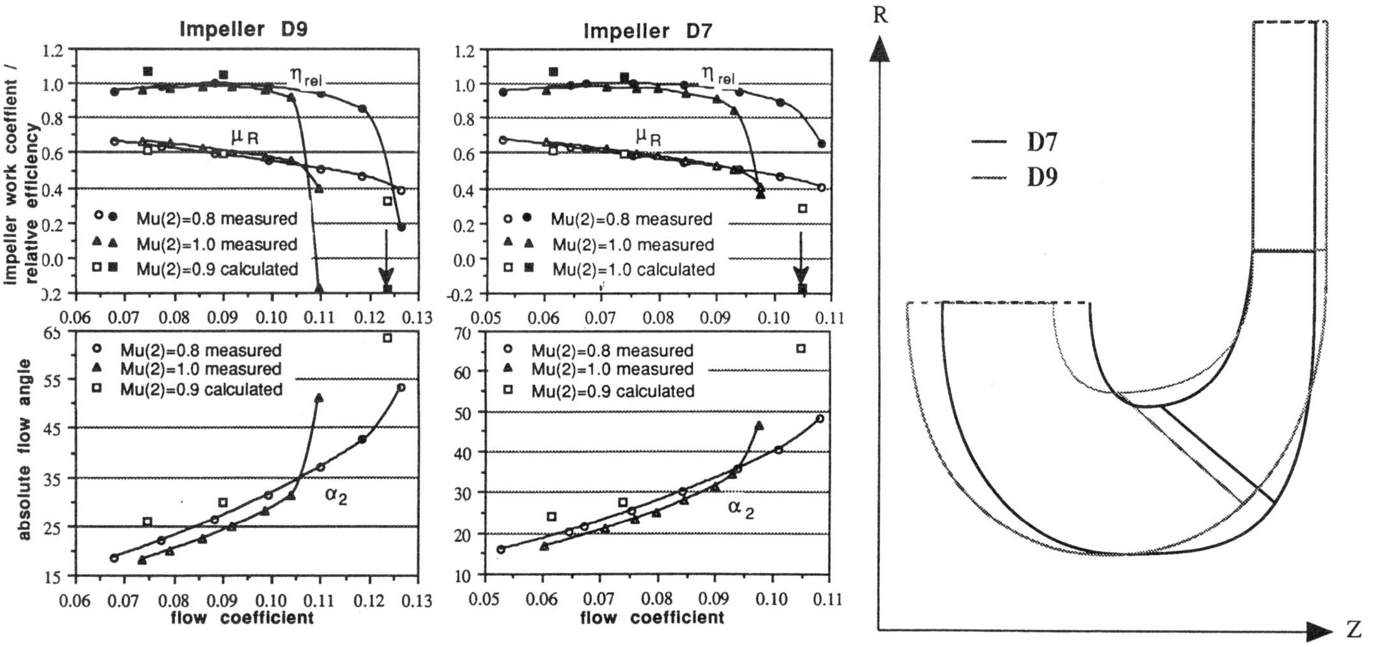

Figure 2. Measured and calculated performance characteristics (μ_R and η_{rel} vs. Φ and α vs. Φ) and meridional channels of the centrifugal impellers D9 and D7.

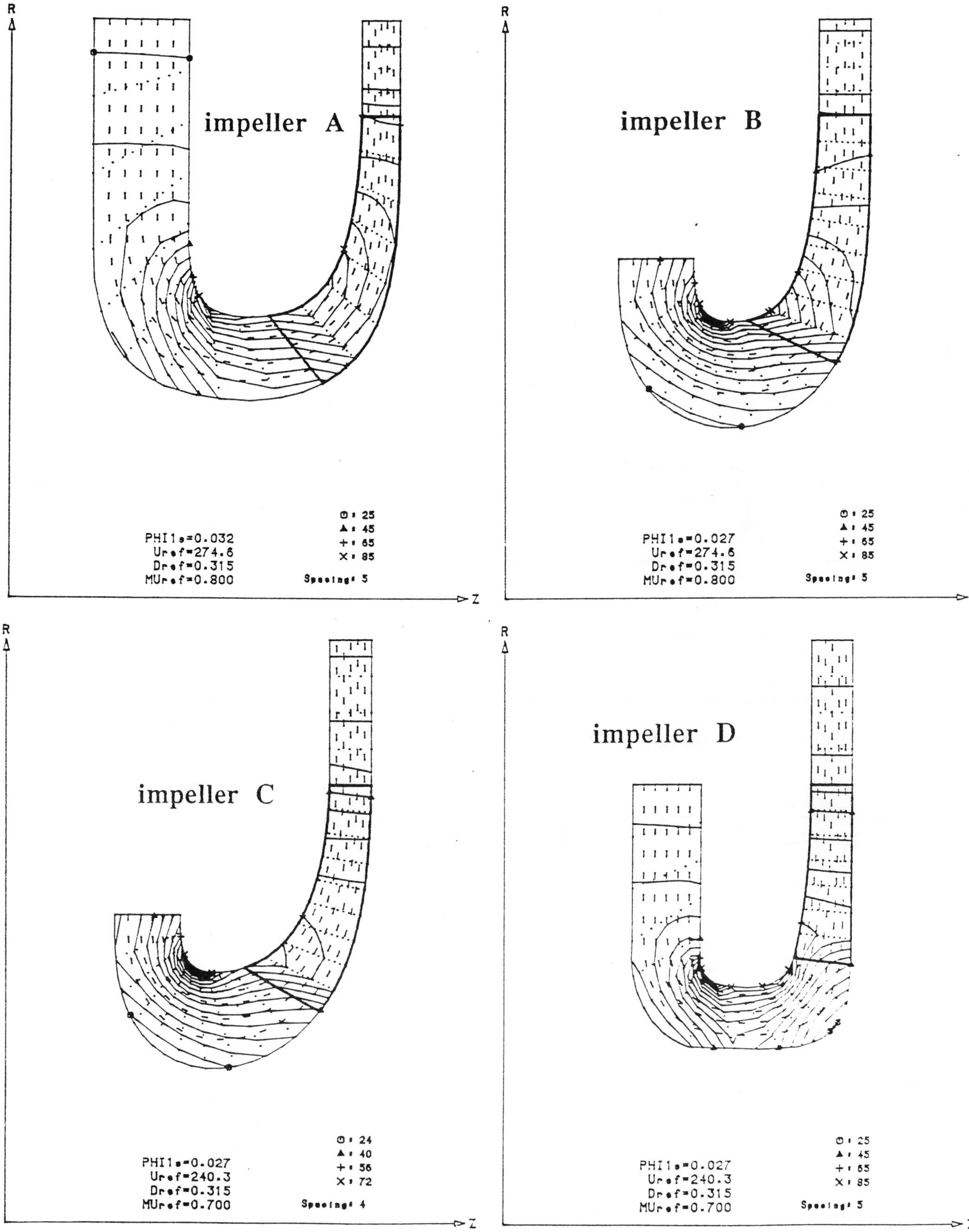

Figure 3. Constant lines of the velocity c_m (WUFLOW) in the meridional plane of the impellers A, B, C and D.

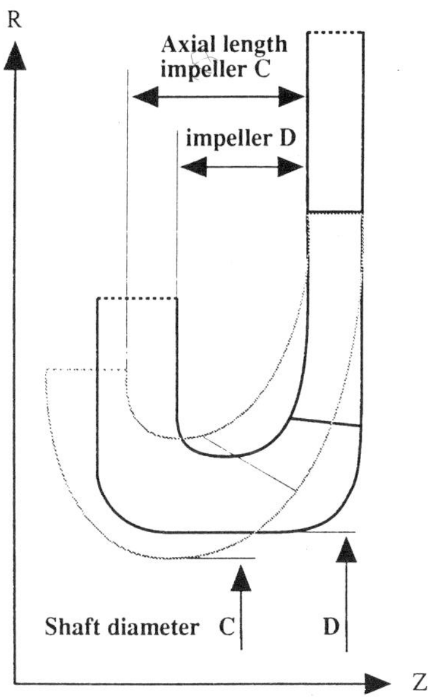

Figure 4. Meridional channel of impeller C and D.

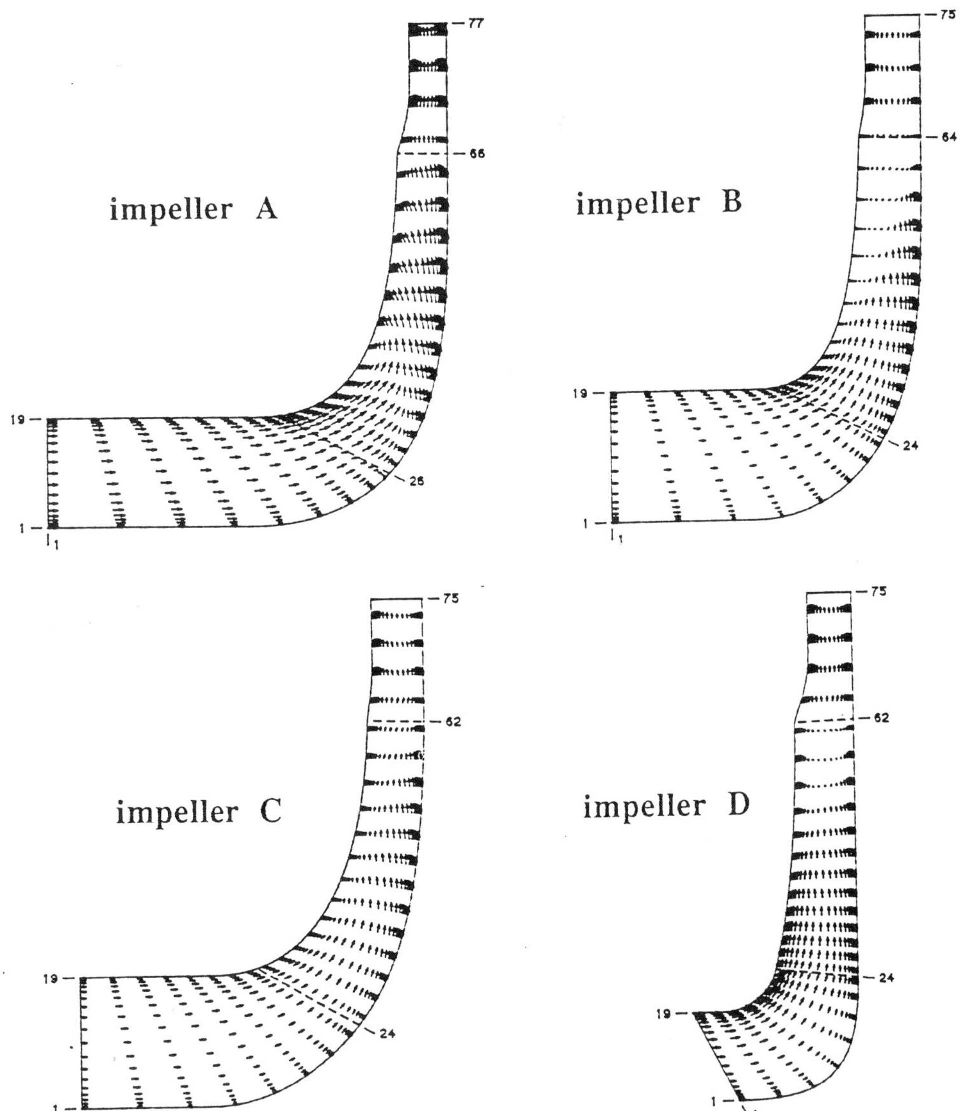

Figure 5. Calculated meridional velocity vectors (BTOB3D) near to the blade suction side of impellers A, B, C and D.

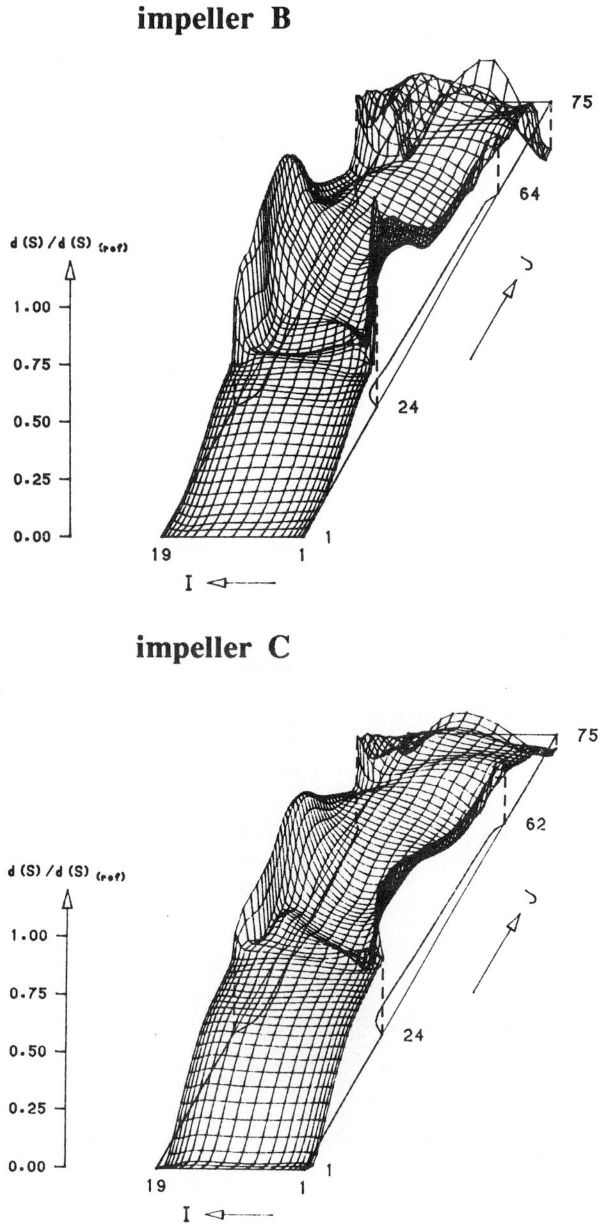

Figure 6. Entropy production near shroud of impellers B and C with index J in streamwise and I in pitchwise direction (BTOB3D).

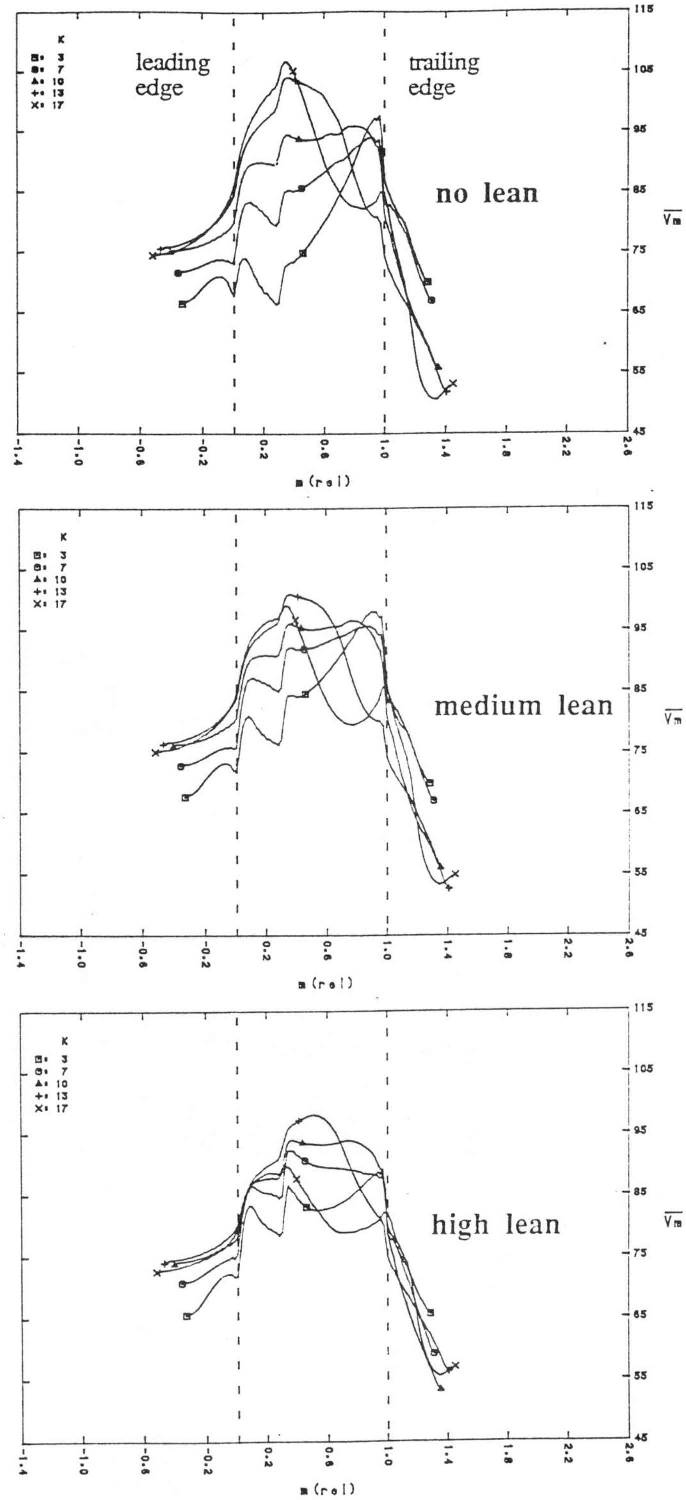

Figure 7. Streamwise development of mean meridional velocity with index K for spanwise distribution: K=3= near hub, K=17=near shroud (BTOB3D).

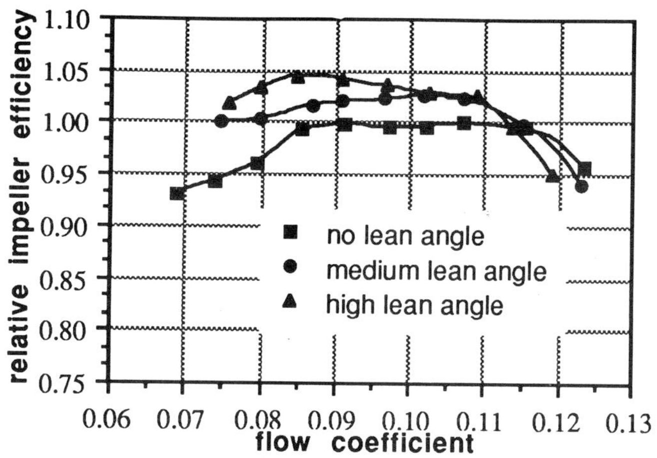

Figure 8. Impeller total to static efficiency for different lean angles.

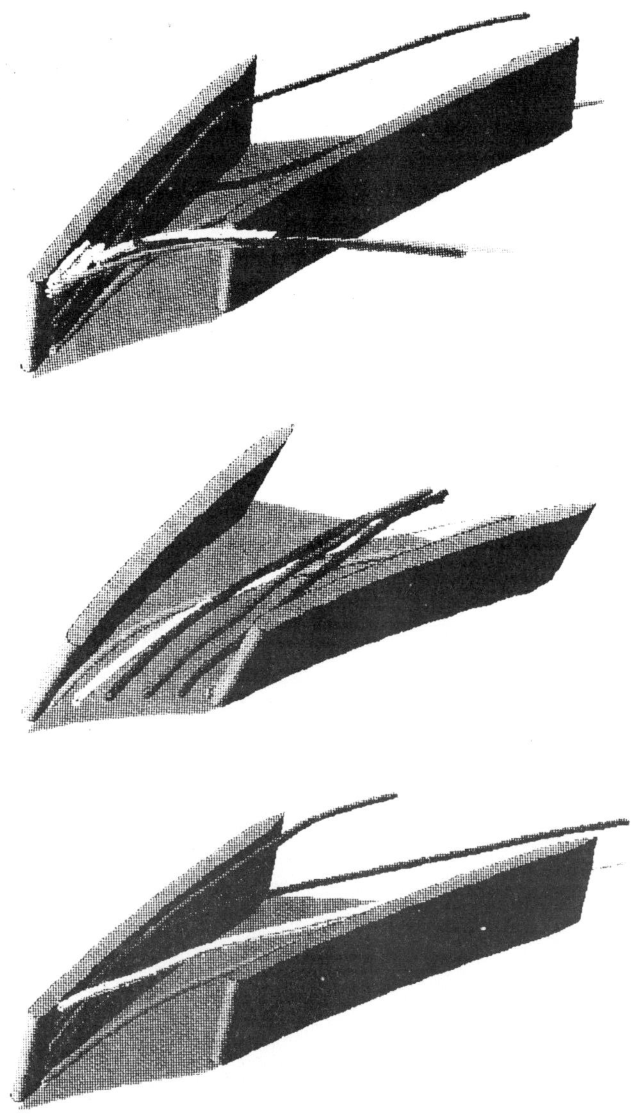

Figure 9. Streamlines in a radial diffuser.
From top to the bottom:
- operating point near surge
- operating point at design flow rate
- operating point near choke

C449/035/93

Development of an axial piston pump for water hydraulic systems

C A BROOKES, PhD, CEng, FIM, M J FAGAN, BSc, PhD, CEng, MBCS,
R D JAMES, BSc, PhD, CEng, MIMechE, MIEE, P R KERRY, BSc, J McCONNACHIE, BSc, MSc and
P W RADCLIFFE, BSc
Department of Engineering Design and Manufacture, University of Hull

SYNOPSIS Progress towards the development of an electrically-neutral subsea sea-water axial-piston pump is described. In particular the design and application of ceramic pistons is discussed, and the prospects for the long term viability of the pump shown to be good.

1 INTRODUCTION

The ever increasing cost of crude oil, the enforcement of more stringent safety standards and an increasing awareness of ecological issues has for some time prompted investigation of alternative hydraulic fluids. For several years now the industry has been using dilute emulsions which offer most of the benefits of oil lubrication with a reduction in both the cost and hazards involved although many risks remain particularly in mining environments (1). However there are certain applications where the use of a pure water hydraulic fluid would be preferable due to its economic and environmental benefits. These applications include subsea diving tools, fire and rescue tools, food processing machinery and nuclear industry robots, and a number of attempts have been made at producing realistic water hydraulic power systems for specialist applications (2,3).

The advantages of using water as an hydraulic fluid are many; its low cost, plentiful supply, contamination tolerance, good heat transfer properties, incompressibility and low pumping viscosity (4). The greatest advantages water has as a process fluid are that it has no flash point and is fire resistant, although care has to be taken to avoid boiling. Water has its disadvantages - corrosion with many metals, poor lubricity, it is heavy and can be prone to bacteriological contamination in closed systems. However it is sufficiently attractive to encourage the development of a specialised pump, particularly at the current time with the ever increasing number of new materials which hold the promise of overcoming many of these difficulties.

The development of advanced polymers has been responsible for much progress towards the realisation of water hydraulic systems which use the water as the lubricant (5,6). Certain polymers are able to withstand the high contact pressures and sliding velocities imposed by boundary and mixed lubrication of the bearing surfaces in pumps. They can also retain their tribological properties for prolonged periods of immersion in water. In combination with treated and/or hardened stainless steels good results have been obtained and pressures up to 100 bar achieved (6). However, since some of the hardest and most wear resistant stainless steels have relatively low corrosion resistance, the ability to use sea-water as the working fluid of the pump requires even more sophisticated and advanced materials and/or coatings whose stability in active water is assured.

2 GENERAL DESIGN CONSIDERATIONS

Early tests with gear and vane pumps showed that these types of pumps were unsuitable for water hydraulic systems because of their high wear rates and internal leakage, and this led the NEL (National Engineering Laboratory) and subsequently J. H. Fenner PLC (Hull) to choose an axial piston pump as the most suitable for water powered hydraulics (1,4,5,6,7,8). Due to the small clearances required to limit leakage between the large number of precision machined components in axial piston pumps, they are generally characterised by high volumetric efficiencies (typically up to 95% for oil) but are also susceptible to wear when the working fluid becomes contaminated.

This current work is concerned with a straight axis axial piston pump with fixed delivery, a schematic layout of which is shown in Figure 1. A cylinder block containing a number of bores is coupled to a drive shaft with each bore containing a piston which is supported by a slipper bearing on a swashplate. The swashplate is inclined at an angle relative to the shaft, and as the shaft is turned the slipper slides on the swashplate causing the piston to reciprocate in the bore. As the piston moves out of the bore, fluid is drawn in through the cylinder port from the inlet port of the port-plate. The situation is reversed as the piston moves into the bore. In this case fluid is forced out of the cylinder port and into the outlet port of the port-plate. A continuous pumping action is established when

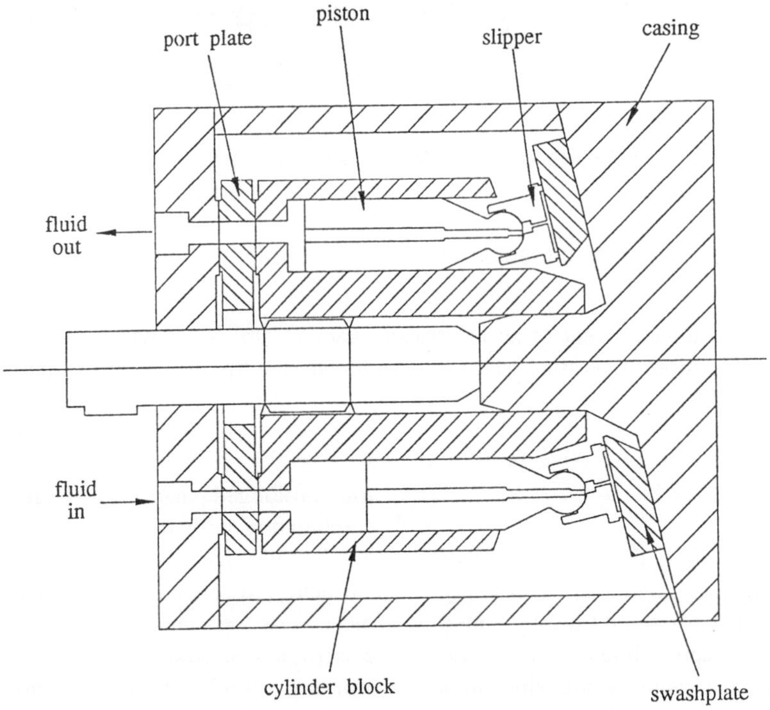

Figure 1. Schematic diagram of an axial piston pump.

all the pistons follow this cycle causing pressure to be raised when the outlet line is subjected to restriction.

The unit as originally developed by the NEL and Fenner used a carbon fibre reinforced polymer (CFRP) and hardenable martensitic stainless steel throughout. These materials provided good tribological performance, however the stringent composition of the alloy required for the hardening process also resulted in a reduction of the corrosion performance of the material and hence its unsuitability for use in aggressive water environments. In order to remedy this it was decided that the new range of advanced engineering ceramics could offer considerable benefits for designers of water hydraulic equipment (9). Generally ceramics are mechanically and physically stable in most environments, and may be produced to very high degrees of accuracy and surface finish. The rapid progress in the development of advanced ceramics has resulted in materials which now have superior strength and Weibull moduli. (The Weibull modulus is a measure of the statistical variation of the properties of a ceramic material and allows design to be conducted on a probabilistic basis rather than a deterministic basis which uses factors of safety (10).)

The reduced viscosity of water means that in a hydrostatic bearing configuration a water-lubricated bearing has approximately one seventh of the film thickness of an oil-lubricated bearing subjected to the same load and flow increasing the possibility of asperity contact. Another disadvantage of the reduced viscosity is that hydrodynamic load generation is smaller when compared to oil, although the frictional drag is less (15). Furthermore, the variation of viscosity with pressure is also two orders of magnitude less than with oil, so that elasto-hydrodynamic lubrication is similarly unlikely to be effective. As a result, mixed or boundary lubrication is probable for water systems, and consequently the bearing materials used are crucial to achieve acceptable performance.

3 CERAMIC COMPONENT DEVELOPMENT

The initial suitability of ceramics for axial piston pumps was tested by the application of a ceramic swashplate. This was tested in unidirectional sliding against a standard polymer composite slipper pad and found to be as effective as hardened stainless steel in tap water. As expected, the ceramic performed well for this non-structural component, and it was decided to employ the ceramic in the most arduous application in the pump - the piston. The piston must have adequate structural strength, fatigue strength, cavitation and jetting erosion resistance and suitable tribological characteristics.

3.1 Materials

The use of ceramic materials has been suggested for several years but has only recently become a practicable alternative to traditional processes of hardening and protective coatings of metals. Indeed, it has been said that the use of harder materials for water lubricated systems is preferable (11). There has already been some limited experience of using ceramics in subsea actuators (3,12) although a high filtration level of 6 μm was used, and in steadily loaded applications such as journal bearings (13,14). The properties of engineering ceramics are radically different from those of their traditional counterparts with much greater toughness and strength and also great improvements in their manufacturability. This has been achieved for some ceramics through transformation toughening whereby a metastable phase in the ceramic expands when a crack touches it thereby putting the crack tip into compression and preventing further growth. The ceramic finally chosen for the pistons was partially stabilized zirconia. This has a high toughness (>10 MPa.m$^{1/2}$), a Weibull modulus of greater than 30 and high flexural and tensile strengths. Although the choice of ceramic was made on the basis of its strength and toughness rather than hardness, it also has a Vickers hardness of typically 1000 kg/mm^2, which is approximately twice that of the hardened martensitic stainless steel previously used in the pump. The thermal properties of the material would not normally recommend its use in bearings, but the localised temperatures are reduced by the high heat capacity and conductivity of water which removes heat twice as effectively as oil (15).

In order to provide the most favourable configuration for the ceramic pistons it was decided that they would be best mated with the polymer composite used elsewhere in the unit. This pairing had proved very successful for the slipper pads and swashplate, and permitted the use of the ceramic components with minimum modification of other parts.

3.2 Design and analysis

Before a detailed analysis of any new pump component can be undertaken, the forces acting in the pump need to be calculated. To this end a series of computer models have been developed, including a complex three-dimensional force analysis of the pump and the generation of a series of parametric finite element models of its components.

The force analysis model is based on a straight axis axial piston pump as shown in Figure 1 and is completely parametric. The parametric form of the program allows the analysis of any equivalent pump, to assess the effects of a change in component size or loading; for example, the changes in load due to altering the bore pitch circle diameter, swash angle, piston length or supply pressure may be determined. The pump is analysed under static and inertial loading thus simulating normal continuous running. All contacts within the pump are assumed to be mechanical except for the port-plate/cylinder block contact. The loads at this contact are calculated by a knowledge of the pressure distributions on the lands, which are then summed over an area in the normal way to give a resultant force (16,17,18). Any of the point contacts may be distributed during the subsequent finite element analysis of individual components to avoid the false effects of the point load. The complete model (19) cannot be described here, but the type of model used is illustrated in Figure 2, which shows the forces acting on a piston and slipper. The driving forces for the piston model are the pressure in the bore above the piston and the inertia forces due to the reciprocating and rotating parts. The bore pressure is determined by an analysis of the timing of the pump (20,21,22).

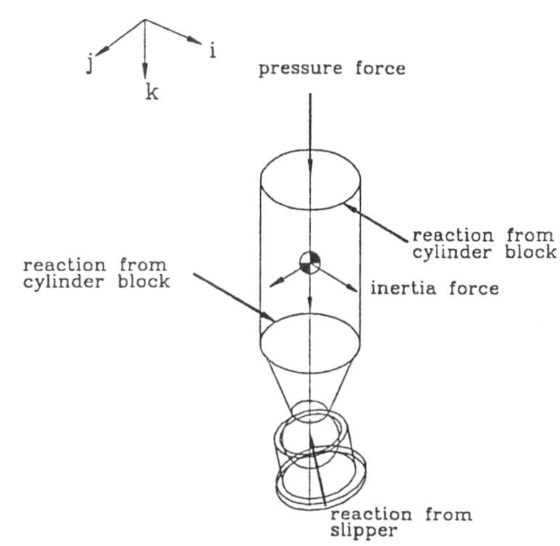

Figure 2. Forces acting on a piston during operation.

The program predicts the reaction forces acting on the piston and the position vectors of those forces which can then be used as input to the finite element models to obtain the stress distributions in the components. However, the contacts between the piston and cylinder bore are in fact distributed over a finite area with the bending loads on the piston causing it to tilt within the bore. The magnitude and distribution of these conformal contact pressures are complex to calculate and depend on the bore pressure, the initial clearances and the materials from which the components are made. To understand how the pressures are distributed in practice, and to gain a deeper insight into the performance of sleeved pistons and lined bores, a series of fundamental contact studies was undertaken. This ultimately allowed a prediction of the

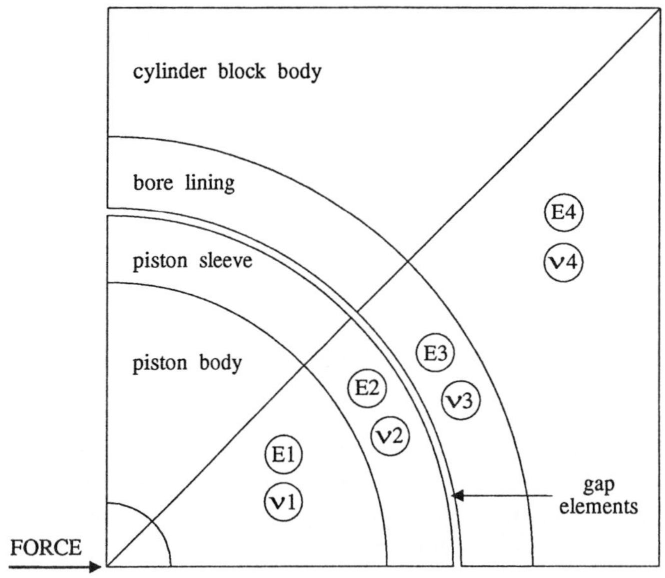

3a. Model representing one quarter of a sleeved piston in a lined cylinder bore.

3b. The Von-Mises stress contours induced by a ceramic piston in a CFRP lined cylinder bore.

stresses in the ceramic pistons and polymer composite lined cylinder block to be made (23).

The simplest finite element models used in the contact studies looked at the general case of conformal contact between cylindrical components, especially where one or other of the components is made up of more than one material as shown in Figure 3a. The models represent a quarter cross-section through a sleeved piston inside a lined bore. Gap elements connect nodes across the clearance between the piston and bore. If the gap element closes, force is transmitted across the gap, if the gap element remains open, no force is transmitted. The force applied at the centre of the piston is obtained from the force analysis. Several series of finite element analyses were undertaken to establish the effects of load, material properties and sleeve or lining thickness on the forces between the components. For example, Figure 3b shows a plot of the von Mises stress contours and the contact size produced between a ceramic 'piston' and a 'cylinder block' with a CFRP lining operating at a pressure of 140 bar.

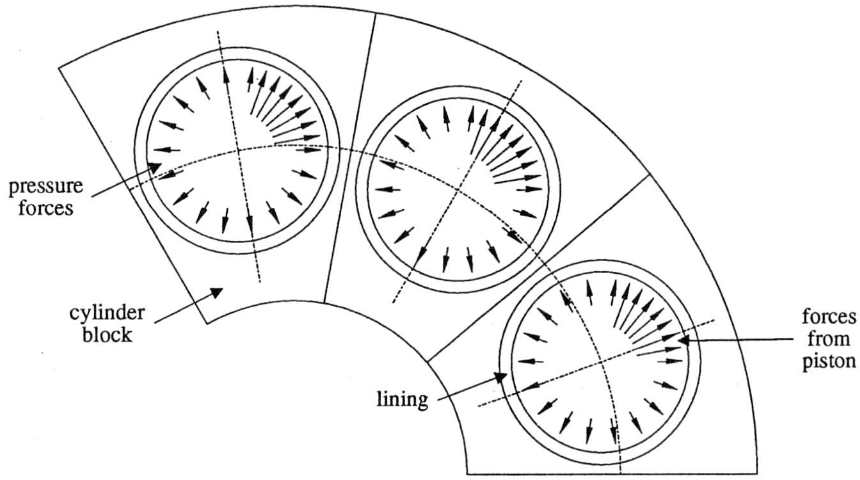

4a. Model of three cylinder bores with the cylinder pressures and piston contact forces.

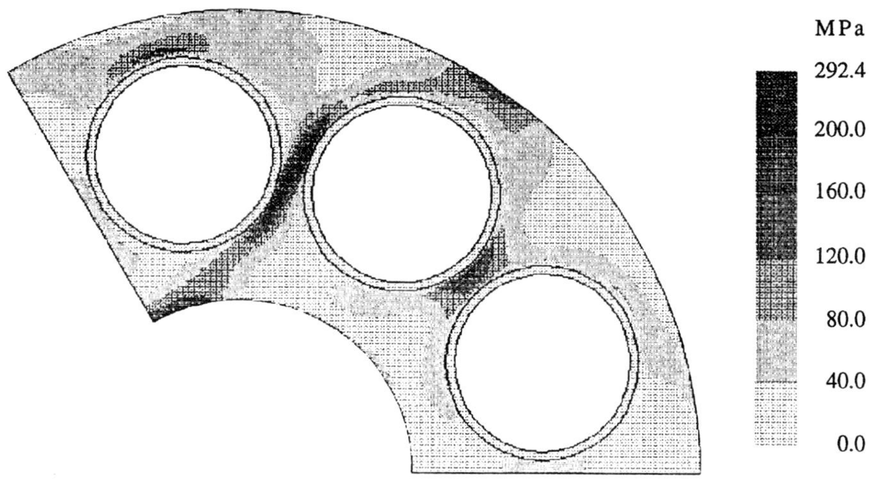

4b. The Von-Mises stress contours induced by ceramic pistons in a CFRP lined cylinder bore, including pressure effects.

The results from these analyses were then applied to a more detailed model of a cross-section through the cylinder block such as that shown in Figure 4a. The model consists of three adjacent lined bores and includes pressure and contact loading which may be different in each bore. As this pressure varies so too does the piston loading, and hence the contact area over which it acts also varies. The contact pressure loading from the pistons is obtained from the analyses above for the appropriate cylinder bore pressure. Figure 4b shows the loading and von-Mises stress contours which result from bore and piston contact pressures for a pressure of 140 bar in two bores and 5 bar boost pressure in the other for a cylinder block with polymer composite liners.

Finally, Figure 5 shows a detail of the neck region of a cross-section through a ceramic piston when loaded with the forces and reactions experienced during its operation (as summarised in Figure 2). A pressure of 140 bar is applied to the end of the piston and the model is constrained at its contact points in the cylinder bore while the load from the slipper is applied on the surface of the piston ball. This produces a stress concentration in the critical neck of the piston which can be evaluated to determine whether it can be expected to lead to failure.

3.3 Final configuration

The results of the detailed analyses on the lined cylinder block and ceramic piston allowed the final configuration and operating limits of the ceramic pistons to be established. The CFRP liners were interference fitted in to

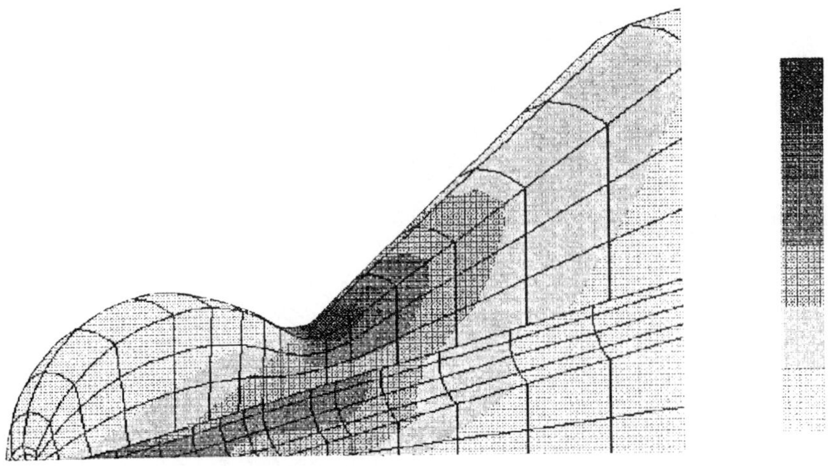

Figure 5. The Von-Mises stress contours in the neck of a ceramic piston during operation.

a standard cylinder block with enlarged bores and held in place with a keep ring. Since the positioning of the bores is crucial to the balance and operation of the pump, each bore was selectively machined to match a particular piston. The pistons themselves were 'green-machined', sintered, ground and polished with 1 μm diamond paste on their shank and ball mating faces to minimise ploughing wear of the polymer composite. A finished ceramic piston (and slipper) is shown in Figure 6.

4 PUMP TESTING

Testing of the unit was conducted on closed-circuit tap water rig boosted by a centrifugal pump with filtration of 25 μm nominal. On-line monitoring of performance was conducted using a programmable logic controller (PLC) and a logging computer which recorded all the operational parameters such as pressure, temperature, leakage and volumetric efficiency. A photograph of the unit is included in Figure 7.

The unit was gradually run-in at increasing rotational speeds and delivery pressures until 45 bar was reached. This process was achieved over a period of approximately 18 hours. Extended running then continued and pressure was raised to the design limit of the standard components which was 140 bar. A graph of the pump's performance is included in Figure 8. To date the pump has achieved a total run time of nearly 650 hours (nearly 60 million cycles), of which 450 hours have been at over 100 bar and 220 hours have been at 140 bar.

The ceramic components have performed extremely well and have even withstood failure of standard components such as slipper retaining rings and slippers which imposed severe excess loads at critical points such as the piston neck. The volumetric efficiency of the unit at 140 bar is approximately 75%. This reduction in volumetric efficiency when compared to an oil pump is due to the increased internal leakage when using water and also the increased clearances associated with the deformation of the low modulus polymer composite. However, the reduction in efficiency due to viscosity and deformation factors is partially offset by the reduced compressibility losses due to the high bulk modulus of water (8). It is important to note that the efficiency remained stable during the pump test and changed only during the first few minutes of a run as thermal and mechanical stability were achieved.

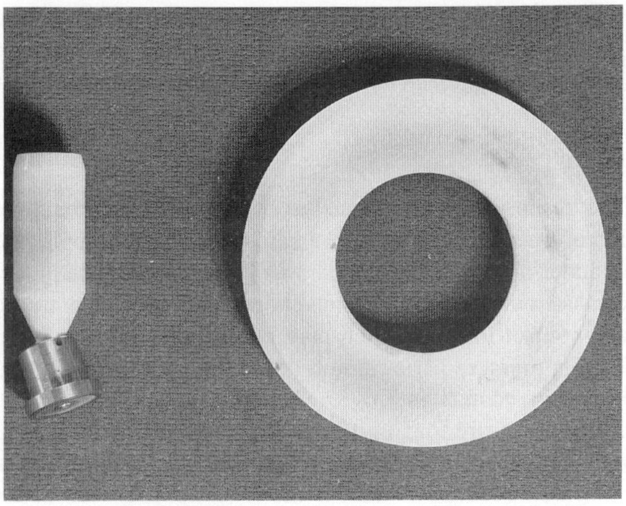

Figure 6. Photograph of a ceramic piston, (metal) slipper and ceramic swashplate.

Periodic examination of components revealed polishing of the polymer surfaces which quickly stabilised after the running in period, and eventually resulted in a 15 μm step in the liners at top dead centre. The mechanism by which this wear occurred would appear to be burnishing rather than abrasion.

Figure 7. Photograph of the test rig, with data-logger.

5 DISCUSSION

The stress raising feature of the neck on the piston would previously have ruled out the use of the more traditional brittle ceramics, however the newer engineering ceramics have proven to be much tougher. The piston is essentially undergoing a continuous bending fatigue test while the pump is in operation, and the results so far of 60 million cycles for each piston shows that fatigue of the ceramic used is unlikely to be a problem. The behaviour of advanced engineering ceramics under fatigue loading is not fully understood, and some researchers believe that they do not have a fatigue life as generally defined for metals. This implies that if a ceramic component survives the first few cycles of operation there are no gross defects in the material, and thus the component should last indefinitely. (No special examination or screening was made of the pistons used in the axial piston pump; they were used as received from the manufacturers.)

The results of the tests on the pump with ceramic pistons and swashplate have proved the viability of ceramics in this type of application. The pistons are one of the most critical components in the pump, requiring both adequate structural and tribological properties. The ceramic piston and polymer composite lined bores appear to offer this combination, and to date have proven successful. It is worth noting that the tests so far have only used tap water, and not sea-water which is the ultimate goal. How the ceramic/CFRP will tolerate debris is to be investigated shortly. However, the same polymer composite running against stainless steel in the original pump led to the wear of the steel rather than the polymer when foreign material entered the system. The precise mechanism of this wear is unknown but it seems likely that foreign particles become embedded in the polymer which then act as a lap.

Having replaced the pistons and swashplate, developments are already underway to completely eliminate the hardened stainless steel from the pump. The other critical interface in the pump is that between the back face of the cylinder block and the port-plate. Ceramic inserts are being developed for this contact, which would allow the use of a more corrosion resistant material for the bulk of the cylinder block; possibly an austenitic stainless steel or metal-matrix composite in the short term. In the longer term however, it is hoped that a complete ceramic cylinder block will be developed.

The cost of ceramics is relatively high compared to the more traditional materials, however economies of scale will reduce the unit cost significantly, as will the development of new and more automated manufacturing methods. There are of course also those applications where the cost of installing, replacing and/or maintaining similar systems (for example, subsea) far outweigh the differential cost of the units, and where significant benefits are likely to be realised.

6 CONCLUSIONS

This paper reviews some of the progress to date in the development of a sea-water pump, and in particular the application of ceramics and polymer composites to the problem. An interdisciplinary approach has been adopted to the problem, and includes an integrated program of material development and testing and advanced component design and analysis. Central to the project is the development of a sophisticated computer modelling and analysis facility, which allows a rapid evaluation of any design and material changes on the performance of the

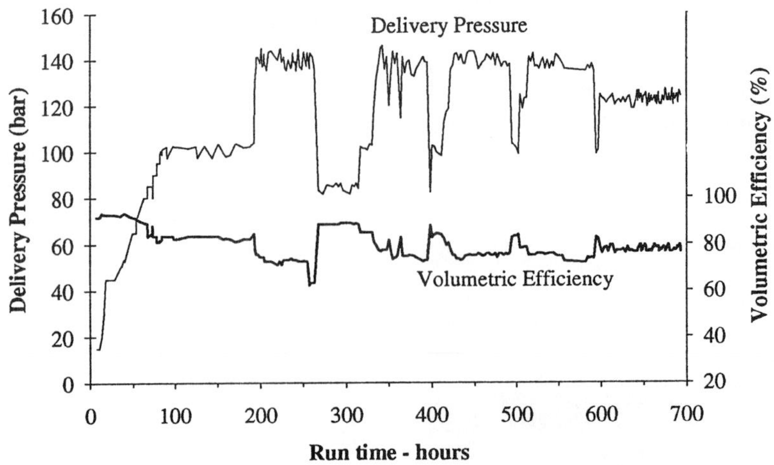

Figure 8. Graph showing pressure and volumetric efficiency of the pump.

units and components. The next phase of the work will address pump testing with sea-water, and component re-design to better exploit the potential of ceramic and polymer materials.

ACKNOWLEDGEMENTS

This work was funded by the DTI - Support For Initiative (SFI) in collaboration with J. H. Fenner PLC.

REFERENCES

(1) KNIGHT, G.C. Water hydraulics - application of water-based fluids to hydraulic systems. *Tribology International*, 1977, 105-107.

(2) TERADA, Y., KOMATSU, A., KANDA, S., INOKUMO, M. and NAGATA, Y. Development of high pressure seawater pump. *Naval Architecture and Ocean Engineering*, 1984, 22, 227-232.

(3) YOSHINADA, H., YAMAZAKI, T., SUWA, T., NARUSE, T. and UEDA, H. Sea-water hydraulic actuator system for subsea manipulator. International Symposium on Advanced Robot Technology, Tokyo, 1991.

(4) MARR, I. M. Tide turns for water fluid power. *Chartered Mechanical Engineer.* 1986, 2, 28-29.

(5) MARR, I. M. The development of hydraulic systems to operate with raw water. *Proc. 39th National Conference on Fluid Power*, 1983, 37, 258-265.

(6) CURRIE, J. A. The development of raw water hydraulics. *Proc. 1st Bath International Fluid Power Workshop*, Bath, 1988.

(7) IFIELD, R. J. The adaptation and development of hydraulic power equipment to operate with 5/95 type non-flammable fluids. *Proc. of Conference on Mining, Quarrying and Tunnelling, I. Mech. E.*, 1974, 157-162.

(8) WYMER, D. G. The use of water hydraulics in machines for mechanized gold mining in South Africa. *Intenational Conference on Mining and Machinery*, Brisbane, 1979, 320-328.

(9) BROOKES, C. A. Tribological application of ceramic materials in water hydraulics: potential and problems. *Proc. 1st Bath International Fluid Power Workshop, Bath*, 1988.

(10) RICHERSON, D. W. Modern Ceramic Engineering. 1982, Marcel Dekker Inc., New York.

(11) LIDGETT, P. J., GOSLIN, D. W. F., RODWELL, C. and RITCHIE, G. S. Hard-on-hard water lubricated bearings for marine applications, *Proc. 13th Leeds-Lyon Symposium on Tribology, Leeds*, 1986.

(12) TAYLOR, P. and MCCULLOUGH, J. Personal communication.

(13) ASANABE, S. Applications of ceramics for tribological components. *Materials and Design*, 1988, 9, 253-262.

(14) KAMELMACHER, E. Design and performance of silicon carbide product lubricated bearings. *Proc. I. Mech. E.*, 1983, 197A, 257-267.

(15) HOTHER-LUSHINGTON, S. Water lubricated bearings. *Tribology International*, 1976, 257-260.

(16) SAITCHENKO, J. S. Force balance conditions of the valve plate and rotor of an axial piston pump. BHRA Report T750, 1963.

(17) SHUTE, N. A. and TURNBULL, D. E. The thrust balancing of axial piston machines: Part 1. BHRA Report RR772, 1963.

(18) SHUTE, N.A. and TURNBULL D. E. The thrust balancing of axial piston machines: Part 2. BHRA Report RR773, 1963.

(19) RADCLIFFE, P.W., FAGAN M. J. and MCCONNACHIE, J. Computer modelling and analysis of axial piston pumps, in: Systems Modelling and Control. (Eds. C. R. Burrows and K. A. Edge), Research Studies Press, 1992.

(20) EDGE, K. A. and DARLING, J. Cylinder pressure transients in oil hydraulic pumps with sliding plate valves. *Proc. I. Mech. E.*, 1986, 200, 45-54.

(21) MARTIN, M. J., and TAYLOR, R. Optimised port plate timing for an axial piston pump. *5th Fluid Power Symposium*, 1978, 51-66.

(22) HELGSTAD, B. O., FOSTER, K. and BANNISTER, F. K. Pressure transients in an axial piston hydraulic pump. *Proc. I. Mech. E.*, 1974, 188.

(23) MCCONNACHIE, J., FAGAN M. J. and RADCLIFFE, P.W. Computer modelling of contacting concentric components, in: Systems Modelling and Control. (Eds. C. R. Burrows and K. A. Edge), Research Studies Press, 1992.

C449/012/93

The myth of cryogenic pumping

Eur Ing A J SEMPLE, CEng, MIMechE
Bechtel Limited, London
P ANSLEY-WATSON
John Brown Engineers and Constructors Limited, London

SYNOPSIS The development of pumps for cryogenic service has proceeded on the basis of modifying an existing design for the new application and then solving the problems as they arise. The "fixes" that have been adopted to solve some of the admittedly difficult problems encountered have been elevated to a proprietary status and used as the basis of a number of myths to justify why certain designs are the only ones suitable for these services. In this paper, the authors seek to expose the myths for what they are thus allowing a clear background for reasoned argument concerning the design of pumps for cryogenic service and their application to varying pumping situations.

INTRODUCTION

The development of any new engineering application tends to be accompanied by teething troubles which are resolved as operational experience is accrued.

In the arcane field of pumping low temperature fluids some of these teething troubles have proved remarkably intractable and have given rise to various myths regarding the optimum solution for the application.

The authors contend that to achieve a successful application it is necessary to understand the nature and limitations of the machine to be used. To achieve understanding it is necessary to dispel the myths.

HISTORICAL BACKGROUND

The development of pumps for cold service on volatile liquids, primarily hydrocarbons but also including the products of air separation units, has featured a number of radically different approaches over the last forty years or so.

The simplest pump known to the authors to be applied was a type of rudimentary barrel pump, comprising a multistage regenerative pump mounted in a suitably sized piece of pipe! The pressure envelope was sealed with gland packing and the rotor was suspended between a pair of grease lubricated deep groove ball bearings. The pump vendor claimed that the soap in the grease settled out at the extremities of the bearing and thus prevented egress of oil or ingress of pumped liquid. Needless to say, this offer was not accepted when made for a ship cool-down duty in the early 1980's.

As early as 1952, pump manufacturers in the United States had recognised the problem associated with the containment of highly volatile low gravity liquids and developed pump designs to handle them. Typical machines borrowed from the oil filled submersible motor technology already applied in down hole and dry dock pumping applications. These units featured an oil immersed induction motor on a common shaft with the pump impeller. The rotor assembly was supported on rolling element bearings. In volatile liquid pumping applications, the drive arrangement was contained within a pressure envelope whose internal pressure was generated by that of the pump and the additional force of the spring loaded diaphragm (fig. 1). Contamination of the motor oil was further prevented by the use of a double mechanical shaft seal between motor and pump.

This design has two significant drawbacks when applied to low temperature pumping. Firstly, the viscosity of the pressurising oil rises considerably when pumping liquids such as propane from storage at approximately -40^{0}C. Secondly, the machine is relatively expensive to build.

The development of the industrialised nations after the Second World and the various Middle East conflicts resulted in an increased demand for pumps to handle large volumes of low temperature hydrocarbons from refrigerated storage, either large shore-side storage tanks or marine tankers. This forced the development of machines that solved the dual problem of the containment of these highly volatile materials and satisfactory operation at low temperatures.

Two generic solutions to the dual problem have been developed, each with its advantages and disadvantages, both of which show a line of descent from the arrangement already described. To some extent, the cause of the problems associated with the pumping of cold liquids can be regarded as a direct consequence of the close relationship in design between the parent machine and these later ones.

In the first solution the electric motor is still contained within the pressure envelope but no longer enclosed within its own pressurised oil bath. Instead it is actually submerged within the pumped fluid. The entire rotating element is carried on two rolling element bearings which are cooled by means of pumped liquid recirculated between high and low pressure areas of the hydraulic path through the machine (fig.2).

In the second solution, advantage is taken of advances made in the development of mechanical seals capable of giving reliable service when operating under ideal conditions of temperature and lubrication. The pressure envelope contains only the hydraulic elements of the pump and the motor is mounted outside. Typically, the pump rotor is supported from an external oil lubricated thrust bearing.

The shaft egress from the pressure envelope is sealed by a double mechanical seal. The seal chamber is separated from the cold part of the pump by a distance piece, maintained at suction pressure, which allows the seal and its barrier fluid to operate close to ambient temperature, and contains a coffer dam to provide the facility to collect any inward leakage of barrier fluid. A self-contained pressuring system for the seal barrier oil is provided which uses the fluid vapour in the distance piece to give the reference pressure needed (fig.3).

In the course of marketing these two competing pump designs, various claims have been made by the manufacturers which have passed into the mythology of pump application engineering in the petrochemical industry. It is worthwhile to examine these claims and put them into context in order to evaluate their worth.

THE "MYTHS"

Effects of Low Temperature

It is often claimed that the problems associated with pumping cryogenic liquids are simply due to the fact that they are "cold". Certainly, the need to operate at very low temperatures does present some problems of material selection but these need be no more intractable than those faced by the more conventional hydrocarbon processing industry when handling fluid streams at around 400^{0}C.

Such high temperatures are regularly encountered as part of the distillation and reforming processes and are not considered troublesome. At such elevated temperatures not only is expansion of the metal itself a problem, but other properties such as creep and other metallurgical aspects combine to make the impact of expansion factors somewhat more arduous than, for example, the mere 192^{0}C temperature differential found in cryogenic duties.

At low temperatures toughness must be considered but there is an extensive range of materials, both austenitic stainless steels and aluminium alloys that have been developed for service under these conditions. It is, perhaps, worth noting that the manufacturer of a particular design of vertical high speed pump driven by an integral gear speed increaser has achieved respectable performance of his machines in low temperature service using only solutions featured in his standard range of process pumps.

Operation over a wide range of temperature can, however, cause problems and this is addressed later in the paper.

Containment

It is often claimed that with the motor submerged in the pumped fluid there is no possibility of any leakage of flammable vapour from within the pump pressure casing to the outside atmosphere. However, risk assessment recognises that gasketted joints such as the closure of the pressure envelope of a pot mounted submerged motor pump can and do fail.

During the engineering of a recently commissioned dense phase ethylene pipeline an assessment was carried out to determine the risk of a major incident occurring as the result of a leakage at an intermediate pump station. While the assessment did find a reduced risk resulting from the use of submerged motor pumps, the higher risk associated with more conventional machinery was still so small as to be acceptable.

Secondary leakage paths

Instances have been reported in which light hydrocarbon vapours have escaped by permeating through the insulation of electric cables and then travelling along between the conducting core and the inside of the insulation before eventually escaping into a non-classified area such as the basement of a plant control room.

With the motor mounted within the pressure containment vessel, it is particularly important to ensure that this does not occur. In recent years considerable care has been taken by the manufacturers to ensure the integrity of their electric penetrations at the pressure boundary. The effect of this effort may be judged from the reports that on one dense phase ethylene pipeline using submerged motor pumps, facilities were provided to detect any such leakage but in the first two years' operation none had been observed(1).

Bearing life

Unfortunately bearing life for submerged motor machines still appears to be a problem since horror stories are frequently retailed as to machines that could barely last a shift without suffering bearing failure. Against these must be set the majority of units about which little is heard but presumably give reliable service. Typically, pumps cited in [1] had achieved some two years uninterrupted service by May 1991.

Although the products handled by the pumps under discussion here range from Butane at 4°C to Liquid Nitrogen at -192°C, the designs of submerged motor pumps applied owe their existence to air refrigeration pumping applications. These units employ ball bearings operating in the pumped fluid and, as indicated above, have acquired an unenviable reputation for short bearing life when applied to fluids exhibiting poor lubricity.

Early applications of the submerged motor machines were for tanker loading and unloading duties where limited duration operations meant that the limited bearing life exhibited by antifriction bearings in fixed registers could be tolerated.

As demand and consequent duration of pumping increased, attempts were made to solve the problem by the application of hydraulic balancing of the rotor.

Obtaining submerged motor machines with a guaranteed bearing life that complies with the API 610[2] specified Bl0 value of 25,000 hours can be difficult. Typically suppliers of these pumps will only predict a life of 8,000 hours and only very recently have been willing to guarantee this. However, it is noted that one manufacturer of submerged motor pumps is about to launch a 16,000 hours bearing life guarantee (under strict terms and conditions).

There is extensive operating experience with other designs of glandless pumps of both the submerged motor and magnetic or eddy current drive type, both of which tend to use ceramic bearings of the hydrodynamic film type. Curiously, no attempt seems to have been made to transplant this hydrodynamic bearing technology already successfully applied on submerged motor pumps handling liquid ammonia, a product with very similar viscosity characteristics.

Hydraulic Balance

Despite this consistent failure to meet the user industry's criteria for successful long term operation of its pumping equipment as expressed in API 610, one of the myths proposed by the cryogenic pump manufacturing industry is its expertise in resolving the problem of hydraulic balance of the impellers. While it is true that improvement has been achieved following the application of hydraulic balancing to reduce axial loadings on the bearings during operation, a myth does appear to have been fostered as a smoke-screen to divert the users attention from the fact that the pumps cannot meet the requirements of bearing life demanded of conventional machinery.

In fact the mainstream pump industry resolved this problem some 30 years ago, as in the balancing of axial hydraulic thrust generated in multi-stage ring section pumps on boiler feed-water service. When pressed on this point the manufacturers have been observed to take refuge behind the effects of the low fluid viscosity of cryogenic liquids.

Comparison of the cross-section of a typical submerged motor pump with that of a conventional alternative reveals a different story, however. In a conventional pump, the residual thrust is maintained in the direction away from the thrust bearing, thus keeping the shaft in tension. In a vertical machine this results in a down-thrust which has to be added to the rotor weight when sizing the thrust bearing from which the rotating assembly is suspended.

In submerged motor pumps the net axial thrust is reversed in order to act upwards against the effect of gravity on the rotor mass. This is achieved by the use of larger wear ring diameters on the hub side of the impellers than employed on the eye side. Any net axial thrust that arises once the pump is at running speed must be limited to that variation in thrust from the value at the rated pumping capacity with change in pump throughput. With appropriate impeller design this variation can be kept within the limits which can be handled by the balance piston arrangement with which these pumps are often fitted and the restoring forces arising in the motor when the squirrel cage is caused to move away from the magnetic centre. Of course, if the magnetic pull is considerably higher than calculated, as could be the case in the event of distorted rotor bars, then bearing life could be compromised.

Thus for a submerged motor pump designed to fully balance its axial thrust loadings, there need be no axial loads for the bearings. Problems that can be foreseen with the bearings are those associated with the ability of the outer race to move quite freely in its housing. Frequent variations in thrust, and hence rotor axial position, can conceivably lead to fretting corrosion between the two while the freedom of the race to move could allow it to rotate leading to the possibility of pick up and severe scoring of the race outer circumference and the housing register in which the race sits. There is, however, little reported evidence of these effects being seen. Instances have occurred where bearing races have cracked suggesting the adoption of incorrect cooldown procedure[3].

As noted in the discussion on the bearings themselves, it seems curious that no attempt has made been to adopt hydrodynamic bearings by this group of pump manufacturers.

Viscosity Effects

Though often quoted as problem, it is not clear why the viscosity of the liquids should present a particular problem, especially since the bearings should not carry any thrust load under normal operating conditions. The concept of this problem becomes more difficult when it is realised that axially split horizontal multistage pumps running at speeds in excess of their first lateral critical in air have been used to pump Ammonia as well as the light hydrocarbons under consideration here for at least the past quarter century.

Such machines rely on the differential pressure across the wear rings to provide adequate shaft damping to control the rotor behaviour and a balance drum to contain the axial thrust generated within the pump and in this respect are no different from their submerged motor counterparts. Unlike the submerged motor machines, however, they do have a long history of reliable service when equipped with rolling element bearings, albeit oil lubricated.

One possible explanation of this discrepancy may be the very nature of ball bearings. In order for a ball bearing to function correctly, it is necessary for the balls and races to be manufactured from hardened materials. Selection of the bearing material for the normal operating temperature is critical to the bearing life of the machine. However, normal bearing materials are not recognised as suitable for low temperature service.

Variable Speed Operation

With increasing emphasis on energy efficiency, variable speed pumping is being employed on pipeline applications to optimise transmission efficiency under all flow regimes. With conventional pumps it is comparatively easy to avoid shaft resonances coinciding with the operational speed. With the submerged motor design, the pump essentially comprises one or more impellers on the end of a motor shaft and thus critical speeds are strongly influenced by motor dynamics and hence are not so amenable to tuning. As a consequence the available operating speed range tends to be defined by the first and second critical speeds of the rotor thus restricting operational flexibility.

To obtain variable speed with submerged motor units it is necessary to use a variable frequency power supply. Experience has proven that the design package of such systems ideally includes the electric motor but this is not always practicable with submerged motor designs. The latter is usually integrated with the pump design and considerable electrical alignment problems have been known to arise with the familiar current source inverter system. In one UK installation, two years after commissioning, it was reported that of the two units installed, only one could be relied upon to run for any length of time [1]. With the advent of readily available voltage source converters this situation can be expected to improve as the necessity to match the inverter characteristics to those of the submerged motor becomes less important.

Specific Speed

Another feature of submerged motor machines is that their design tends to employ a lower specific speed for the impeller than is found in conventional machinery, as it is desirable to reduce the number of impellers hung from the submerged motor shaft. While this results in a pump with a shorter axial length, reducing site construction costs, this feature has tended to result in a lower hydraulic efficiency which, coupled with the electrical losses from the motor, increases the heating of the fluid during its passage through the pump. However, this is unlikely to be a significant problem on the low power pumps generally used in transfer of liquid between refrigerated storage facilities.

Internal Recirculation

All centrifugal pumps recycle a portion of their throughput round each individual impeller. Normally, however, the resultant rise in the temperature of this recycled fluid does not have any significant impact. In the submerged motor pump a considerable rise in temperature can occur in that portion of the pumped fluid which is recirculated through the motor bearings. Eventually this fluid is returned to the pump suction by which time it may well be above its boiling temperature for the specified suction conditions and its appearance in the proximity of the impeller eye can harm the pump's NPSH performance. This is resolved most ingeniously by installing an inducer to ensure that the pump can handle the suction conditions it creates for itself rather than those the user believes he has created. Inducers, by their nature, only operate effectively over a relatively narrow flow range, thus limiting the turndown capability of the pump. This has been thought to contrast poorly with the turndown performance of the conventional machine.

In recent years, however, the better understanding of the behaviour of fluid passing through centrifugal pumps achieved by Fraser[4] and others suggests that conventional pumps have less turndown ability than was previously thought.

In the area of turndown capability there is probably little to choose between the two types of machine.

Seal leakage

It is generally acknowledged that mechanical seals leak. Certainly, with any mechanical seal the potential for leakage, both inwards as well as outwards, exists.

However, once that potential has been acknowledged, it is relatively easy to design a system which not only prevents the flammable liquid from leaking outwards, but which also restricts to an absolute minimum any contamination of the pumped fluid by seal barrier fluid.

Such systems have been described in detail before[5] and in essence comprise a double mechanical seal with the barrier fluid pressurised from the pump suction via some form of servo device which also serves to segregate the two fluids. This allows the inboard seal to run with the minimum differential pressure required to maintain the liquid film equilibrium between the faces. In more sophisticated designs, a distance chamber is provided between the pump suction chamber and the mechanical seal which allows the seal to run at or approaching the atmospheric ambient temperature rather than that of the cold fluid. This eliminates cold starting problems associated with ice formation at the seal faces and ensures an ideal operating environment for the mechanical seal.

It is not unknown for pumps featuring such seal arrangements to run without attention and without failure for as long as eight years, approximately 64000 hours (in excess of twice the B10 bearing life demanded by API 610).

The use of the distance chamber facilitates the collection of any inward leakage by causing it to be flung outward away from the pump shaft and allowing it to be collected in the lower extremity of the chamber. In a particular installation pumping cold ethylene, facilities were provided to permit blowdown of any such leakage but during the first two years operation no leakage had been noted.

CONCLUSIONS

It is the authors opinion that considerable effort has been invested by a limited number of manufacturers in attempting to persuade users of cryogenic pumping equipment that the submerged motor pump is the only solution. The authors contend that, given the problems that are known to occur during the operation of these machines, the case is not proven.

In fact the application of conventional pumps with purpose designed mechanical sealing systems can provide more reliable and more economic solutions than the submerged motor pump in its present state, apart from such applications as in-tank pumps, both shore-based and on board liquified gas tankers, where the submerged motor pump offers particular advantages in maintenance handling.

It is hoped that this discourse has identified those of the "myths" expounded by the industry which are real from from those that are less clearly so.

Those "myths" which can be dispensed with are:

- Low Temperature
- Containment
- Hydraulic Balance
- Seal Leakage

The "myths" which are in reality problems needing to be addressed at the enquiry stage are:

- Secondary Leakage Paths
- Bearing Life
- Viscosity Effects
- Variable Speed Operation

It must be stated that the observations made in this paper cannot all be laid at the door of the equipment manufacturers. To some extent the continuing status quo in the design of pumps for this type of application has to be the result of the failure of the end users to actively encourage improvements in design, especially in those areas affecting pump life.

Improvements in the operating life of these machines will have a dramatic effect on operating costs and hence on the profitability of the process.

THE FUTURE

It is the authors opinion that the future for the sealless pump is assured, due primarily to the drive from legislation[6]. To this end, and to meet the challenge from alternative technologies the submersible pump manufacturers will have to take note, at least in part, of the comments made in this presentation.

Review of the up and coming technologies would indicate that the potential for magnetic drive pumps will be a serious challenge to the present strong position of the submersible motor pumps. Available power of magnetic drive machines has been steadily increasing in recent years, and the authors have been party to a novel pumping configuration which would overcome a number of complaints presently levelled at this type of drive system.

The double mechanical seal arrangement must be regarded as having reached the pinnacle of its development and pumps featuring it will be supplanted with one of the alternatives mentioned above.

The authors would like to suggest the following areas for future development:

Bearing Design

Hydrodynamic bearings and alternative bearing materials should be actively investigated, especially the use of modern materials technology. Ceramics and advanced polymers would appear suitable candidates for this particular application.

Hydraulics

The present range of hydraulic coverage from both the submerged motor and external motor manufacturers can restrict the operation of the plant. In addition, planned LNG operations would appear to be increasing in size; in a recent study it was necessary to run twelve pumps in parallel to meet the design throughput. This must be considered excessive.

Variable Speed

The use of variable speed operation should be more actively considered as this can significantly improve the operating range of the plant, and reduce operating costs. Manufacturers need to investigate this area to ensure compatibility between their machines and the speed control package.

With development along the lines suggested, the submerged motor pump is the the way forward for cryogenic pumping applications.

REFERENCES

1. J.W.Wem & D.M.Cullen, "Supercritical Ethylene Pipeline Boosters, Pumps or Compressors?", I.Mech.E., 23 May 1991.

2. API Standard 610, "Centrifugal Pumps for General Refinery Service", 7th. Edition, February 1989.

3. R.H.Watkins, "Experience gained from Submerged Motor Pumps in Cryogenic Service", I.Mech.E., C156/80, 1980.

4. W.H.Fraser, "Recirculation in Centrifugal Pumps", Worthington Group, McGraw Edison Company.

5. P.Ansley-Watson, "Mechanical Seals for Flammable Liquids", I.Mech.E. Seminar, "Shaft sealing in Centrifugal Pumps", 12 February 1992.

6. P.C.Barnard, "Process Pumps - The Way Forward", I.Mech.E. Seminar, "Process Pumps - The state of the art", 1991.

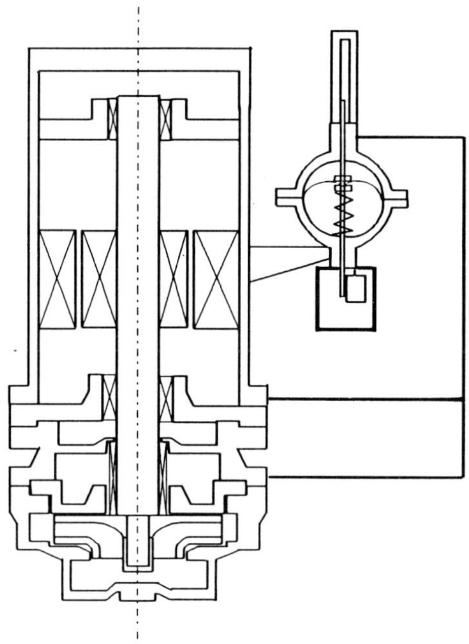

Fig.1 Oil filled motor-pump assembly

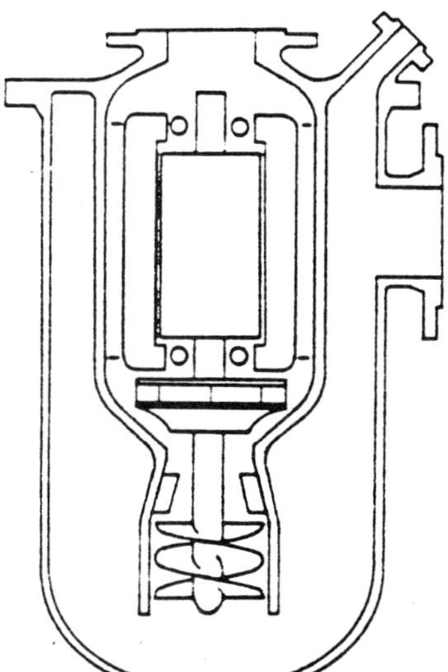

Fig.2 Submerged motor-pump assembly

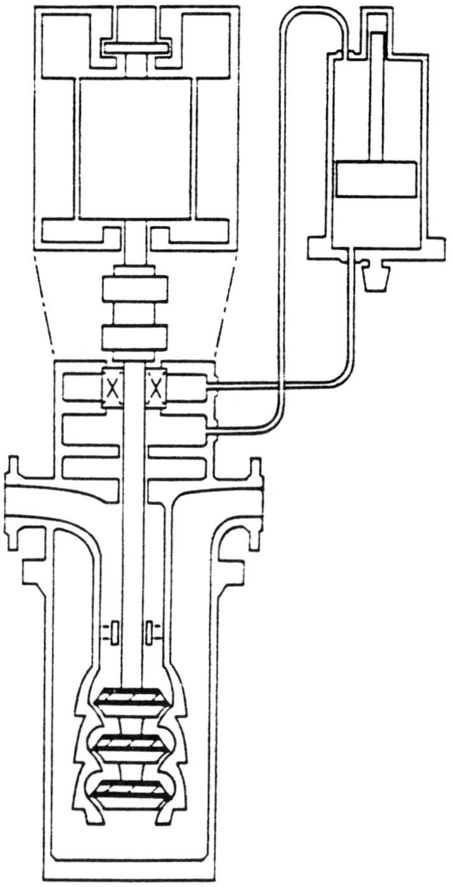

Fig.3 Canned pump with mechanical seal

C449/032/93

Wear resistant materials for oil-field fluid machinery – a selection procedure

Eur Ing R J K WOOD, BSc, PhD, CEng, MIMechE, MICorr, D COOPER, BSc, PhD, CPhys, MInstP and F A DAVIS, BSc, PhD, CEng, MIM
BP Research, Sunbury-on-Thames

SYNOPSIS Fluid machinery used in BP operations is subjected to extremely hostile environments which impose a variety of wear mechanisms on components. Down time due to wear and component failure can be very costly and the need to select materials based upon an understanding of tribological properties is essential. The growth in the transfer of tribo-technology is slow. It is often assumed that selecting materials on the basis of suitable mechanical, thermal and physical properties will automatically lead to their possessing wear mitigating properties. This is often not true. This paper describes a suite of laboratory test methods and a procedure to interpret wear data developed by BP to aid material selection.

1 INTRODUCTION

Oil-field fluid machinery can be subjected to extreme forms of adhesive, abrasive and erosive wear. To overcome these wear problems, critical components are hard-faced or surface treated. The selection of these surfaces is complicated and is hindered by the absence of the relevant wear data. To provide such tribo-technology transfer BP have developed a suite of wear test methods and a procedure to interprate results to aid material selection and highlight potential tribological failures in oil production fluid machinery (1-2). Materials investigated included examples currently in use, such as the cobalt based "Stellite" alloys, tungsten carbides, silicon carbides and silicon nitrides.

To illustrate the use of the material selection procedure this paper considers three typical components of oil-field fluid machinery; rotary pump seals, multi-phase pump impellers and gate valve surfaces. These components are exposed to a range of wear mechanisms in service but usually a different dominant mechanism will lead to failure. Mechanical seals on rotary pumps suffer from abrasive type wear in operation which down-rates the seal performance. Figures 1a and 1b illustrate abrasive wear of tungsten carbide seal ring face. A multi-phase pump impeller is subjected predominantly to sand erosion wear mechanisms. Typical eroded surfaces of a pump impeller are shown in Figures 2a and 2b. Adhesive wear/galling is a catastrophic wear process occurring in gate valves as seen in Figures 3a and 3b.

2 EXPERIMENTAL

2.1 Test Methods

Three test techniques were developed to investigate the wear characteristics of the materials exposed to the three wear mechanisms. Wear volume was measured using a linear variable differential transformer or deduced from mass loss. Dry sliding, adhesive wear tests were carried out using a pin-on-disc machine (Figure 4a) coupled to a microcomputer for data-logging. The tests were set up so that a flat ended cylindrical pin sample was loaded against a horizontal rotating disc sample. Test loads between 5 and 50N and a sliding speed of $1ms^{-1}$ were used. The same pin-on-disc machine was employed for the abrasive tests.

Initial tests were carried out with a 180 grit silicon carbide (SiC) abrasive paper bonded to a steel support disc. Further tests were made using a synthetic sandstone disc. Test loads of 10N and 50N, and sliding speeds of $0.01ms^{-1}$ and $1ms^{-1}$ were used. Erosion resistance was evaluated using a sand/well water slurry erosion rig (Figure 4b). Two conditions were investigated: low energy (particle size $120\mu m$ and a jet velocity of $16.5ms^{-1}$ at 25°C), and high energy ($235\mu m$ particles at $30ms^{-1}$ and 50°C).

2.2 Materials Evaluated

Table 1 lists the materials evaluated and typical properties at room temperature.

3 RESULTS

An example of dry sliding, adhesive wear data for Stellite sliding against itself or a ceramic is shown in Figure 5(a). A more useful way of presenting this data is to express it as a specific wear rate (m^3/Nm); this has the advantage of highlighting fundamental aspects of the wear process, in particular, transitions. Figure 5(b) shows the above results expressed as specific wear rates; the upper bar represents wear of the pin and the lower that of the disc. For Stellite sliding against itself, the transition from mild to severe wear is clearly illustrated and can be confirmed by metallographic analysis. Table 2 lists the specific wear rates for the material combinations investigated and transition loads where appropriate.

In the abrasion tests it was found that against 180 grit SiC paper the wear rate decreased with sliding distance. Data expressed as specific wear rates revealed a constant 2-body abrasive rate for the first revolution of the disc which then decreased due to abrasive degradation. Sandstone discs offered an abrasive which did not degrade with time and a constant wear rate resulted. 2-body abrasive wear rates obtained with both SiC paper and sandstone are listed in Table 3.

Analysis of the erosion results show that the scar profiles fall into two categories: Ductile materials show a W-shaped scar while brittle materials exhibit a U-shaped scar. From the overall volume loss measurements the volume loss per particle impact was calculated (3) as shown in Table 3.

4 DISCUSSION

Pin-on-disc tests show the characteristic transition from mild to severe wear for Stellite sliding against either itself or a ceramic disc. However, against the ceramic counterface a higher transition load was recorded. For ceramic pins against a Stellite disc, severe wear at high load occurred due to thermal spalling which resulted in cutting/abrasive wear of the disc. In the case of silicon carbide sliding against itself, wear occurred by the combination of polishing (fine abrasion) and spalling. Above the transition load spalling predominates. For silicon nitride against itself abrasive wear occurred along with deposits which increased with load. Tungsten carbide, either sliding against itself or a ceramic counterface exhibits mild wear performance and offered an acceptable material counterface. The best combination to date was, however, silicon nitride against tungsten carbide.

Against SiC paper, the softer Stellite material exhibited wear rates which fell with increasing load. This was due to increased sliding of the abrasive, rather than cutting. No such effect was observed for the hard tungsten carbides or ceramics. Variations in sliding speed had no effect on wear for the conditions studied; this enabled the time duration of tests to be optimized and results from a range of distances to be compared. Against SiC paper, Stellite exhibited the highest high wear rate. This was considered to be excessive and not related to the practical situation where Stellite is used successfully in many abrasive applications. The difference in the laboratory test and field experience was considered to be due to the hardness of SiC abrasive (~2600Hv). Sandstone, with a hardness range 550-800Hv provided a more representative abrasive and enabled more useful discrimination to be made between materials.

In the erosion tests, the slurry jet produces a range of particle impingement angles over the target surface, decreasing from normal incidence at the centre of the jet. Ductile materials exhibit maximum damage at an angle of 20-30°, to generate a W-shaped profile. Brittle materials exhibit maximum wear at normal incidence, giving a U-shaped scar, see reference (1). The W-shape scar profiles obtained for silicon carbide and stellite indicate ductile behaviour while silicon nitride and tungsten carbide exhibit brittle behaviour. The materials ranking in order of increasing erosion resistance, considering both high and low energy erosion results, is stellite, silicon nitride, tungsten carbide and silicon carbide.

Transfer of laboratory data to engineering design is complex due to the combination of wear mechanisms that occur in practice. To facilitate this technology transfer, a rating system might be envisaged in which the impact of each of the relevant types of wear is expressed using a rating system. Table 4 illustrates the basis of such a system.

Using this rating system, wear data (Tables 2 & 3) could be presented as material performance charts (Figure 6). Selection of materials/surfacing processes would be then based upon a weighted appraisal of the influence of relevant wear mechanisms specific to an application. Table 5 gives an example of how such an approach might be used to select a pump rotor material in combination with various casing materials.

Summation of the product of the test rating and the weighting factors would give a Stellite rotor a weighted average rating of 5.30 compared to 2.40 for silicon carbide. Silicon carbide would therefore be the preferred material on wear resistance grounds. The difference in performance between Stellite and silicon carbide represents over 40 times improvement in wear resistance.

Similar guidance to material selection for mechanical seals and gate valves are given in Tables 6 and 7. Despite the poor rating for Stellite by the authors it is recognised that this material is successfully used in many gate valve applications and problems only occur when conditions become very severe, requiring a more costly material solution. These tables are guide to aid designers if problems occur.

For mechanical sealing faces it can seen that silicon carbide sliding against itself is superior to a silicon nitride combination. Such results agree with the findings of Komiya et al (4).

5 CONCLUSIONS

5.1 Adhesive wear

Tungsten carbide, either sliding against itself or a ceramic counterface exhibits mild wear performance and offered an acceptable material counterface. The best combination to date was, however, silicon nitride against tungsten carbide.

5.2 Abrasive wear

Against SiC abrasive, sliding speed had no effect on wear and no significant differences were detected between the materials tested. This was thought to be due to the high hardness of the SiC abrasive. Sandstone abrasive offers better discrimination between the materials.

5.3 Erosive wear

From the W-shape erosion scar profiles of silicon carbide and stellite these materials show ductile behaviour while brittle scars result for silicon nitride and tungsten carbide. The materials ranking in order of increasing erosion resistance is stellite, silicon nitride, tungsten carbide and silicon carbide.

5.4 Selection Procedure

A proposed weighted rating appraisal scheme is presented which provides the design engineer with a material selection methodology based upon a fundamental understanding of wear mechanisms.

ACKNOWLEDGMENTS

The authors would like to thank Messrs M Bunte, M Binfield and J Lewis for carrying out much of the experimental work and preparation of the figures. Permission to publish this paper has been given by the British Petroleum Company plc.

REFERENCES

(1) COOPER D, DAVIS F A & WOOD R J K, Selection of wear-resistant materials for the petrochemical industry, *J.Phys.D:Appl.Phys.*, 25, (1992), A195-A204.

(2) DAVIS F A, COOPER D & WOOD R J K, Engineering of wear-resistant surfaces for the oil industry, *presented at Nordtrib 1992 Helsinki , 5th Nordic Symposium on Tribology, Vol 2, published in Tribologia, Finnish Journal of Tribology*, Vol 11, No.3, 1992, p 107-116.

(3) MOORE A J and WOOD R J K, Erosive wear mapping of pipeline materials, *Plastic Pipes VIII conference*, Koningshof, Sept. 1992.

(4) KOMIYA M, MATSUI S & HIRABAYASHI H, Investigation on surface failure of hard materials for mechanical seals, *Proc. of 13th International Conference of Fluid Sealing*, Brugge, Belgium, April 1992.

TABLE 1 Materials evaluated and typical properties at room temperature

	PHYSICAL	MECHANICAL							THERMAL		
Material	Density (ρ) kg/m^3	Flexural Strength MPa	Compression Strength MPa	Ultimate Tensile Stress (σ_f) MPa	Yield Stress (σ_y) MPa	Hardness H_{v30}	Young's Modulus (E) GPa	Fracture toughness (K_{IC}) $MPa\sqrt{m}$	Coefficient of thermal expansion (α) $\mu m/mK$	Specific Heat $Jkg^{-1}K^{-1}$	Thermal Conductivity (k) W/mK
Stellite	8440	-	1680	900	540	415	210	-	11.4	418	15
Tungsten Carbide	14950	2640	5690	-	-	1600	630	9.8	4.6	-	100
Silicon Carbide	3100	550	3900	160	-	2800	410	4.6	4.0	670	126
Silicon Nitride	3275	190	550	240	-	1320	315	4.0	3.0	800	16

TABLE 2 Dry sliding, adhesive wear results

Material	Combination	Wear Rates x 10^{17} (m^3/Nm)		Transition Load (N)
Pin	Disc	Pin	Disc	
Stellite	Stellite	550-650	560-3700	5
Stellite	Silicon Carbide	200-350	50-140	20
Stellite	Silicon Nitride	550-1000	400-820	20
Tungsten Carbide	Tungsten Carbide	12-30	20-35	
Tungsten Carbide	Silicon Carbide	25-40	100-350	
Tungsten Carbide	Silicon Nitride	30-45	25-90	20
Silicon Carbide	Silicon Carbide	260-400	750-825	20
Silicon Carbide	Stellite	150-360	510-10 500	
Silicon Carbide	Tungsten Carbide	20-2000	10-610	
Silicon Nitride	Silicon Nitride	4000-7500	44 000-120 000	
Silicon Nitride	Tungsten Carbide	2-5	30-90	
Silicon Nitride	Stellite	200-350	1400-7500	

TABLE 3 Abrasive and erosive wear results

Material	Abrasive		Erosive	
	SiC ($10^{-12}m^3/Nm$)	Sandstone ($10^{-15}m^3/Nm$)	Low Energy (μm^3/Impact)	High Energy (μm^3/Impact)
Stellite	15.4-35.9	34.7-64.7	0.105-0.147	1.34
Tungsten Carbide	4.1-5.0	2.8-6.7	0.0117-0.0143	0.52
Silicon Carbide	6.0-7.9	0.52-5.00	0.0024-0.0029	0.13
Silicon Nitride	9.5-12.0	7.6-9.0	0.032-0.057	1.76

TABLE 4 The wear rating system

WEAR TEST	WEAR UNIT	TEST RATING					
		0	2	4	6	8	10
Dry sliding, adhesive	m^3/Nm	10^{-17}	10^{-16}	10^{-15}	10^{-14}	10^{-13}	10^{-12}
Dry abrasive on silicon carbide	m^3/Nm	10^{-14}	10^{-13}	10^{-12}	10^{-11}	10^{-10}	10^{-9}
Dry abrasive on sandstone	m^3/Nm	10^{-16}	10^{-15}	10^{-14}	10^{-13}	10^{-12}	10^{-11}
Low energy sand erosion	μm^3/impact	10^{-4}	10^{-3}	10^{-2}	10^{-1}	1	10
High energy sand erosion	μm^3/impact	10^{-2}	10^{-1}	1	10	10^2	10^3

TABLE 5 Material selection for pump rotor

Pump Rotor Material		Silicon Carbide		Stellite	
Pump Case Material		Stellite		Stellite	
WEAR TEST	Weighting Factor (WF)	Test Rating (TR)	WF x TR	Test Rating (TR)	WF x TR
Continuous dry adhesive	0.00	3	0.00	6	0.00
Intermittent dry adhesive	0.05	4	0.20	7	0.35
Dry abrasive on SiC	0.00	5	0.00	6	0.00
Dry abrasive on sandstone	0.15	2	0.30	5	0.75
Low energy sand erosion	0.50	2	1.00	6	3.00
High energy sand erosion	0.30	2	0.90	4	1.20
Weighted Average Rating Σ (WF x TR)			2.40		5.30

TABLE 6 Material selection for mechanical seal faces

Seal Ring Face		Silicon Carbide		Silicon Nitride	
Stationary Seal Face		Silicon Carbide		Silicon Nitride	
WEAR TEST	Weighting Factor (WF)	Test Rating (TR)	WF x TR	Test Rating (TR)	WF x TR
Continuous dry adhesive	0.05	4	0.20	7	0.35
Intermittent dry adhesive	0.05	5	0.25	10	0.50
Dry abrasive on SiC	0.40	5	2.00	6	2.40
Dry abrasive on sandstone	0.35	2	0.70	3	1.05
Low energy sand erosion	0.15	2	0.30	5	0.75
High energy sand erosion	0.00	2	0.00	4	0.00
Weighted Average Rating Σ (WF x TR)			3.45		5.05

TABLE 7 Material selection for gate valve surfaces

Gate Valve Face		Tungsten Carbide		Stellite	
Gate Valve Seal		Silicon Nitride		Stellite	
WEAR TEST	Weighting Factor (WF)	Test Rating (TR)	WF x TR	Test Rating (TR)	WF x TR
Continuous dry adhesive	0.30	2	0.60	6	1.80
Intermittent dry adhesive	0.30	3	0.90	7	2.10
Dry abrasive on SiC	0.00	5	0.00	6	0.00
Dry abrasive on sandstone	0.20	2	0.40	5	1.00
Low energy sand erosion	0.15	4	0.60	6	0.90
High energy sand erosion	0.05	4	0.20	4	0.20
Weighted Average Rating Σ (WF x TR)			2.70		6.00

FIGURE 1a Abraded 4.5" diameter mechanical seal ring of width 0.5"

FIGURE 2a Eroded 4.5" pump impeller

FIGURE 1b Example of an abraded tungsten carbide surface

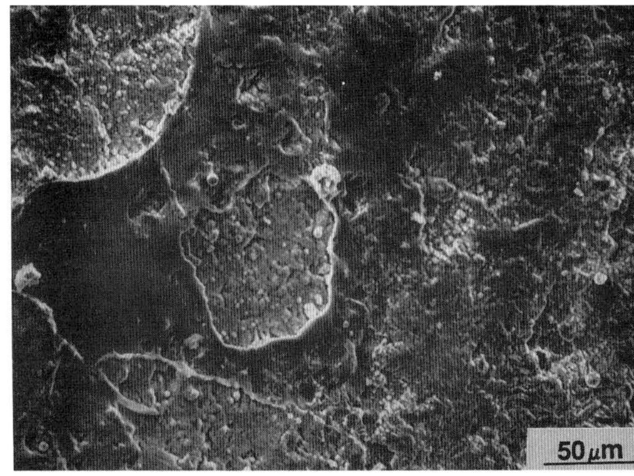

FIGURE 2b Example of an eroded high crome white cast iron surface

FIGURE 3a Galled 4" gate valve components

FIGURE 3b Example of a galled Stellite surface

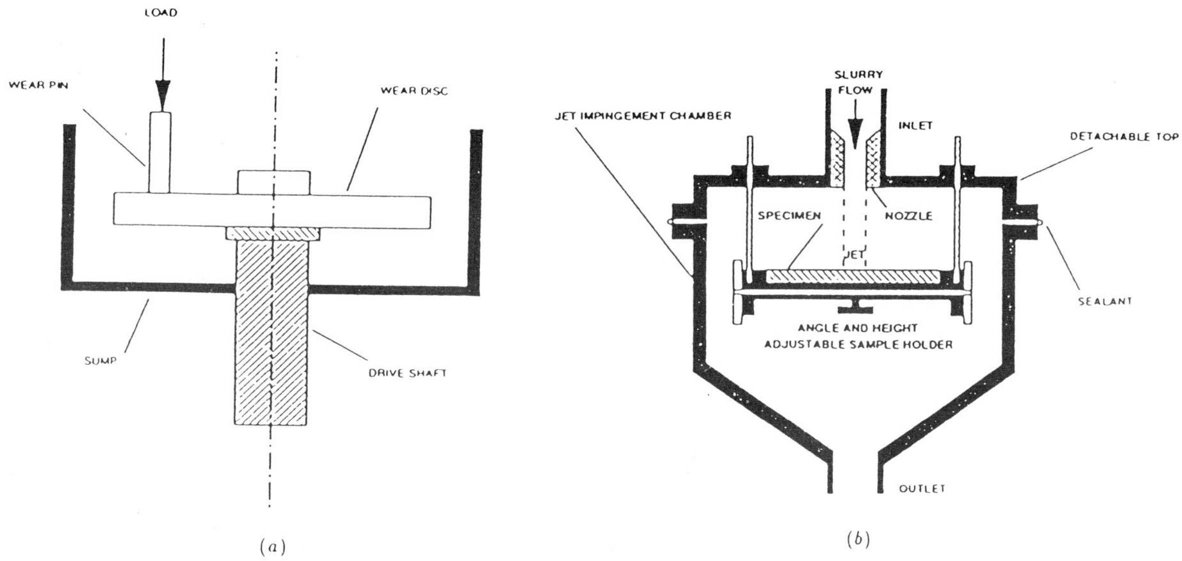

FIGURE 4 Apparatus: (a) pin-on-disc machine; (b) slurry impingement erosion rig

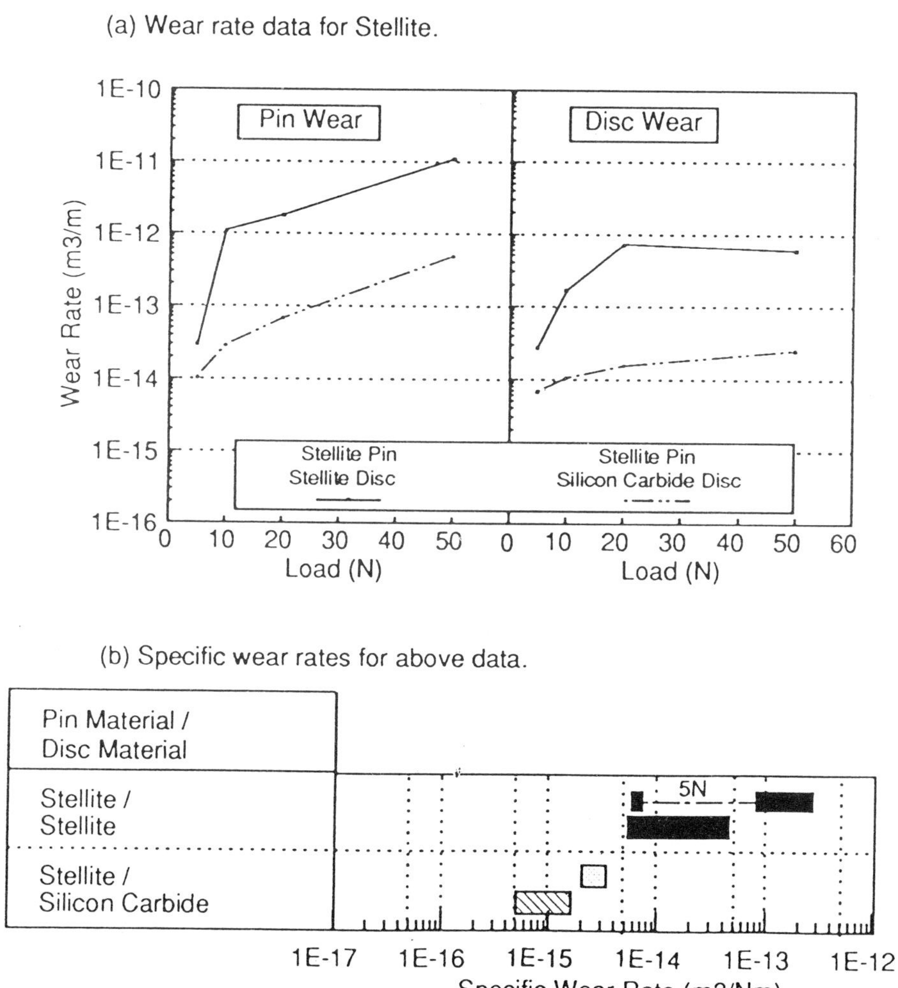

FIGURE 5 Typical dry sliding wear results

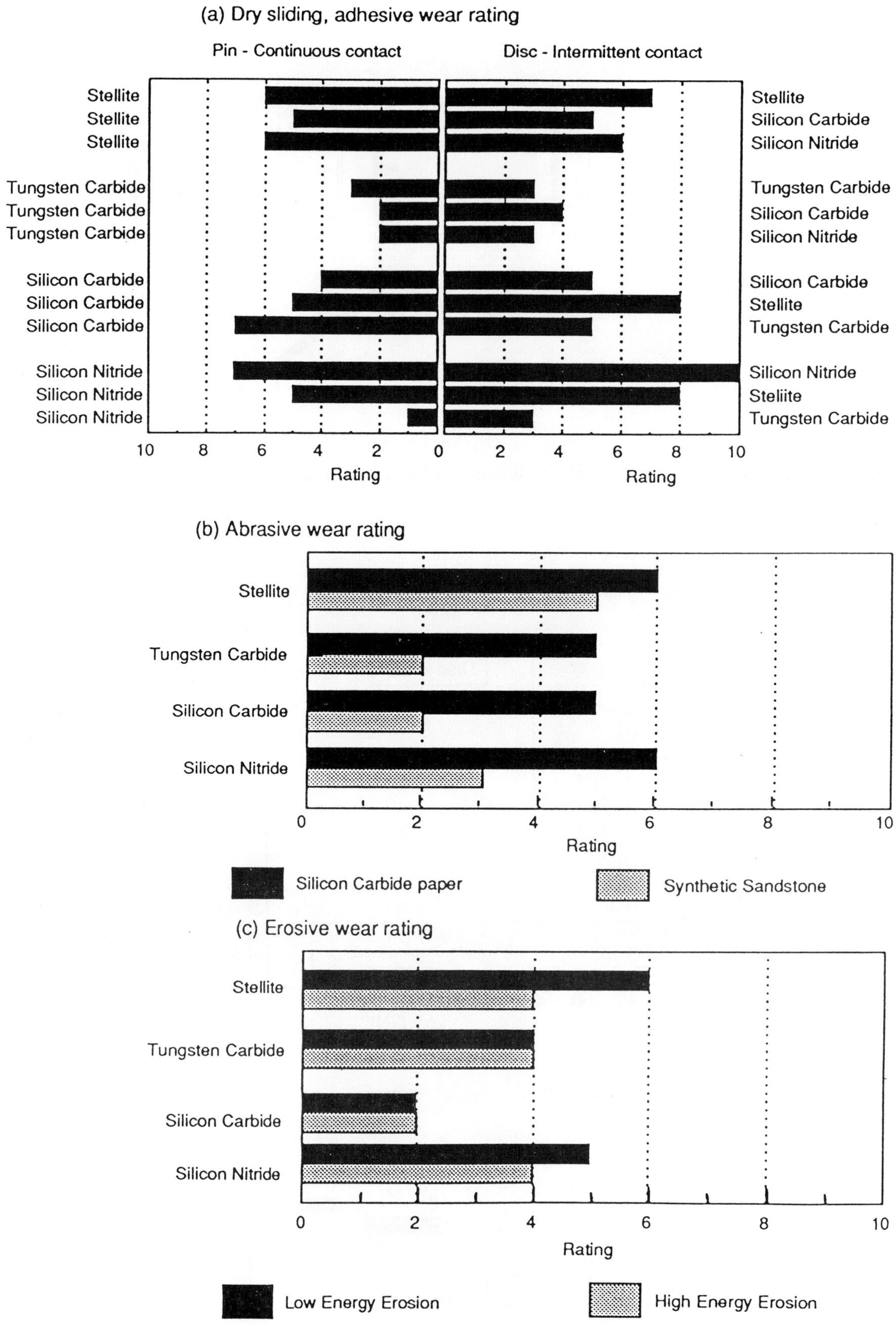

FIGURE 6 Material performance charts

Seal oil reclamation – a different approach

J M FORESTER
Wellman Furnaces Limited, Warley, West Midlands

SYNOPSIS An outline of the development and operation of a simple, robust steam based system for continuous stripping of contaminants from seal oil used in compressors and pumps for oil and gas processing, particularly when located on offshore installation.

1. INTRODUCTION

In centrifugal pumps and compressors, pressurised mineral oil often provides the active barrier in the labyrinth systems used as a seal between the shaft and the main housing.

By design, these seals restrict and control any flow outward along the shaft to prevent leakage of the process media. Inevitably contact takes place at the interface between the two systems leading to contamination of the seal oil by the process media.

If these process fluids are immiscible or easily separated from the oil then no further action may be required. Where they are hazardous or detrimental to the system then consideration should be given to their possible effect on the seal and/or lube oil. In particular, any changes in the physical properties of the oil, such as flashpoint, viscosity or levels of corrosive contaminants which may adversely effect the mechanical function or safe operation of the equipment must be looked at in order to guarantee safety of operation.

2. EFFECTS OF CONTAMINATION

Numerically, systems moving hydrocarbon compounds are a predominant group effected by the problem of contamination, with oil from the seal traps becoming entrained with components from the process. Gaseous products such as methane tend to diffuse quite rapidly from the oil, particularly as the pressure drops on being discharged from the seal traps. It is the mid range hydrocarbons between C_3 and C_{14} which generate the problems, as they cannot be simply separated from the oil. The presence of these fractions alters the performance characteristics of the oil significantly.

In addition, the presence of hydrogen sulphide and carbon dioxide in some gases and oils, can further compound the problem by generating corrosive products which can further damage equipment.

The data in Table 1 gives a typical hydrocarbon distribution for oil discharged from the degasser tank on a natural gas compression skid. In this profile, those hydrocarbons in the range C_3 to C_{12} produced an imbalance in the composition of the oil, as the typical profile for new oil has no measurable hydrocarbons below C_{14}.

Table 1 Sample of ISO 46 grade mineral oil collected from the outlet of a standard API 614 degasser tank on a natural gas compression skid.

Composition	Wt %
C_{1-2}	0.01
C_3	0.02
C_4	0.06
C_5	0.09
C_6	0.17
C_7	0.47
C_8	0.89
C_9	0.54
C_{10}	0.44
C_{11}	0.30
C_{12}	0.18
C_{13}	0.09
C_{14}	0.06

The result is a reduction in flashpoint and viscosity to a level where disposal becomes necessary on performance and safety grounds.

The standard degasser tank, designed in accordance with API 614 has little effect on this oil as the fractions concerned will not easily "weather off", even when subject to extended periods of agitation and heating.

3. PRINCIPLES FOR ADOPTION OF STEAM STRIPPING

The original impetus for the project stemmed from comments made regarding the performance of existing oil recovery treatments during an informal meeting with various interested parties. Basically there appeared to be an opportunity for a simple, reliable low maintenance system which would be capable of processing oil continuously in an online situation and return the cleaned oil for immediate reuse.

Continuous single pass processing was a desirable feature as it avoided the need to install storage facilities for the hazardous contaminated oil generated when using a batch type system.

Effective online reclamation involves a continuous stream of oil flowing through the unit, extracting the contaminants as it passes through the process environment, discharging the waste products in a suitable form for easy disposal, whilst returning the reclaimed products for further use.

The close similarity between the contaminants and the structure of the base oil dictates that selective extraction of these impurities which adversely effect the properties of the oil would be necessary.

Due to cost and service requirements, the use of chemical extraction or adsorption techniques can be eliminated, leaving variations in temperature and pressure as the simplest process variables to manipulate, in order to drive off the contaminants.

Extraction of these detrimental fractions can be achieved effectively by heating a thin distributed film of oil, allowing them to diffuse out as gases or vapours which are flushed away for disposal with any waste process fluids.

In this situation, the operating temperature must be strictly controlled to avoid overheating and possible carbonisation of the oil, but must be sufficient to ensure efficient heat transfer and rapid heating of the oil.

Reduction of the pressure above the oil during heating can improve the diffusion rate but as this involves passing a thin film of oil over a large heated surface in an evacuated chamber, continuous processing can be difficult to achieve. As control depends on maintaining a constant pressure for a given level of extraction, the deviations in flow as the various fractions diffuse off can impair the efficiency of the system.

In many cases the oil has to be recirculated until a constant pressure and temperature profile has been achieved before it can be returned to the main seal system, effectively turning what should be a continuous process into a batch operation.

Thus evaluation of the options available indicated that the use of a temperature controlled process, where a hot fluid comes into intimate contact with a well distributed film of oil, would offer the best chance of success in providing an effective single pass continuous process.

4. DEVELOPMENT OF PROCESS AND HARDWARE

The basic characteristics of the system were established using a model based on laboratory glassware and involved a series of tests using various hardware configurations.

The potential of steam as both a heating medium and a process fluid became evident as contamination levels in the test samples were increased; steam stripping maintained effective levels of extraction.

This approach has simplified design requirements as steam raising can be achieved quite easily, the small quantities of water involved being condensed as a waste product, and fed to drain with the extracted liquid hydrocarbons and any corrosive products present.

In order to establish that the process characteristics determined in the laboratory would be maintained as the system was scaled up, an all metal prototype was constructed for an oil flow of 200 litres/day.

In practice this unit proved to be more efficient than the laboratory model, allowing the steam to oil ratio to be reduced so that a final process flow of close to 400 litres/day was achieved without any reduction in system performance.

The prototype proved that the process was reliable, simple to maintain and that with a relatively simple temperature based control system, fully automated processing was possible.

By isolating the electric immersion heaters in a separate vessel, heating potable water to raise steam for use as the process media and heat source, temperature control in the process chamber will effectively be self regulating.

The maximum attainable temperature would be restricted to 100°C so overheating and carbonisation often associated with direct heating can be avoided.

Interaction between the base oil and the steam should be negligible as these oils have been formulated for use with steam turbines. Under the temperature conditions experienced during the treatment cycle no measurable degeneration of the oil should result from the reclamation process.

In testing the development unit, various configurations of process chamber packing were tested, maximising the surface area available, improving oil distribution and streamlining the flow of steam through the system.

The basis configuration for the process was established as represented by the flow diagram Fig.1.

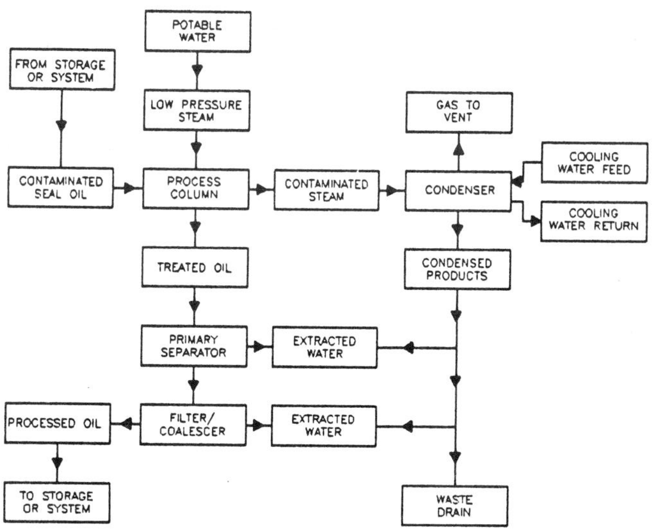

Fig 1 Simplified Process Flow Diagram

Performance date for the system was established using samples of the base mineral oil mixed with hydrocarbon solvents to produce the desired hydrocarbon profile. By varying the type of solvent a range of profiles were tested, the bias being shifted from C_3 to C_5 through to C_7 to C_{11} in order to determine the performance envelope for the system.

In addition, when subjected to repeated contamination and clean up cycles, the oil maintained its physical properties and analytical profile throughout the trial.
The evaluation programme was backed up by tests on samples taken from various compressor installations as part of a series of trials carried out for potential end users.

Many of these samples were collected during routine maintenance when the oil was being changed, so it was heavily contaminated with a wide range of compounds. The results obtained were very encouraging, mirroring the improvements achieved during the trials with solvent contamination.

Table 2 shows the results from processing a 205 litre sample of contaminated ISO 32 grade mineral based turbine oil from the seal oil system on an offshore natural gas compression skid.

Table 2 Properties of a sample of ISO 32 grade mineral turbine oil before and after processing.

	Before	After
Kinematic Viscosity @ 40°C mm^2/S	18.1	30.0
Flash Point (°C) (Pensky Martin Closed Cup)	<20	181
Water Content (mg/kg) (ASTM D1744)	540	100
Component (Wt %)		
C_1	<.01	<.01
C_2	0.05	<.01
C_3	0.35	<.01
C_4	0.87	<.01
C_5	1.08	<.01
C_6	1.07	<.01
C_7	1.49	0.02
C_8	1.23	0.02
C_9	0.38	0.02
C_{10}	0.13	0.02
C_{11}	0.06	0.02
C_{12}	0.02	<.01
$C_{13}+$	93.27	99.9

The oil was in extremely poor condition having values for flashpoint, viscosity and water content well outside those acceptable for normal operation. In a single pass we were able to restore this sample to an operationally acceptable condition, capable of immediate re-use.

One area which was seen as a potential problem was the effect of mixing steam with the oil, particularly with regard to possible additive depletion during processing. However to date from information available, the action of the reclamation process does not adversely effect the performance of the oil nor measurably reduce additive levels. The application data sheets for these oils provides further support for this statement, recommending their use in steam generators as they have additives which are resistant to the action of steam.

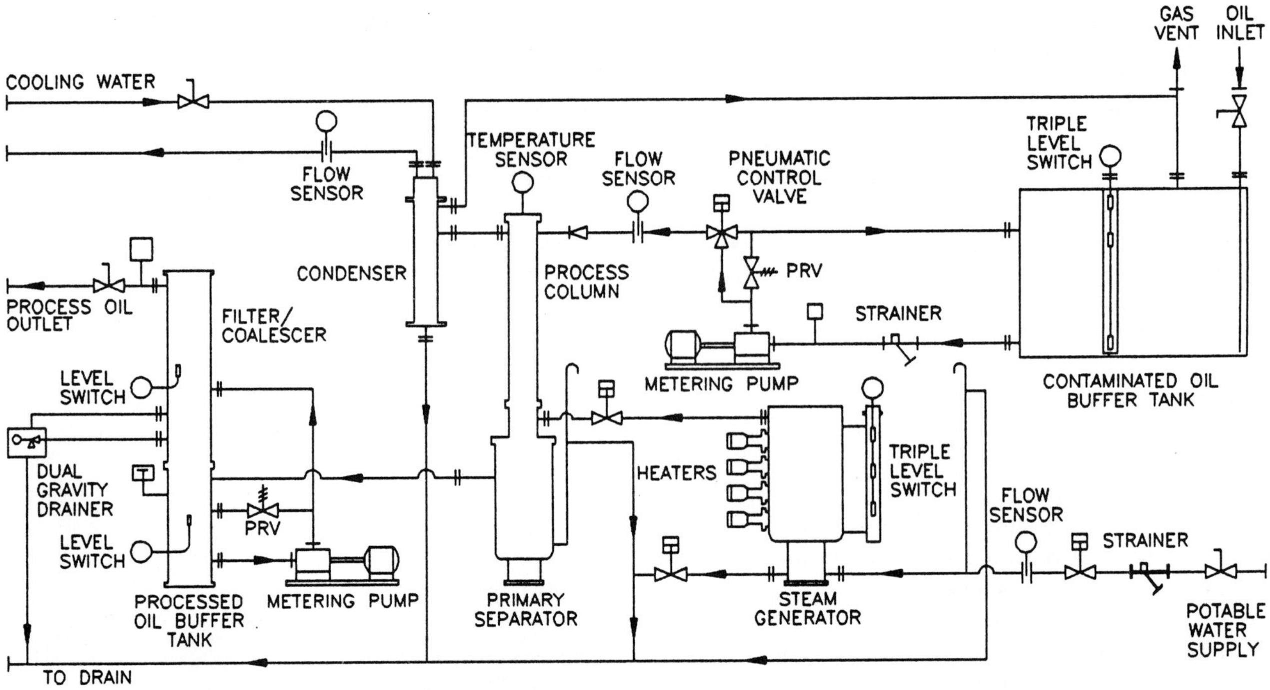

Fig 2 Schematic layout of Seal Oil Reclamation System

5. PRE-PRODUCTION EVALUATION AND TESTING

In view of the rapid progress made with the prototype system, development was taken one stage further by producing a pre-production unit.

This encompassed the design and construction standards required to install and operate the system in a hazardous area, located either land based within a refinery/terminal or on an offshore production platform.

The object of producing an uncommitted system was to allow the unit to be placed on a trial basis with various interested parties. This would have allowed potential end users to test the system on their equipment and establish the performance of our process in their particular environment.

However just as the construction phase had been completed, an opportunity arose to install the system on an offshore natural gas production platform where a new compression module was in the process of final commissioning.

The installation consisted of two gas turbine driven, multi stage, export compressors, which had run into difficulties over the quality of the oil being returned from the seal oil system to the main reservoir.

In normal operation, the oil discharged from the seal traps was passed to a heated degasser tank, where a series of baffles ensured that it was retained for an extended period of time to allow the volatile components to "flash off" to vent.

In practice the degasser tank proved ineffective, the oil which it discharged was still contaminated and over a relatively short period of time the physical properties of the oil in the main reservoir would deteriorate to a point where they were unusable.

The main cause for concern was that the flashpoint of the main reservoir had been found to fall from a nominal 195°C for 'new' oil to below 160°C, the imposed minimum operating limit, in as little as four days. This situation resulted from oil with a flashpoint, on occasion as low as 40°C, being returned to the reservoir from the degasser tank.

Timing dictated that the unit was despatched fully certified but not fully commissioned, to be installed on the open deck space next to the compression module. Temporary pipework was installed, allowing contaminated oil to be drawn from tanks, processed and returned to storage for reuse.

The skid, nominally one metre square by two metres high, complete with hazardous area control panel, was able to process between 350 and 400 litres of contaminated ISO 46 grade mineral turbine oil per day, raising the flashpoint from nominally 50°C to 190°C, with similar improvements in viscosity and hydrocarbon profile.

By carrying out the initial commissioning and operation of the system in conjunction but without commitment from the end user, a free flow of information and recommendations regarding future design and operation was obtained.

From the data collected during these production runs, a larger unit was proposed with detail changes to the design to improve serviceability, extend the operational range and reduce maintenance requirements.

6. THE FINAL STAGE

The design, evolved from the result of these development stages, has culminated in a practical low maintenance system.

The schematic Fig. 2 details the layout of the current system which encompasses all the basic components of the early systems refined to produce a fully automated skid based unit, a typical example being illustrated in the photograph Fig. 3.

Even with the optional buffer tank, this skid nominally two and one half metres by one metre by two metres high, would pass through a standard accessway, allowing the unit to be fitted into existing installations.

Fig.3 Typical Seal Oil Reclamation System

Control of the system can be integrated into an existing distributive control system or be a standalone system by incorporating a small programmable logic controller mounted in either a skid mounted hazardous area rated control box, or a separate sheet metal control panel for location in a safe area.

Systems are pre-commissioned at our works prior to despatch, providing the user with the option of a full test run, using either samples of contaminated oil from the actual installation or fresh oil contaminated with controlled levels of solvent.

Operationally, when monitored over a period of twelve months, one system maintained the flash point of a seal and lube oil reservoir at nominally 190°C, thus easily satisfying the limit of 180°C set by the compressor manufacturer.

7. CONCLUSION

The application of seal oil reclamation can be justified on economic, safety and environmental grounds, providing an opportunity to maximise the service life, minimise consumption and avoid wasteful disposal of a valuable natural resource.

The installation of a simple steam based reclamation process has proved to be an effective, reliable system which is proven in the offshore production environment. The unit as shown in Fig. 3 can be manufactured to a standard specification in order to reduce delivery time and minimise production costs and so maximise profitability to the client.

The design has also been geared to individual clients requirements so that changes to specification can easily be incorporated.

REFERENCE

1. Arco British Ltd - Operational data regarding the performance of a Seal Reclamation System - Private communication 1992.

Film stiffness characteristics of a spiral-groove gas face seal

P BASU, PhD and Z WILLIAMS, ScB
EG & G Fluid Components Technology Group R & D, Cranston, Rhode Island, USA

SYNOPSIS The operating gas film stiffness is a critical design parameter for trouble-free operation of noncontacting gas face seals in centrifugal compressors. The higher the stiffness, the more robust is the design against various uncontrollable factors such as shaft angular misalignment, vibration, transients, dimensional tolerances and the like. The fluid-mechanical model for the flow through a spiral groove gas seal and the calculation of film stiffness are discussed here. The parameters influencing the gas film stiffness are identified and can be optimized for maximum film stiffness using the discussed computer code.

NOTATION

d	diameter
f	Fanning friction factor, $\tau_w = f\ (\rho u^2/2)$
f_{open}	seal opening force
f_{close}	seal closing force
h	film thickness
h_{avg}	average film thickness over the hydrodynamic section, $(h_o + h_j)/2$
h_g	groove depth
k	film stiffness
n_g	number of grooves
M	Mach number
N	speed
p	pressure
r, θ	radius and circumferential coordinates
$\hat{r}$,$\hat{\theta}$	Unit vectors in the radial and circumferential directions
S	non-dimensional spiral span, $(d_o - d_j)/(d_o - d_i)$
T	temperature
u	average fluid velocity in radial direction
W	width ratio of land to groove
α_g	spiral angle
β	seal face coning (+ive if converging film)
γ	specific heat ratio
μ	viscosity
ξ, η	transformed coordinates
ω	angular velocity
τ_w	Wall shear stress
Λ	compressibility number, $6\mu\omega r_o^2/p_o h_{avg}^2$
$()_b$	value at seal balance diameter
$()_g$	groove parameter
$()_i$	value at seal inside diameter (i.d.)
$()_j$	value at spiral inside diameter, (also called junction diameter (j.d.))
$()_o$	value at seal outside diameter (o.d.)

1 INTRODUCTION

The past ten to fifteen years have seen an increase in application of noncontacting gas seals in centrifugal compressors, replacing buffered oil film seal systems. The selection or conversion to a dry seal system saves many thousands of dollars in operating costs by reducing energy consumption and maintenance downtime [1,2]. A schematic diagram of such a gas seal assembly is shown in Figure 1. The face geometry design of such gas seals is based on combined hydrodynamic and hydrostatic principles. The hydrodynamic grooves are machined on the hard rotor face. Among various types of hydrodynamic grooves used, the spiral groove pattern is known to have superior film stiffness characteristics. As shown in Figure 2, the spiral groove hydrodynamic section is followed by a hydrostatic sealing dam. At higher compressibility numbers 'Λ', the spiral grooves act to pump the gas to a higher pressure at the spiral/dam junction than what is obtained statically. The seal is designed to have a small but safe operating gap to minimize leakage and high film stiffness for better reliability and life.

A comprehensive computer code has been developed at the authors' company over the last several years to design the subject gas seals for the present as well as more demanding future compressor applications [3,4]. Given a set of operating conditions, geometrical design parameters and material properties, the code predicts the average operating film thickness and its detailed profile, operating film stiffness, leakage, opening force and power consumption. Various other detailed information such as pressure, Mach number and temperature distributions over the seal face may also be obtained in graphical as well as printed forms, if desired by the seal designer. The mathematical model underlying the code considers pressure, thermal and centrifugal distortions of the stator and rotor faces. An axisymmetric finite element routine is integrated with the code to iteratively calculate the seal face pressure distortions. A viscous flow model is used over the spiral groove section; whereas fluid inertia is also included in describing the flow over the dam section. The analysis allows for "choked flow" condition at the seal outlet which is the seal i.d. for an outside pressurized seal.

The predicted leakages for various designs and applications have been correlated with the experimental results over the last several years. In reference [4], the measured leakage is about 60% higher than the predicted value at the design point. This correlation was performed in 1991. Since then the code has been upgraded and the subseqent correlations have been showing agreement within 10-30%. Since the leakage is a strong function of the average film thickness, it is hypothesized that the film thickness prediction would be even better although no direct correlation has been made as yet.

In the following sections, the flow model and film stiffness calculation are briefly presented. The parameters affecting the film stiffness are varied and their influences shown graphically. Also some of the isssues pertaining to the stiffness optimization are discussed.

2 MATHEMATICAL MODEL

The iterative calculation of flow parameters over the seal face and the face distortions are discussed in [3,4] in detail. However, the flow model is briefly presented in this section to introduce some new studies done subsequent to the above publication.

2.1 Hydrodynamic section

The flow over the hydrodynamic section is assumed to be governed by the compressible Reynolds equation which in vector form is given by :

$$\vec{\nabla}.\left\{ph^3\vec{\nabla}p - 6\mu phwr\hat{\theta}\right\} = 0 \qquad (1)$$

This equation is solved over a sector consisting of a land and groove pair (Figure 3) to obtain the two dimensional pressure distribution. Since no angular misalignment is assumed, the pressure distribution over the other sectors will be identical.

In order to solve the above equation, fixed pressure boundary conditions are imposed on the two radial boundaries e.g. $p(r_o) = p_o$ and $p(r_j) = p_{j+}$. 'p_{j+}' is the junction pressure on the hydrodynamic side, there being a loss of pressure across 'r_j' due to sudden contraction. However, this pressure loss is not quite appreciable under most conditions. Periodicity conditions are applied along the two circumferential boundaries. The Reynolds equation is then solved with both finite element (FE) and finite difference (FD) methods. The reasons for using the two different methods are firstly to validate one analysis against the other. Secondly, over the ranges of variation of the spiral angle 'α_g', one method functions better than the other in terms of accuracy, convergence and mass conservation. Hence, the user has the option of choosing either method depending on the case under study. Figure 3 shows a typical computational grid on a 15° spiral sector. A special case of spiral groove seal is a radial groove seal when '$\alpha_g = 90°$'. This case has been analyzed for a round-bottomed seal with both methods in [5]. Since then, the methodologies have been extended to general spiral case i.e. 'α_g' ranging between 0 to 90°. With smaller values of 'α_g' such as less than 10°, the computational meshes for both FE and FD become highly distorted and the discretization errors tend to increase. Since the groove geometry does not follow the cylindrical coordinate lines for a general spiral angle, the following transformation is made [6] for the finite difference method,

$$\xi = r \qquad (2)$$

$$\eta = \theta - \cot\alpha_g \ln\frac{r_o}{r} \qquad (3)$$

With this transformation, the new coordinate lines 'ξ and η' would follow radial and circumferential lines respectively. In effect, the spiral lines are straightened out to be radial lines and a spiral sector becomes a pie-shaped sector. This transformation also changes the governing equation (1) but it can be rearranged in a similar conservative vector form so as to adopt the same finite difference discretization technique discussed in [5].

With either of the methods (FE or FD), a set of nonlinear algebraic equations is obtained which is then solved with Newton-Raphson linearization and 'successive over relaxation' methods.

2.2 Hydrostatic dam section

The compressible flow over the dam is described with a one-dimensional treatment [3]. The model includes fluid radial inertia, viscous friction, cross-sectional area change and entrance losses. The "choked flow" condition at the outlet can be predicted, as this often happens with a high pressure seal. The Reynolds number dependent expressions for friction factor are used for laminar and turbulent flow conditions in the radial direction. The governing equation for an outside pressurized seal is given as follows :

$$\frac{dM^2}{dr} = -\frac{2\left(1+\frac{\gamma-1}{2}M^2\right)M^2}{1-M^2}\left(\frac{1}{r}+\frac{1}{h}\frac{dh}{dr}\right) - \frac{2\gamma M^4\left(1+\frac{\gamma-1}{2}M^2\right)}{1-M^2}\frac{f}{h} \qquad (4)$$

The above equation is numerically integrated with proper boundary conditions to obtain Mach number distribution along the dam. Since a sharp gradient exists at the seal outlet, a variable grid is used with finer grids near 'r_i' and progressively coarser grid towards 'r_j'. The pressure and temperature distributions are next determined. However, for low pressure applications with '$M_i < 0.1$', a compressible viscous flow model [5] is used instead of the above one.

2.3 Junction condition

At the junction radius, the non-isentropic pressure loss at the entrance to the dam section is taken into account as follows [7] :

$$\frac{p_{j-}}{p_{jo}} = \left[1 + \frac{\gamma - 1}{2}\left(\frac{M_{j-}}{C_L}\right)^2\right]^{-\frac{\gamma}{\gamma-1}} \quad (5)$$

where, 'p_{j-}' is the pressure on the dam side of the junction and 'p_{jo}' is the stagnation pressure at the junction obtained from 'p_{j+}' and 'M_{j+}' by isentropic relationship. The entrance loss coefficient 'C_L' is taken as 0.6.

The junction pressure 'p_{j+}' (and hence, 'p_{j-}') is iteratively obtained so as to match the mass flow rate through the hydrodynamic and the hydrostatic sections.

2.4 Film thickness distribution

The film thickness distribution '$h\ (r)$' is an input to this fluid model. In the design code, '$h\ (r)$' comes from the face distortion calculation modules which in turn requires '$p\ (r)$' from the fluid module. Hence, these two distributions are iteratively obtained. However, in the film stiffness study described in the following section, the '$h\ (r)$' is assumed to be a linear function as follows for simplicity sake :

$$h(r) = h_i + \beta(r - r_i) \quad (6)$$

2.5 Example pressure distributions

For presenting examples, the chosen baseline parameters are given in Table 1. If any parameter is set at a different value, it will be so indicated.

Table 1 Baseline parameters

d_o	150	mm
d_i	115	mm
h_i	3.75	μm
β	0	μrad
p_o	7	MPa
p_i	0.1	MPa
N	16000	r/min
T	150°	C

Figure 4 shows the pressure distribution over a spiral sector for $h_i = 2.5\ \mu$m. The pressure distribution is mapped on a '$\xi - \eta$' plane. The radial and circumferential coordinate lines indicated on the plot are actually 'ξ' and 'η' lines respectively. As seen, for this compressibility number (Λ) of 30, the pressure at the junction 'p_j' is higher than that at the o.d. i.e. 'p_o'. Therefore, a spiral groove seal not only develops hydrodynamic pressure due to the step effects in the circumferential direction but also creates an elevated level of pressure, sloping upwards towards the junction due to the spiral pumping effects against the sealing dam. The bi-directional seals, in general, lack this pumping ability and hence, have weaker film stiffness characteristics than the conventional spiral groove seal. With a proper choice of spiral parameters, the film stiffness of a spiral-groove seal can be optimized for further reliability and life.

Figure 5 shows the one dimensional pressure distribution over the entire sealing face. The 2D pressure over the hydrodynamic section is converted into an equivalent 1D pressure profile based on equal opening force considerations. The pressure at 'r_j' is clearly seen to be higher than that at 'r_o'. It may be noted that the pressure drops very sharply near the exit at 'r_i' and the "choked flow" condition exists for the case under consideration. The exit pressure 'p_i^*' is 0.19 MPa which is higher than the applied back pressure 'p_i' of 0.1 MPa.

2.6 Film stiffness

After the pressure distribution is determined, the seal opening force is obtained by integrating it over the face area.

$$f_{open} = \int_A p\ dA \quad (7)$$

$$f_{close} = \pi(r_o^2 - r_b^2)p_o + \pi(r_b^2 - r_i^2)p_i + f_{spring} \quad (8)$$

where 'f_{spring}' is the spring force at the back of the stator (refer to Figure 1). The opening force versus film thickness curve, commonly known as '$f - h$' curve is shown in Figure 6 for the baseline case. Although the 'X-axis' shows 'h_i', the distorted face profile '$h\ (r)$' is accounted for in the calculation. At the operating point, the opening force is equal to the closing force. The film stiffness at the operating point is the negative slope of the '$f - h$' curve there, given by the following expression :

$$k = -\frac{df_{open}}{dh_i} \quad (9)$$

3 FILM STIFFNESS CHARACTERISTICS

In designing a gas seal, the 'spiral parameters' e.g. 'r_j, W, α_g, n_g and d_g', are first nominally chosen based on experience. Then, the design code is run several times with various other seal geometrical parameters continuously adjusted until the desired level of operating '$h\ (r)$' is achieved. Next 'h_i' and an average 'β' are obtained from '$h\ (r)$' for this base design. In a subsequent film stiffness maximization study, the five above 'spiral parameters' are optimized subject to various constraints, discussed later. This study is done with an optimization code, different from the gas seal design code. This code consists of only the fluid module of the gas seal code and the necessary optimization routines and loops built around it.

3.1 Dependence on film thickness and coning

Before going into discussion on stiffness optimization, let us examine how the 'base stiffness' is influenced by 'h_i' and 'β'. Figure 7 shows the dependence of 'k' on 'h_i' when coning is zero. As seen, a very high film stiffness can be obtained by running the seal at the lowest feasible film thickness. This minimum permissible film level is basically dictated by considerations such as full-film lubrication, interface heat generation at the operating conditions, starting torque and initial rubbing etc. With increase in film thickness, the stiffness rapidly decreases.

Figure 8 shows the influence of 'β' on film stiffness at a fixed 'h_i'. Referring to the top curve, the film stiffness increases considerably with negative coning. This is due to the fact that the spiral section comes closer to the other sealing face and hence, the compressibility number increases rapidly, thereby enhancing the hydrodynamic effect and stiffness. With increase in coning in the positive direction, however, the film stiffness rapidly decreases and finally, the hydrodynamic action fades into the background. In this limiting case, the film stiffness mainly comes from the hydrostatic effect due to the convergent film in the radial flow direction. The second curve pertains to the same seal under static condition and highlights the purely hydrostatic effect. As seen, at the baseline system pressure of 7 MPa the seal develops a reasonable hydrostatic stiffness and retains it even up to a modest amount of negative coning due to the presence of the groove step in the radial direction. Of course, the stiffness will eventually become negative with further decrease in coning. Because of this hydrostatic stiffness, the seal is likely to stand a certain amount of reverse rotation at higher pressures until the negative hydrodynamic component of the stiffness overcomes the hydrostatic component. The third curve is for reference purpose only and for a seal without any hydrodynamic groove. As seen, the seal starts to develop negative stiffness even at a small amount of positive coning. In summary, looking at Figures 7 and 8, it is evident that a thinner operating film and a small amount of negative coning could improve the film stiffness substantially.

3.2 Dependence on 'spiral parameters'

Once suitable operating levels of 'h_i' and 'β' are determined, the film stiffness 'k' can be further optimized by adjusting the five 'spiral parameters', as mentioned earlier. The following functional relationship can be written

$$k = k\left(r_j, W, \alpha_g, n_g, d_g\right) \tag{10}$$

The stiffness is to be optimized subject to several constraints (some are quite subjective). A partial list is given below.

1. $f_{open} \leq (f_{open})_{base\ design}$. For many combinations of the above parameters, an increase in 'k' is attendant with a proportionately higher increase in 'f_{open}'. Under this circumstance, the seal operating 'h_i' will increase and hence, 'k' will decrease appreciably (refer to Figure 7), thereby offsetting any gain in stiffness. On the other hand, for certain combinations, 'f_{open}' will decrease with an increase in 'k'. For these cases, the operating 'h_i' will decrease by a small amount (because of the steep gradient of the '$f - h$' curve with decrease in 'h_i') and results in further increase in stiffness [8].

2. Maximum allowable leakage. Sometimes leakage increases with changes in 'spiral parameters'. It is to be ensured that the leakage stays within the allowable limit during the optimization process.

Once an optimium spiral design is obtained, the gas seal code is run with these parameters to ensure that 'h_i', average 'β', leakage and film stiffness are within the intended limits.

A preliminary study is presented next to show the influence of the parameters on the film stiffness. These parameters are varied over five discrete levels. Figures 9 and 10 present the stiffness variations with the five individual parameters. The level of only one parameter is changed at a time while keeping the others at the baseline levels. The 'X-axes' of these plots indicate the parameter levels.

The study presented here is by no means a full optimization study, which is beyond the scope of this paper and hence, the optimum values shown here are not the true ones. The intention here is to present the features and the capabilities of the computer codes developed for design optimization of gas seals. In an actual optimization study, all the spiral parameters, except 'n_g', are varied continuously between their set upper and lower limits whereas 'n_g' is varied discretely over integer values. Furthermore, the variation of parameters are simultaneous instead of one parameter at a time because of the presence of the interaction effects among parameters.

4 SUMMARY

Film stiffness is a critical design parameter for high speed, high pressure gas seals. A mathematical model for calculating the film stiffness of a spiral groove gas face seal is presented here. It is shown that a thinner operating film and a small amount of negative coning could improve the film stiffness substantially under certain operating conditions. The film stiffness can be further optimized by suitably adjusting the spiral parameters. An optimization study, using the discussed code, can indicate the ranges of design parameters for high film stiffness and insensitivity to manufacturing tolerances.

ACKNOWLEDGEMENT

The authors like to thank EG&G Fluid Components Technology Group R&D, Cranston, Rhode Island for permission to publish this paper and Gas Seal Group, Industrial Division, EG&G Sealol for providing experimental correlation data. The authors also express thankfulness to Dr. A. Datta, Messrs. A. Deamer, J. Gardner and C. Kowalski for their valuable suggestions, and to Ms. D. Andrade and Ms. J. Jackson for correcting the manuscript.

REFERENCES

[1] CARTER, D. R. Application of Dry Gas Seals on a High Pressure Hydrogen Recycle Compressor. The Seventeenth Turbomachinery Conference, Dallas, Texas, USA, 1988, 3-7 (Texas A&M University).

[2] SHAH, P. Dry Gas Compressor Seals. The Seventeenth Turbomachinery Conference, Dallas, Texas, USA, 1988, 133-139 (Texas A&M University).

[3] GARDNER, J., & BASU, P. Analysis, Design and Laboratory Testing of a 1500 psig (103 bar), 15500 rpm Tandem Gas Seal in a Centrifugal Compressor. ROMAG '91 : Magnetic Bearings and Dry Gas Seals Conference and Exhibition, Washington D. C., USA, 1991.

[4] GARDNER, J., BASU, P., & MURAKI, R. Analysis, Design, Laboratory Testing of a High Pressure, High Speed, Reduced Axial Length Gas Seal. The 13th International Conference on Fluid Sealing, Brugge, Belgium, 1992, 389-412 (BHR Group Ltd/Kluwer Academic Publishers).

[5] BASU, P. Analysis of a Radial Groove Gas Face Seal. *Tribology Transactions*, 1992, Volume 35, 1, 11-20.

[6] JAMES, D. D., & POTTER, A. F. Numerical Analysis of the Gas-lubricated Spiral-groove Thrust Bearing compressor. *Journal of Lubrication Technology*, ASME, 1967, 89, 4, 439-444.

[7] ZUK, J., LUDWIG, L. P. & JOHNSON, R. L. Compressible Flow Across Shaft Face Seals. The 5th International Conference on Fluid Sealing, Warwick, England, 1971, H6-81-108 (BHRA).

[8] KOWALSKI, C. Suggestions given in private communications. Gas Seal Group, EG&G Sealol, Cranston, Rhode Island.

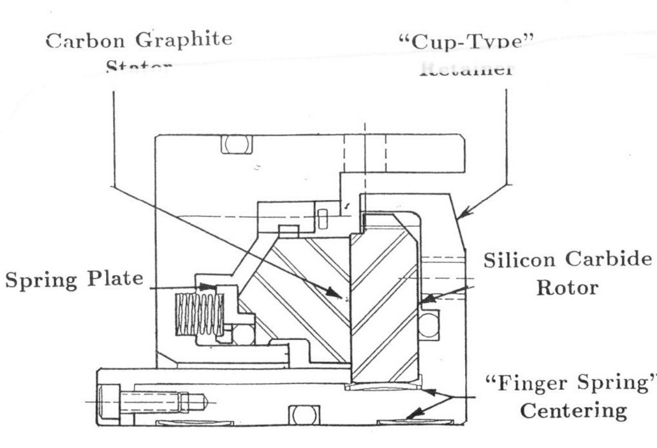

Fig 1 Schematic diagram of a gas seal assembly

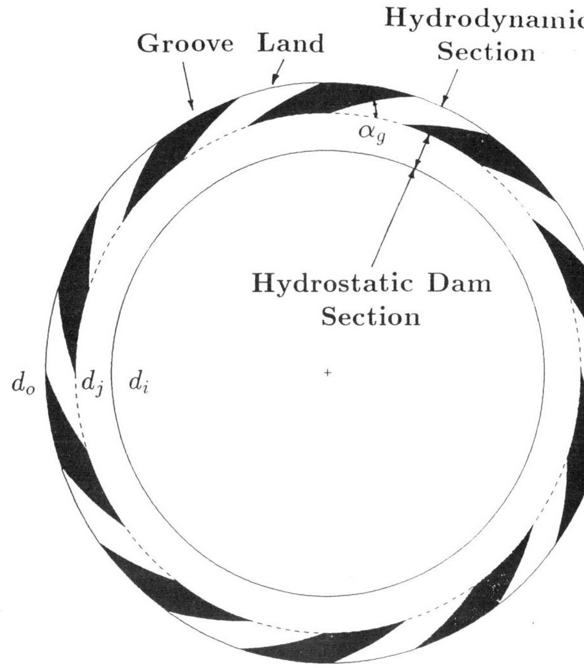

Fig 2 A spiral groove seal face

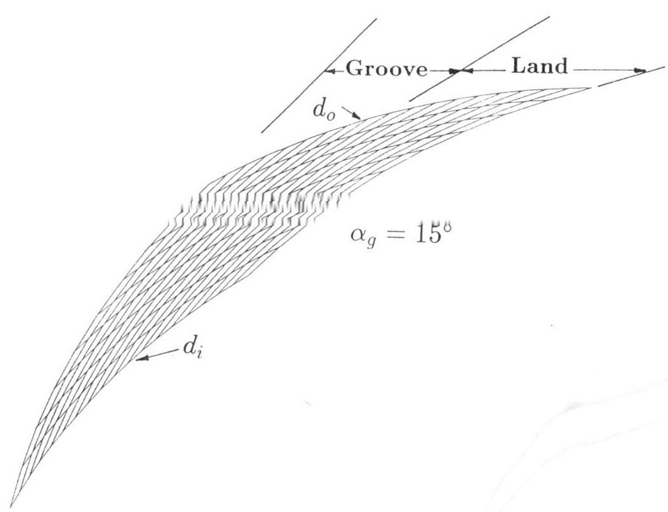

Fig 3 Computational mesh on a spiral groove sector

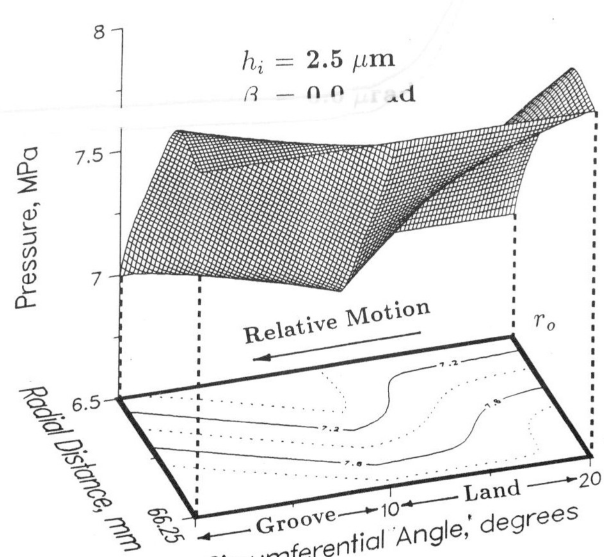

Fig 4 2D Pressure distribution over the hydrodynamic section

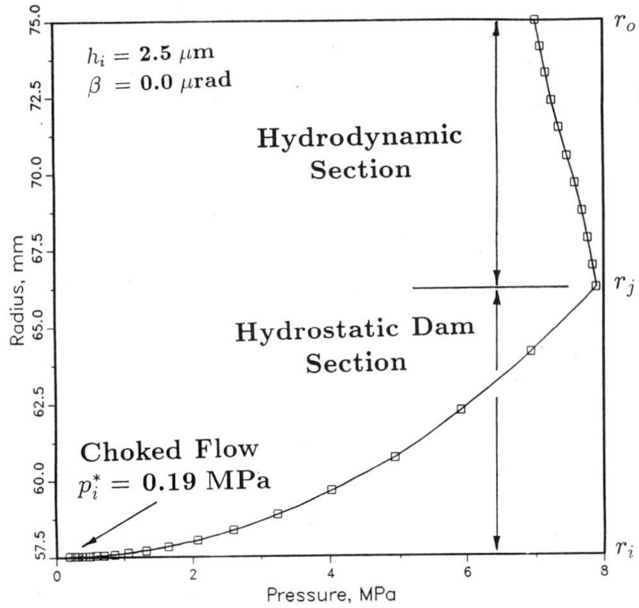

Fig 5 1D Pressure distribution over the seal face

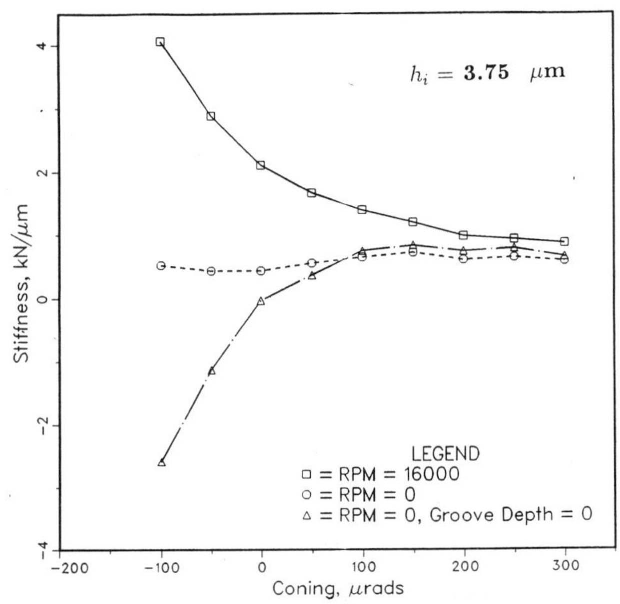

Fig 8 Stiffness variation with coning

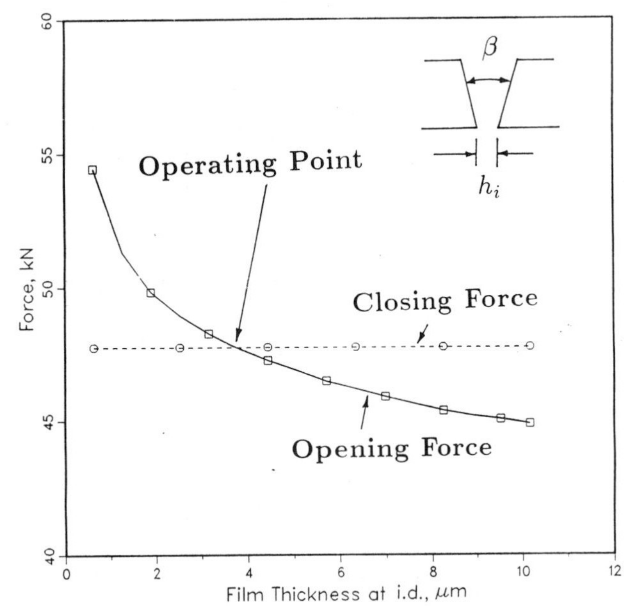

Fig 6 F-h plot

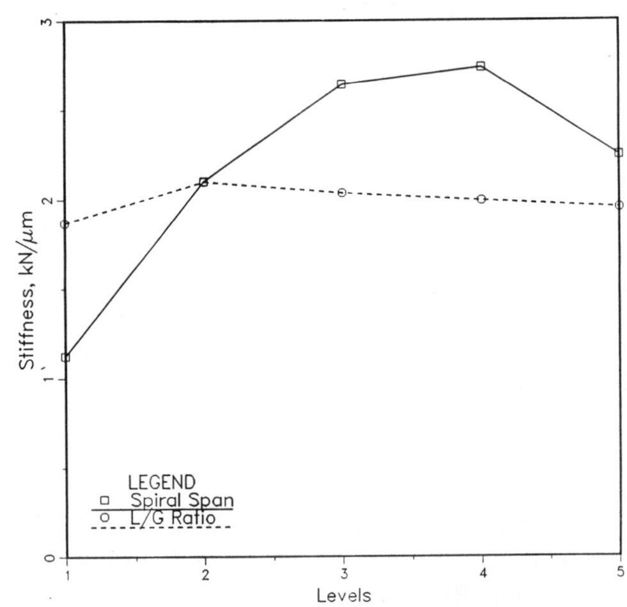

Fig 9 Stiffness variation with spiral span and land-to-groove width ratio

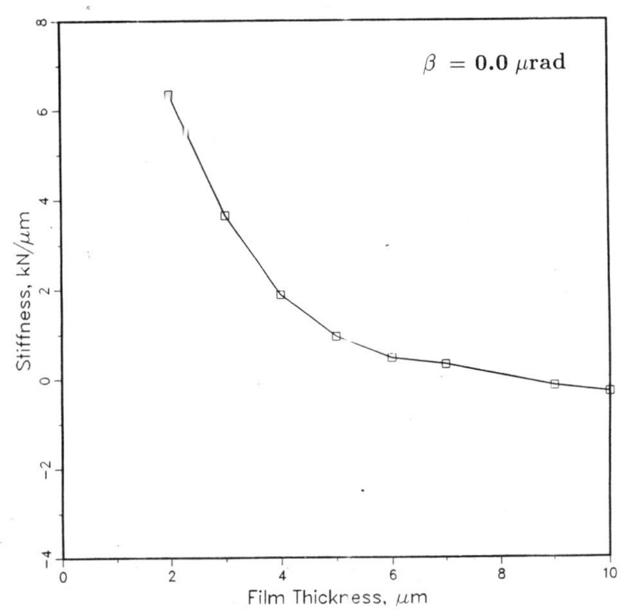

Fig 7 Stiffness variation with operating film thickness

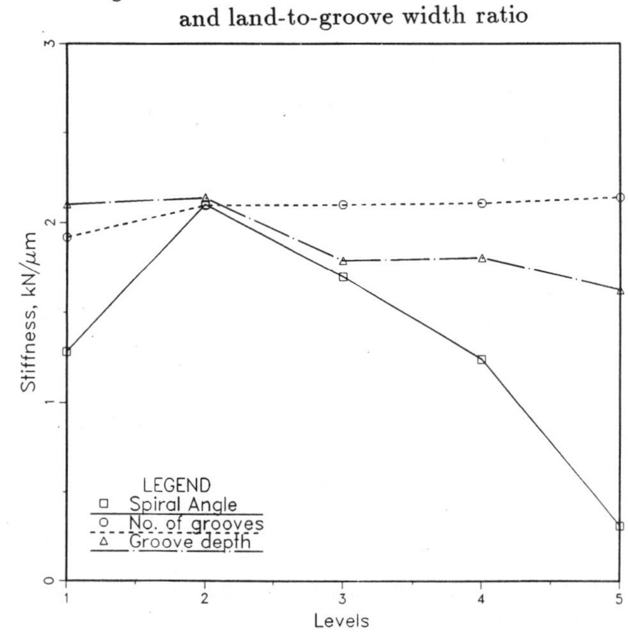

Fig 10 Stiffness variation with spiral angle, number of grooves and groove depth

C449/004/93

A new approach for rotating systems torsional vibration analysis using an energy method

L G M KOOP, Ing
Thomassen International BV, Rheden, The Netherlands
J MOLENAAR, PhD
Eindhoven University of Technology, The Netherlands

SYNOPSIS

To guarantee successful operation of a rotating system it is necessary to carry out a torsional vibration analysis.
The torsional behaviour of machines with reciprocating mechanisms requires thorough investigation. With the application of Variable Speed Drive E-motors a large speed range and start-up torsional behaviour has to be investigated. In this case it is not always possible to meet the required separation margins from natural frequencies over the entire speed range. Therefore it is essential to describe the components and the applied torques in the system as accurately as possible.
In this article a new approach of calculating the torsional vibrations and a comparison of calculated and measured results is presented.

INTRODUCTION

In a rotating system different types of elements may be present, coupled by means of torsionally elastic components.

1. Purely rotating components such as flywheels, shafts, and rotors of turbines or E-motors
2. Connecting rods that exhibit rotating as well as oscillating motion.
3. Purely oscillating components such as crossheads, pistons and pistonrods.

Figure 1 shows a sectional view of a reciprocating compressor in which all above mentioned parts are present.

The calculation of torsional vibrations in a system consisting of a set of reciprocating components is very complex. The difficulty is that the equations of motion are non linear, originating in the oscillating masses and the connecting rod.
Usually these equations are linearised by reducing the inertias of the oscillating masses to the radius of the crank which they are connected to, and by dividing the connecting rod into a rotating and an oscillating part.
In this way the reciprocating masses are replaced by discs with a constant moment of inertia, which is basically incorrect but a common simplification. This problem is not new and has been investigated in the past more or less successfully by scientists in Europe (see lit. 2 and 3). The solutions have never been resulted into a practical method for executing computations.

In coöperation with the Eindhoven University of Technology in the

Netherlands, a computerprogram has been developed to simulate the real torsional system exactly.In this program elements of type 1,2 and 3 have been incorporated.

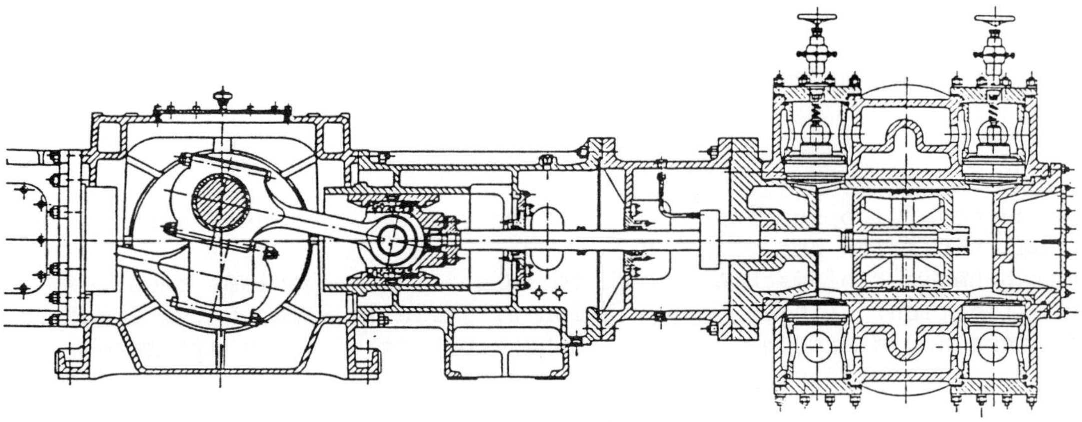

fig. 1 Sectional view of a reciprocating compressor

COMPARISON OF THE CLASSICAL APPROACH VERSUS THE NEW APPROACH

New	Classical
High accuracy.	Less accurate.
Simulation of start-up behaviour as function of speed.	Not possible.
Possibility of testing the influence of different types of damping including non-lineair.	Generalized damping only.
The simulation of unstable regions.	Not possible, quasi-static.
Natural frequencies to be determined from the response calculations.	Calculation of natural frequencies and mode shapes.
Requires a powerful computer.	Faster calculations.

NOMENCLATURE

I	Inertia [kgm²]
m	mass [kg]
n	element number
R	Crankradius [m]
a,b	coordinates [m]
Ø	crankangle [rad.]
t	torsional stiffness [Nm/rad]
d	damping [Nms/rad]
T	kinetic energy [Nm]
Q	coefficient of work [Nm]
U	deformation energy [Nm]
C	damping energy [Nm]

THEORY AND COMPARISON WITH THE CLASSICAL METHOD

The used theory is based upon using Lagrange's formalism. To apply this it is necessary that the energy of all elements is calculated. The three types of elements have different kinetic energies and different angular momenta. The resulting equations consist of a system of ordinary equations with initial conditions. In general it is not possible to solve this system analytically. Numerically it is however possible. It is solved by means of a set of routines from ODEPACK. (see lit. 1)

Lagrange's equations are based on expressions for the different types of energy present in the system. These energies are :

1. Kinetic energy
2. Potential energy/deformation
3. Damping/dissipative energy
4. Work (Applied torques and forces)

1.KINETIC ENERGY

a.The rotating element

The kinetic energy of a purely rotating body (disc) depends on the inertia and the angular velocity and is described in the formula in figure 2.

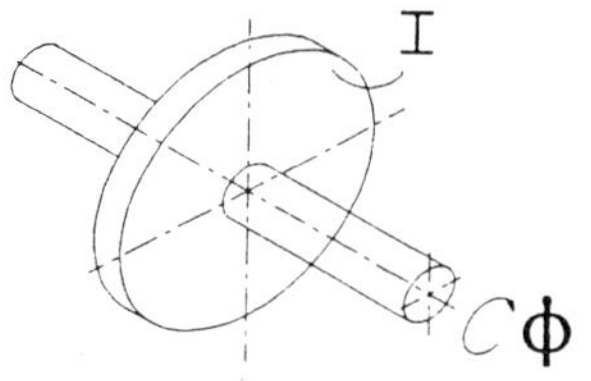

$$T = \tfrac{1}{2}\, I\, \dot{\phi}^2$$

fig. 2: Kinetic energy of a disc

b.The connecting rod

The orbit of the centre of gravity of the connecting rod is eggshaped. The connecting rod generates massforces in the horizontal and vertical planes.
The correct solution is to determine the mass, the centre of gravity and the moment of inertia around the centre of gravity and to describe thereby of the centre of gravity.
In figure 3 the kinetic energy of the connecting rod is shown.
In the classical approach the mass of the connecting rod is divided into a reciprocating and a rotating part. This approach ignores the pendulum inertia about the centre of gravity.

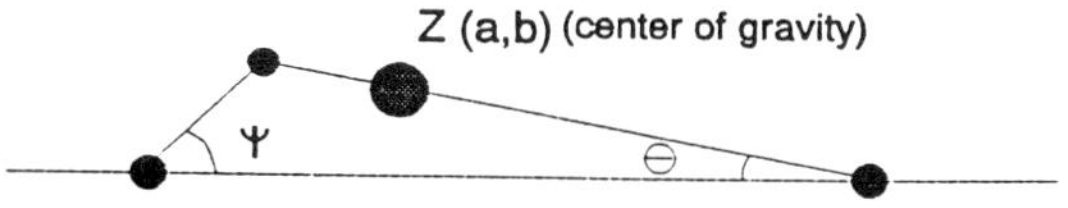

$$T = \tfrac{1}{2}\, m_{rod}\, (\dot{a}^2 + \dot{b}^2) + \tfrac{1}{2}\, I_{rod}\, \Theta^2$$

fig.3: Kinetic energy of a connecting rod

c.The reciprocating masses

The moment of inertia of reciprocating masses is not constant over one revolution. Its instantaneous value depends on the

crankangle. When the crankangle is 90 degrees the moment of inertia is maximum. When the connecting rod is in line with the web the moment of inertia is zero. Figure 4 shows how this problem is solved in the classical approach. The mean value of the inertia is added to the inertia of the crank and the forces caused by the acceleration and deceleration of the masses must be calculated individually.
The kinetic energy of the masses is a function of mass and instantaneous velocity. This velocity is a function of rotating speed and crankangle. In the newly developed method the massforces need not to be calculated separately or individually but are automatically coupled to the speed. The formulas for calculating the kinetic energies are presented in fig. 3 and 5.

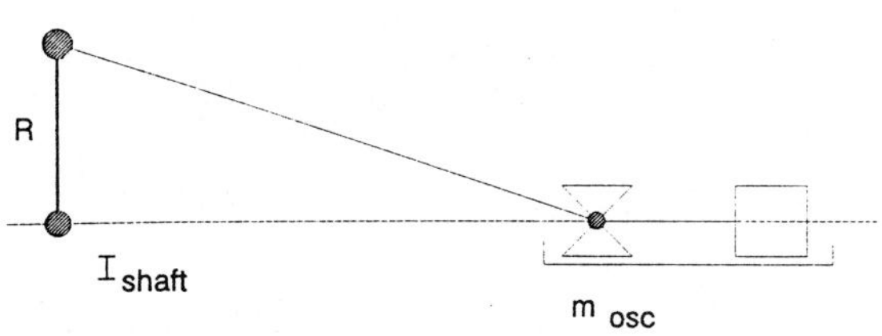

$$\text{Inertia} = I\,\text{shaft} + m_{osc} * R^2$$

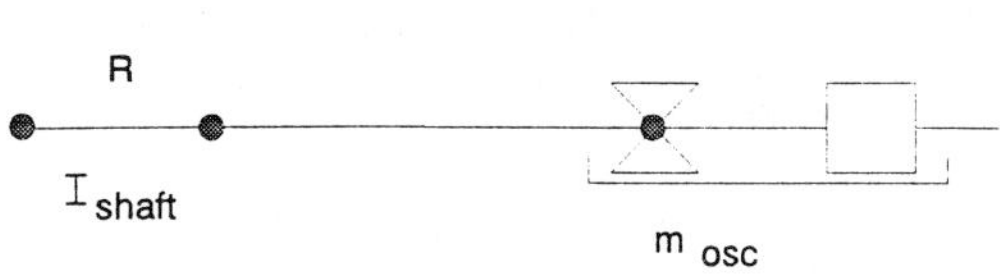

$$\text{Inertia} = I\,\text{shaft}$$

$$\text{Mean Inertia} = I\,\text{shaft} + \tfrac{1}{2}\, m_{osc} * R^2$$

fig. 4: Simplification of inertia of reciprocating masses.

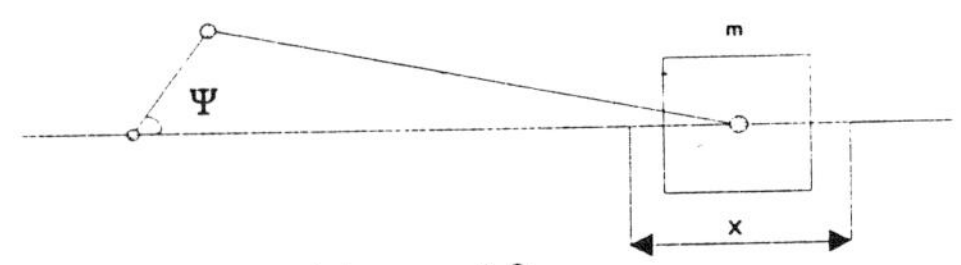

$$T = \tfrac{1}{2}\, m\, \dot{x}^2$$

fig. 5:Kinetic energy of reciprocating mass

2.POTENTIAL ENERGY/DEFORMATION

The deformation energy is the energy stored in the elastic components of the system. These components can be the shaft itself or elastic couplings. The deformation energy is given in fig 6.

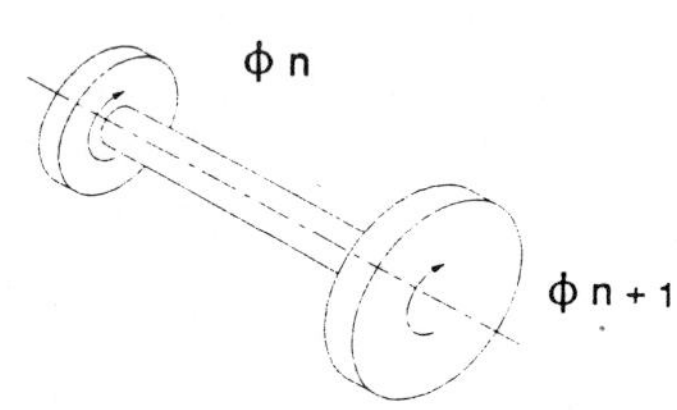

$$U = \tfrac{1}{2} t\,(\phi_n - \phi_{n-1})^2$$

fig. 6: Deformation energy

3.THE COEFFICIENT OF WORK

The coëfficiënt of work is the work which is done by external forces or torques as a function of change of crankangle.
Figure 7 demonstrates the coefficients of work of the three types of elements.

1. Purely rotating bodies: Q equals the applied torque
2. Reciprocating bodies: Q equals the force on the piston times a term depending on the crankangle.
3. Systems with a constant speed such as rotating electric fields : Q equals zero.

The work that is performed at an angular change $\delta\phi$

1. Rotating bodies

$Q = Torque\ (time,\ \phi,\ \dot{\phi})$

2. Pistons

$Q = Force_{piston}\ (\phi\ .\ \dot{\phi})\ *\ coëff\ (\phi)$

3. Rotating field (constant speed)

$Q = 0$

fig. 7: Coëfficiënt of work

4.THE DAMPING

The dissipation energy caused by damping has been assumed to be a function of the speed (viscous damping).
In figure 8 the several types of damping which can be used in the program are demonstrated.
Three types of damping have been included:

a. Reciprocating parts
When the actual speed is determined, it is possible to introduce the damping caused by the reciprocating parts. This means that this damping is not constant over a revolution.
The damping is a function of friction coefficient, weight and non constant velocity.

b. Rotating parts
Damping between rotating and nonrotating surrounding

c. Relative damping
The damping between two adjacent elements
This damping is a function of angular speed difference and damping coefficient between two elements.

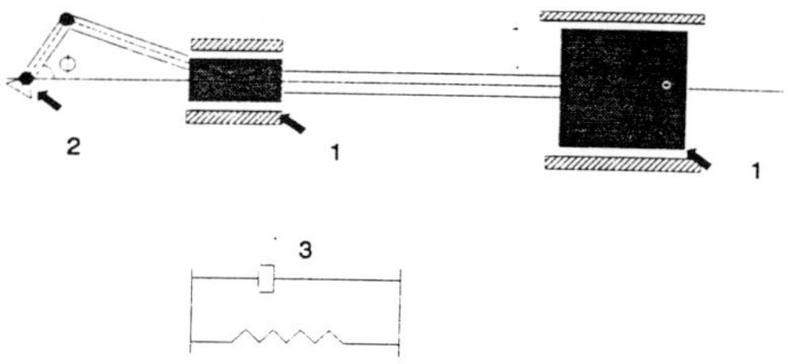

1. Friction of oscillating parts
$C = f_{osc}\ *\ (\dot{\phi}\ *\ coëff\ (\dot{\phi}))$

2. Friction of rotating parts
$C = f_{rot}\ *\ \dot{\phi}$

3. Friction between elements
$C = f_{shaft}\ *\ (\dot{\phi}_i\ -\ \dot{\phi}_{i-1})$

fig. 8: Damping (C)

From the relationships in the above items 1 through 4 the equations of motion can be derived. The result is a set of n+2 second order differential equations.
The advantages of using this method are:

1. The movements of the connecting rods and oscillating elements are now exactly taken into account.
2. Possibility of applying external torques as a function of speed or crankangle such as starting torques of E-motors
3. Possibility to investigate the influence of the damping generated by the piston.
4. Possibility to calculate the effect of the torsional behaviour on the current fluctuations of the E-motor.
5. Non-linear effects of couplings can be simulated.
6. The massforces (as used in the classical approach) are automatically generated

Comparison of results

Torsional models of a 6-cylinder and a 4-cylinder compressor have been made using the "classical" and the "real" approach where the cranks in the classical model have been modelled by means of a disc. The torsional stresses have been calculated with the new program using the same amount of damping.

The table below shows the ratio real/classical results for the major harmonic frequency components and for the 0-peak overall level.

	Harmonic components						overall 0-pk
	3	6	7	9	10	11	
6-cylinder							
374 RPM	1.0	1.0	*	1.3	-	*	1.16
339 RPM	1.0	1.0	-	0.83	0.5	2.22	1.12
Harmonics	1	11	12	13	14		
4-cylinder	1.0	2.14	2.36	1.95	*		1.36

* This is negligible in the disc model
- this component is not present in both models

TABLE:Ratio of calculated torsional vibrations REAL/CLASSICAL model

From the table above it can be concluded that the calculation of the real model produces higher torsional vibration levels. This cannot only be explained by the higher contribution of the harmonic components which are near or in resonance, but also by the contribution of components in the real model which do not show up in the results of the disc model.

The most important one in the 6-cylinder model is the 11th harmonic. This component is also found in the measurements on the real compressor. This phenominum appears also in the 4-cylinder model but in this case as the 14th harmonic. These additional resonances are caused by the time-dependance of the inertia of the reciprocating parts.

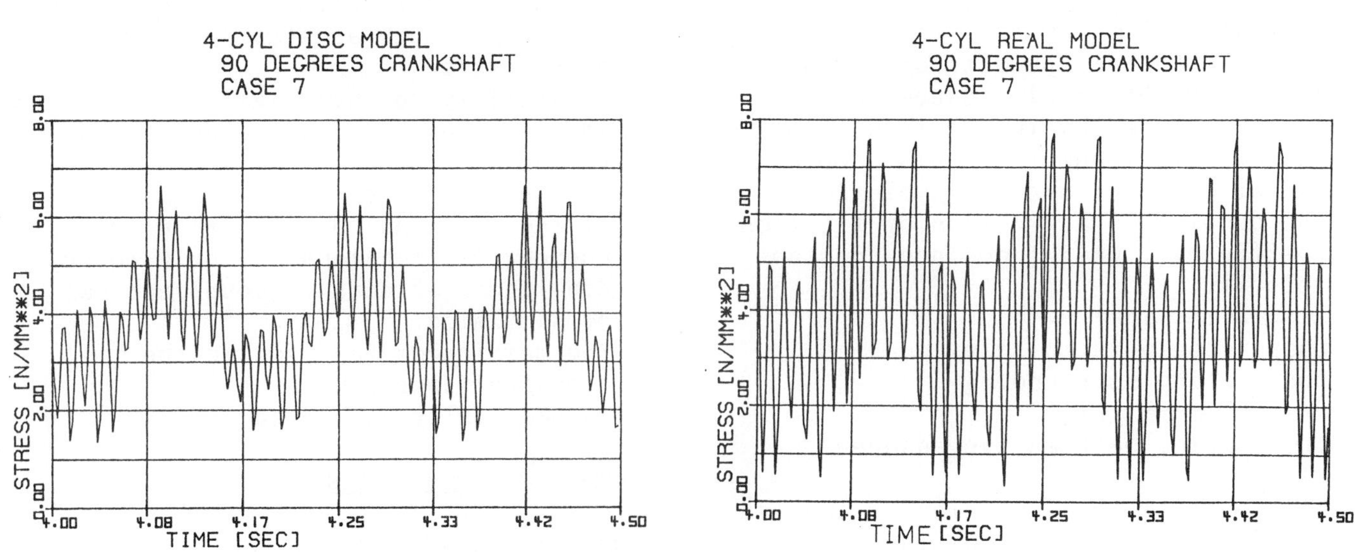

fig.9a: Calculated torsional vibrations of a 4-cylinder compressor

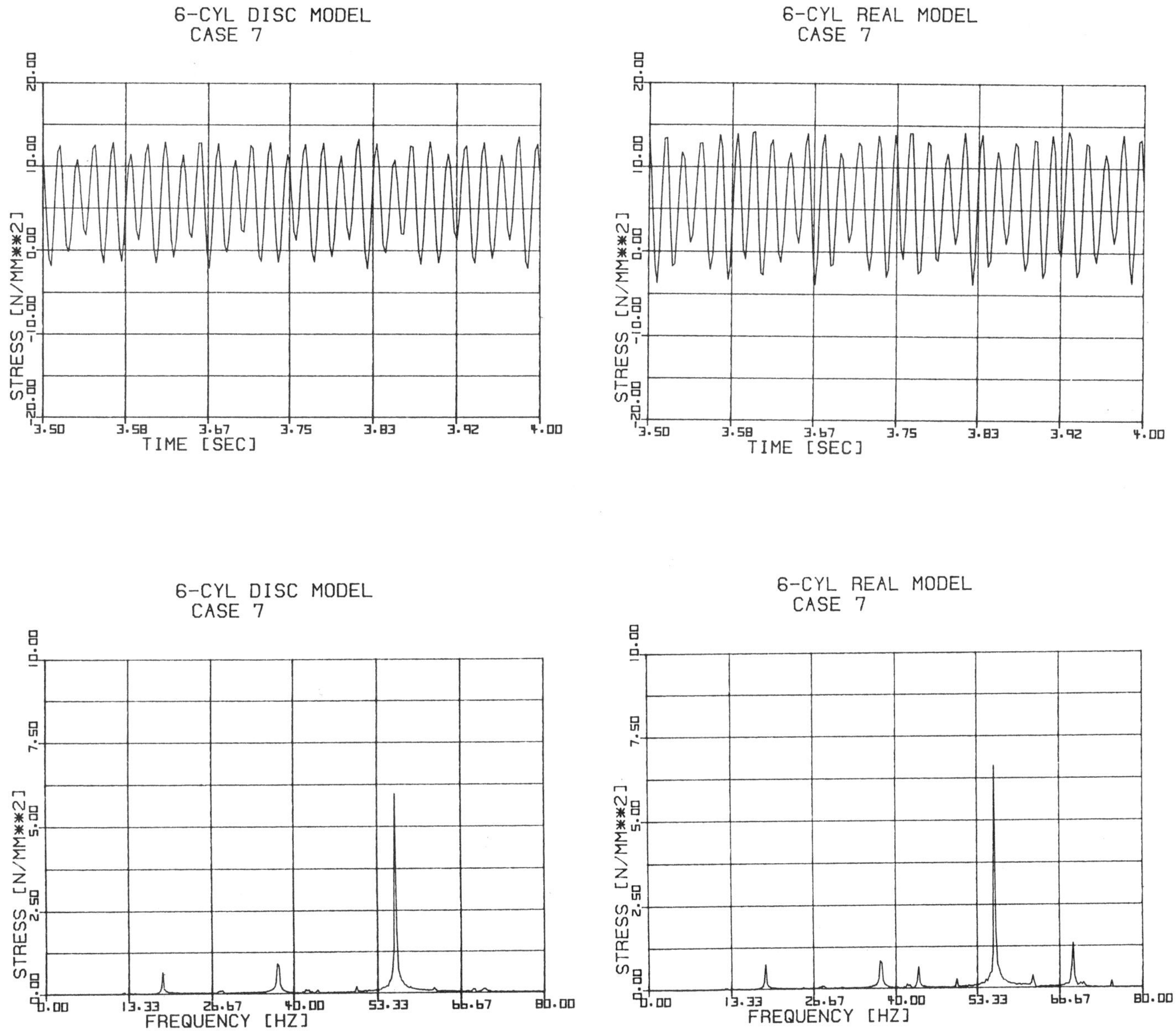

fig.9b: Calculated torsional vibrations of a 6-cylinder compressor

Transiënt conditions (such as speeding-up or running-down) can be calculated using an external moment on the driver unit which causes the system to accelerate, or by using a variable driving frequency. An example of this is shown in figure 14.
This option is also valuable for finding the critical speeds in the system. By varying the driving speed of the system the maximum torsional vibrations can be found within the speed range or within a margin of the operating speed.
When the mathematical formula of the reswitching phenomina of an E-motor driver is known, the torsional implications of such an occurance can also be calculated.

The influence of the damping

The influence of the different types of damping has been numerically investigated.

a. Damping type a (piston friction damping).
It appeared that the absolute damping according to type a has a minor effect on the vibration amplitudes. If only this type of damping is applied in the system the vibration amplitudes remain significantly above the measured values also when the power losses introduced in this way are higher than the actual values. The figure 10 below shows the influence of the mechanical losses on the magnification of the 1st node resonance frequency.

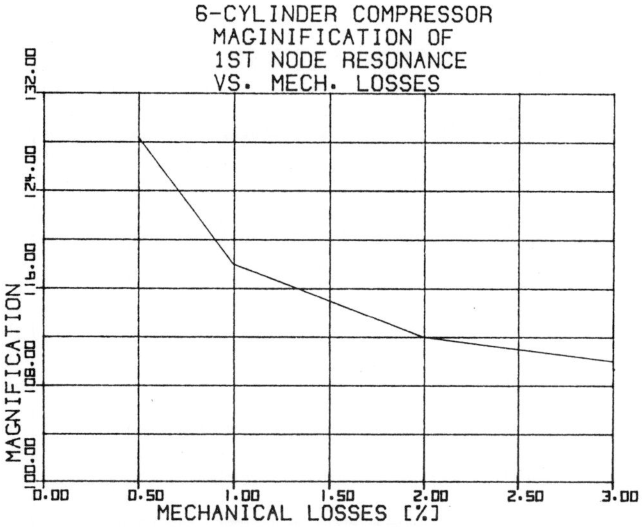

fig.10 6-cylinder compressor. Magnification of 1st node resonance vs. mech. losses.

b. Damping type b (absolute damping)
The investigation of the absolute damping of the rotating parts shows comparative results as for type a. It has not been investigated for large damping coefficiënts

c. Damping type c (relative damping)
This type of damping is generated by the squeeze film effect in the bearings of the reciprocating and rotating parts. Measurements as described by others (ref.2) show that there is a relationship between the maximum vibration amplitudes and the number of bearings. In literature the damping has been expressed as:

$$Ccyl = k * \omega Jcyl$$

Ccyl = damping per cylinder
k = a factor to be determined
J_{cyl} = Cylinder inertia
ω = system natural frequency

The term Ccyl has been determined with the help of the results of the measurements and the numerical results. In figure 11 a graph of the amplification factor versus the factor k has been plotted. The calculation results appeared to match the measurements when k has the value of 0.5.
It must be pointed out that this approach is based on the classical theory and that is it only valid for one node resonance conditions. When a system is running at a non-resonant condition the amplitudes are mainly caused by the forced vibrations and not by damping.

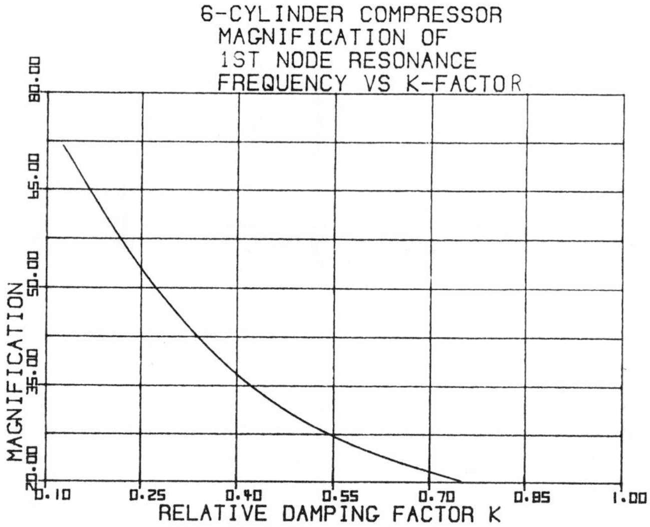

fig.11 Influence of relative damping on magnification factor.

Examples and measurements

We have measured torsional vibrations on the six-cylinder compressor driven by a Variable Speed E-motor of 4.5 MW.

In so doing we were also able to test our new program.
The speed could be varied between 185 and 375 RPM. By varying the speed it was possible to find the resonance frequencies and to measure the amplitude of frequency components in the spectrum of the torsional vibrations at these particular speeds.

One problem is that when running at resonance speed the magnitude of vibration is determined by the damping.

We introduced the amount of damping in the system equal to the mechanical losses present using damping type a. Then damping type c was used to match the amplitude of the resonant harmonic component with the measurements. After analysis it proved that there was a remarkable coincidence between the calculated and measured values not only for the resonance condition but also for other operating speeds. In particular the frequency analysis shows much information on this comparison. (see fig 12 and 13).

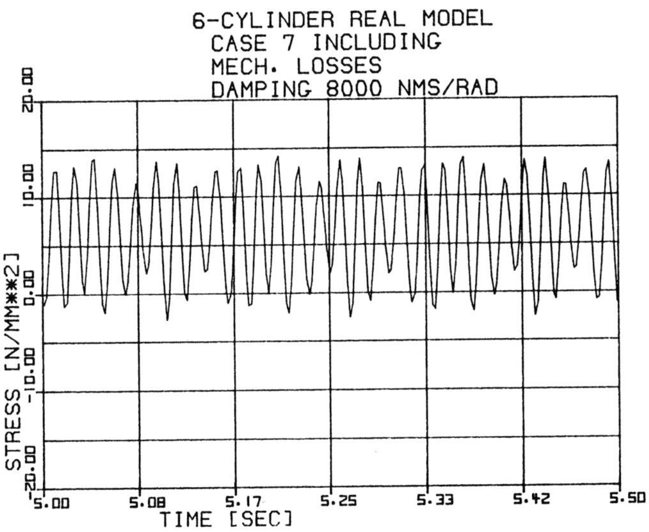

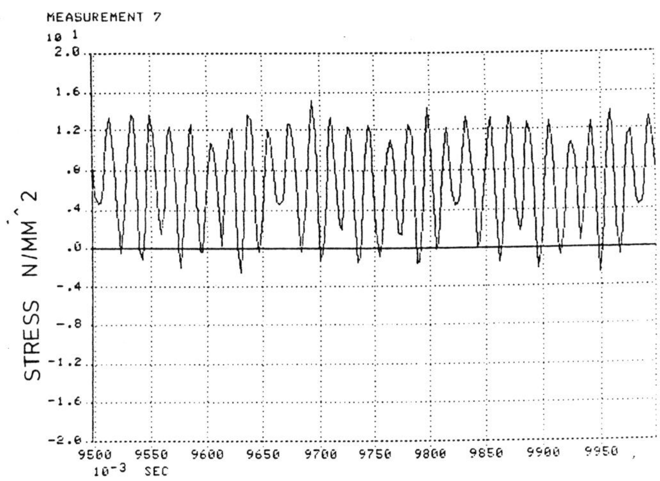

Measured torsional vibrations

Fig. 12: Comparison of measured and calculated torsional vibrations
Time/stress plot

The analyses of the stress shows the frequency components.
All the measured and calculated frequencies are present in the same order of magnitude.

If a sweep of the rotating velocity is performed the resonance frequencies can be found. This is done in practice and has also been simulated (fig 14)

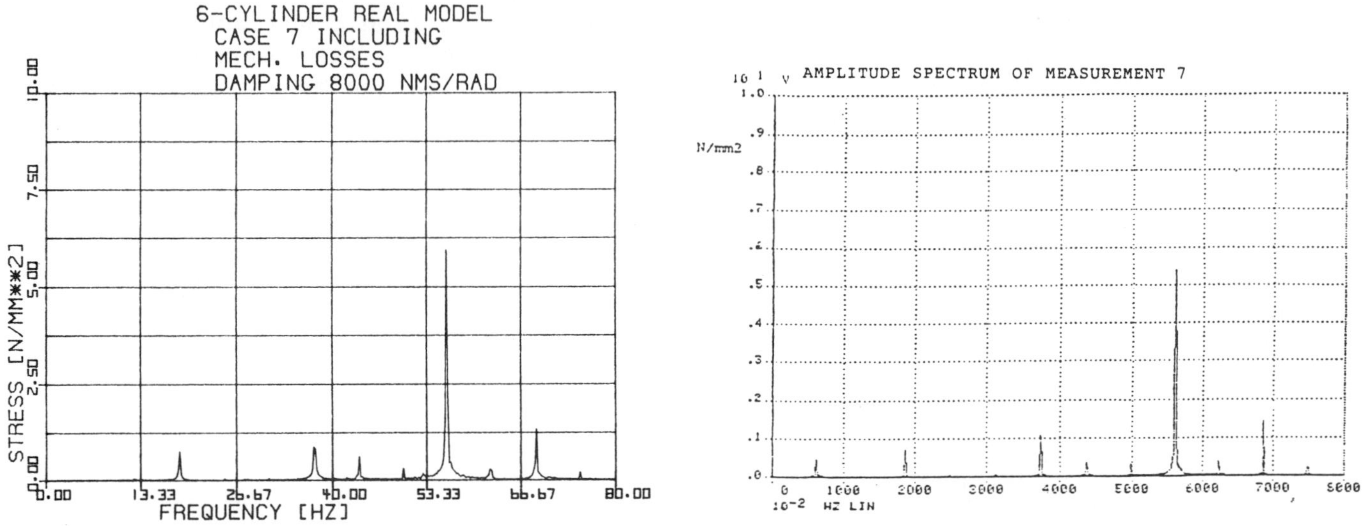

Measured torsional vibrations

fig. 13:Comparison of measured and calculated torsional vibrations
Frequency/stress plot

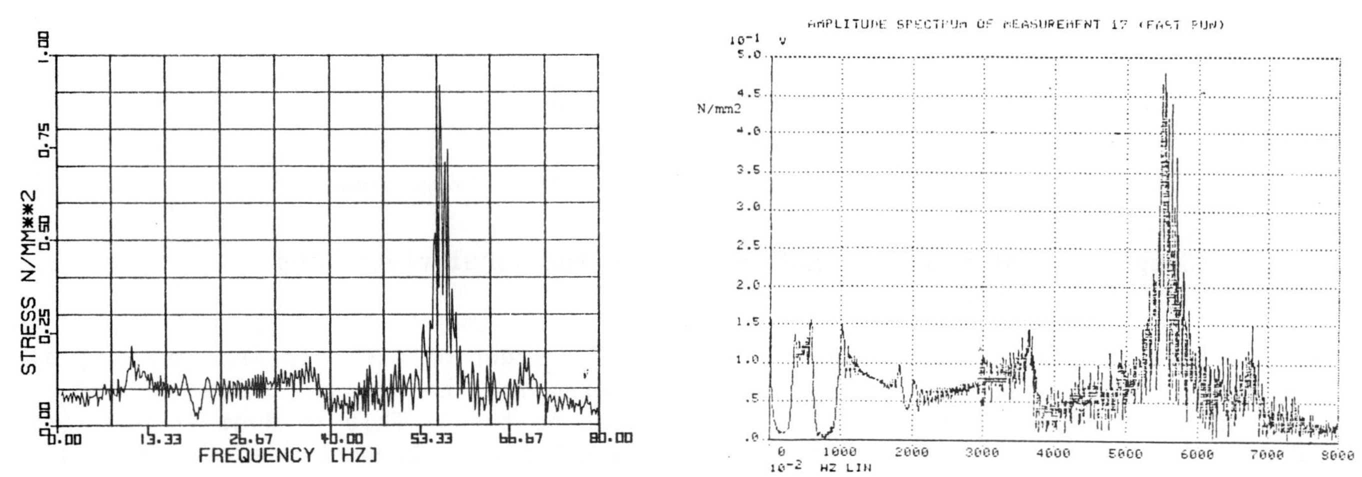

fig.14: Comparison measurements and calculations of torsional vibrations
Frequency analysis of speed sweep.

Conclusion

When the measurement results are compared to the results of some system analysed by the classical approach, notable differences appear, demonstrating the advantage of the newly developed method. Analysis for complex systems can now be performed with increased confidence, and "non-explainable" difference in outcome between measurements and computations employing the classical approach no longer occur. The shortcoming of the classical method, ie its inability to predict accurately transiënt conditions, is now no longer a limitation.

When compared to the outcome of the classical methods, significant divergencies can be noted.

1. The peak-to-peak level of the overall dynamic torques for a six-cylinder compressor are 16% higher and 36% higher for a four-cylinder compressor with the cranks at 90°.
2. The amplification of the resonant harmonic component is approximately 25% higher.
3. Remarkable is the occurrence of a significant second resonance condition at 11 times compressor speed in the six-cylinder model which is totally absent in the outcome of the disc model, and which shows up in the measurements on the real compressor.

References

[1]ODEPACK,A systemized collection of ODE solvers in Scientific Computing, R.S. Stepleman et al.(eds),North-Holland, Amsterdam,1983

[2]Hafner K.E., Maas H., Torsionsschwingungen in der verbrennungskraftmaschine, 1985, Springer Verlag Wien,ISBN 3-211-81793-X

[3]J.Molenaar, Torsional vibrations in a Crankshaft, Proceedings of the 4th European Conference on Mathematics in Industry,pp 337-345,1991,editors H. Wacker and W. Zulehner, B.G. Teubner Stuttgart and Kluwer Academic Publishers.

C449/009/93

Liquid ingestion into reciprocating compressors

J F C ARTHUR, BA, CEng, MIMechE and H B CARRICK, MA, PhD, CEng, MIMechE
ICI Engineering, European Region, Billingham, Cleveland

SYNOPSIS Experience of liquid ingestion by a large user of reciprocating compressors is used to quantify the risk of loss of containment. The possible sources and protection available are reviewed and check lists to assist assessment developed. Protection measures during design and in service are included. Two case studies of incidents are described.

1 INTRODUCTION

Reports from the period 1965 to mid 1980 of incidents ascribed to liquid ingestion into reciprocating compressors were collected from within the ICI group. The number of incidents is related to the population, and the incident rate and probability of loss of containment derived and put into context.

The possible sources of ingestion, consequences, and protection methods were considered and finally recommendations for design, operation and maintenance procedures to avoid ingestion damage are made.

2 REPORTS OF INCIDENTS

Instances of liquid ingestion for which basic information was available are listed (Appendix 2) and those involving loss of containment are identified. All the details of incidents have not been included but all details available were considered in developing the sources, effects and protection methods.

3 SOURCES OF INGESTION

A guide list has been drawn up (Appendix 3) to show the sources of ingestion and protective measures available.

A summary and illustrative line diagram are provided in Figures 1 & 2

4 THE POPULATION OF COMPRESSORS

Group I. A total of 250 reciprocating compressors, 52 on hazardous duties.

Group II. A total of 380 compressors

5 THE INFLUENCE OF COMPRESSOR DUTY

The duty of a reciprocating compressor is relevant to liquid ingestion because of possible loss of containment of dangerous gas, and the higher incidence of liquid ingestion on duties associated with evaporation of refrigerant and similar situations where there is normally liquid present near the machine inlet. The following types of duty need particular consideration:

a) Hazardous gas handled
(usually Flammable or Toxic)
b) Low hazard gas handled in a confined space
(potential asphyxiation hazard)
c) Refrigeration duties.
(saturated gas normally present)

6 CONSEQUENCES OF LIQUID INGESTION

6.1 The effect on the Machine

Where the quantity of liquid exceeds the clearance volume above the piston at top dead centre (this is typically 5 to 15% of the swept volume but can be higher, particularly on low pressure ratio applications) severe damage is likely to be caused to the following:

a) valve cages
b) motion work (possibly limited by weak link such as crosshead wedges)
c) cylinder head and liner
d) big-end bearings

Where the quantity is less than this effects are mainly confined to valve plates and piston rings. It is rare for an event to be proved conclusively to be due to ingestion because the evidence may evaporate or drain away.

Additional damage as a result of the machine continuing to run after motion work damage is a distinct possibility.

One case of continuous satisfactory operation with liquid present by design and with no knockout pot is known. The inlet cooler is horizontal and the inlet connection to the horizontal cylinder is at the top. The quantity of liquid is similar to the quantity of lubricant fed to the cylinders and it mixes/dissolves in the lubricant without apparently excessively decreasing the lubrication effect.

6.2 Injury to Personnel, damage to Nearby Equipment

Most incidents do not result in a loss of containment and the consequences are limited to the machine. However where loss of containment does occur the risk of fire or explosion is higher than for many leakages as the motion work is likely to be running in a damaged state, giving rise to sources of ignition such as hot surfaces or incendive sparks.

The consequential effects of loss of containment are likely to be injury to personnel in the area from the gas released, and, where ignition occurs, damage to personnel and nearby equipment from blast, missiles and fire.

6.3 Repair cost

The repair cost of the machine is likely to be higher than for normal routine repairs as all components have to be examined for evidence of damage or distortion. Significant ingestion puts abnormal stresses on nearly all components and the examination should take this into account and not be limited to that required to check for typical failures.

7 FREQUENCY OF OCCURRENCE

7.1 Incident Rate

From specific machine incidents of damage due to liquid ingestion, reported from ICI UK sites, it is possible to make an approximate estimate of their frequency:-

Table 1 Liquid ingestion incident frequency

Site	Incidents Reported	Machine-years in Operation	Incident Frequency (per machine -year)
A	2	748	0.003
B	7	412	0.017
C	3	345	0.009
D	3	575	0.005
E	4	2397	0.002
Total	19	4477	0.004

Machine-years in operation is derived from

R x T x L

Where:

R = Running period = 350 days per year

T = Proportion of time running = 0.66 (assumes average 50% spare capacity - one spare machine for every two duties - with equal running time for each machine)

L = Machine lifetime
15 years average for sites A B D E
30 years for site C

7.2 Sensitivity Analysis

The overall incident frequency per machine-year for possible variations in individual estimates made for factors in the analysis above is as shown in Figure 3. As a result a range of 2×10^{-3} to 8×10^{-3} incidents per machine-year is likely

7.3 Loss of containment

Two incidents in nineteen caused a significant loss of containment:-

No 12 - 300 kg of fluorocarbons
No 28 - 300-400 kg of Carbon dioxide

The probability of loss of containment is therefore 0.1 per incident. However, due to under reporting, it is estimated that up to two more loss of containment incidents could have occurred, so the upper limit probability is taken as 0.2.

The frequency of loss of containment is therefore:-

incident frequency
x loss of containment probability

a range of 2×10^{-4} to 16×10^{-4} per machine-year with a mean value of 6×10^{-4} per machine-year

8 PROTECTION AVAILABLE

Possible means of prevention, protection, or mitigation of damage are listed below. These include items from specifications and operational incidents. A chart of sources and available methods for dealing with them is provided in Appendix 3, a summary in Figure 1 and an illustrative line diagram in Figure 2.

8.1 General protection

a) Liquid coolant pressure below process pressure
b) Cooling by gas to avoid the possibility of inward liquid coolant leakage.
c) Frost precautions on coolant system to avoid cylinder or exchanger leaks.

8.2 Inlet system protection

a) Adequately sized knock-out pots or effective separators to separate liquid and retain it until removed from the system
b) Cyclone and/or low velocity separation.
c) Drainage by automatic trap or level control.
d) Gauge glass or level indicator, high level alarm and trip with separate connections on inlet condensate removal vessel. Five minute capacity between alarm and trip. Note that dirt can cause sticking of a float operated level switch, ultrasonic switches are potentially better, having no moving parts.
e) Drains on pulsation dampers.
f) Moisture removal device in inlet. Knitmesh pads require steady flow at specified rate and are therefore not recommended as pulsation can cause re-entrainment.
g) Drain points of adequate size (minimum 25 mm).

8.3 Avoidance of liquid accumulation

a) Piping falling to drain points to avoid hold-up.
b) Automatic isolation valves:
    (i) to act as barrier to migrating vapour from inlet system
    (ii) to supplement NRV preventing discharge system contents under pressure being transferred into compressor
c) Drain points of adequate size (minimum 25 mm).
d) Manual bypass round condensate removal level control valve or trap.
e) Heating to prevent condensation.

8.4 Protection within the Machine

a) Top inlet bottom exit horizontal cylinder.
b) Unloading facilities, either inlet valve lifting or additional volume, these provide additional capacity to accomodate liquid or less resistance to escape of liquid from the cylinder. These features could assist at start up when the risk of ingestion is higher than usual but may not be in use at other times
c) Cylinders maintained above condensation temperature.
d) Spring relief of cylinder head.
e) Bypass to inlet operated by excessive pressure.
f) Small capacity, limited rate, cylinder lubrication supply with timed cut off on machine stopping. Topping up of lubricant monitored
g) Pressure testing of cooling circuit components.

8.5 Ancillary Equipment

a) High integrity heat exchangers (particular care in; selection of material, weld details, tube support)
b) Separator for continuous liquid removal.
c) Use of inert gas to purge condensing gas.
d) Control of process liquid inventories to avoid over filling.

8.6 Operational measures

a) Check no liquid present by draining via visible drain point.

c) Check traps operating.
d) Check traps clear.

9 SUMMARY OF GUIDANCE

9.1 General

The approach selected does not consider the probability of ingestion causing machine failure as once a risk of ingestion has been identified it is likely that it cannot be discounted and therefore the reasonably practicable prevention measures identified should be employed.

Use the methodology of this paper to review the hazard and decide what measures are appropriate. A cost versus benefit assessment may be required for cases where the risk to equipment and production is high but that to personnel is low.

Note that one of the times of higher risk is when a machine is started. At this time there may be more potential for ingestion, protective measures may be temporarily bypassed, and personnel are more likely to be in the vicinity. This will require attention when defining operating procedures, see Appendix 4.

9.2 Assessment of Risk

Decide whether there is a risk of liquid ingestion by considering the sources of ingestion (Appendix 3) in conjunction with the illustrative diagram (Figure 2).

9.3 Determine Action

Consider appropriate methods of prevention, protection and mitigation, from Figure 2 and Appendix 3 and decide which are relevant in the particular circumstances. All relevant measures should be taken unless the risk to individuals and of major accident can be shown to be acceptably low. In the latter case consider economic factors.

9.4 Operation & Maintenance

The possible continuing actions available to maintain good operating conditions are listed in Appendix 4. These should be considered in the light of local circumstances and those appropriate incorporated into operating and maintenance instructions.

APPENDIX 1. TERMINOLOGY

Knock out pot, or Catchpot
: A vessel at compressor inlet whose function is to intercept liquid which may occasionally be present in the gas stream.

Separator
: A device designed for continuous removal of liquid from a gas stream.

Damper
: A vessel placed close to cylinder inlet or exit to reduce pressure pulsation levels.

Trap
: A device to permit liquid to pass while not permitting gas to pass.

APPENDIX 2. INCIDENTS SUMMARY

* denotes loss of containment

	Compressor	Comments
1	Halls, Pump-down units	'88; Damaged pistons, rings, valves, bearings
2	Halls, Refrigeration	early '80's; Bent connecting rods
3	Bryan Donkin	'88; Cylinder liner damage
4	Stal-Levin, refrigeration	Many times; Low stage valves damage
5	Consolidated Pneumatic	'85; Condensation when shut down. Valves damage
6	Bellis & Morcom	'89; H2SO4 carryover. Piston disintegrated, rod broken, head cracked
7	Recycle refrigeration	'72; Piston ring breakage
8	Alley No 3 transfer	some years ago; Broken piston rod and head
9	Recycle vent gas	many times; Filled with water, no failure
10	Burckhardt	many times; Fracture of distance piece
11	Sulzer (labyrinth)	'88; Aftercooler failure/water inwards. Flame trap blew, crosshead wedges protected machine
*12	Phillips Cryogenic	Sept 85; Liquid logged, soldered joint failure- process side of head
13	York Refrigeration	Many times; Faulty level control. Broken connecting-rods, worn crankshaft
14	Hall's Refrigeration	88; Valve damage
15	Sabroe Refrigeration	'89; Valve seals failed
16	Cooper Bessemer	Cracked crosshead
19	Halberg	'71; valve & piston problems, cylinder block cracking while running
20	Esslingen	1973/1974
21	Plant Air	1965
22	Sulzer boil off	piston, piston rod keys crosshead damage, no valve damage. 6 incidents in one year. 3 high level trips installed on K O pot
23	Carrier gas	no damage, high level in inlet system
24	Ingersoll Rand	'83; loose piston nut due to possible carry-over '84; valve break up, '88; piston pick up
25	verbal report	occasional reports of liquid in valve chambers
26	Sulzer	'81; failed white metal bearings, '84; cracked piston & valve plate starting with liquid in cylinder, piston

		rod/crosshead wedges loosened
27	Halberg	'70; process carryover 600 valve faults (broken plates) in 7 years. Ring wear & failure
*28	Peter Brotherhood	Feb '89; Intercooler leak on start up, cylinder fracture
29	verbal report	valve plate breakage at start up, drainage incomplete
30	Sulzer	Oct 73; wedges loose, exchanger leakage on inlet. Oct 73,77, Feb84; piston damage.
31	verbal report	build up on valves, mist/particulate carryover once/year
32	Ammonia	fouled valves, piston ring failure mist/particulate carryover several times per year
33	Ingersoll Rand	intercooler leak, heavy knock, no damage
34	Linde	cooler coil in sump fracture due to overfilling & turbulence, liquid into cylinder. Crankshaft broken, cylinder head cracked
35	Peter Brotherhood	numerous incidents of water leakage due to vibration of intercooler tubes at baffle plates
36	Nuovo Pignone	July '89; possible ingestion leading to delayed mechanical damage
37	Linde	Mar.90; Intercooler leak off line (o-ring), extensive cylinder head damage

APPENDIX 3. POSSIBLE SOURCES OF LIQUID & METHODS OF PROTECTION.

3.1 Applicable to a quick re-start

3.1.1 Inlet

a) Source

Leak of coolant into gas line
i) from pre-coolers or intercoolers
ii) from bypass coolers
iii) from jacketed pipes

Liquid trapped in inlet pipe above machine inlet valve

From process side of upstream equipment
i) scrubber or evaporator overflow
ii) catchpot overflow
iii) condensation (gas migration plus build-up)

b) Protection

Use air cooling
Suitable size knock out pot
Cyclone disengagement in knockout pot
Suitable drainage from knockout pot
Level indicator, alarm, trip, trap,
Level control
Control of filling systems
Inventory measurement & control
Pipework
i) no hold up in inlet, bypass
ii) suitable drains for checking piping and damper
iii) trace heating
iv) thermal insulation
Immersion heater to boil off liquid
Operating procedures required as appropriate
Cylinder configuration
bottom outlet for horizontal cylinders

3.1.2 Machine

a) Source

Cylinder oil supply continues after machine stopped
Inward coolant leak into cylinder
Cracked liner after freezing
Failed gasket
Condensation in cylinder due to cold coolant

b) Protection

Cylinder oil
i) small inventory
ii) top up procedures
iii) feed rate setting

Cylinder pressure relief
i) spring loaded cylinder head
ii) bursting disc
iii) internal relief valve

3.1.3 Discharge

a) Source

Reverse flow through defective non-return system including machine valves
Leaks in after cooling or heating equipment
Delivery system overfilling
Condensation in discharge pipework

b) Protection

As inlet 3.1.1.b
Automatic isolation valve

3.2 After Valve Isolation

3.2.1 Inlet

a) Source

As above 3.1.1.a but with the possibility of liquid being trapped behind isolation valve. This may advertise its presence by external effects such as frosting of piping.

b) Protection

As inlet above 3.1.1.b
Tracing/insulation
Non condensing gas pressurising
Coolant below process pressure

3.2.2 Machine

a) Source

As above 3.1.2.a

b) Protection

Timed cylinder lubrication cut out
Non-freezing coolant
Keep coolant flowing
Lower pressure coolant
Detect loss of coolant from closed system
Bar over before start
Heat coolant to avoid condensation

3.2.3 Discharge

a) Source

As inlet side above 3.1.3.a

b) Protection

As inlet above 3.1.3.b
Non return valves

3.3 After slip-plating
As above but possible draining of accumulation when de-slip-plating

3.4 While Operating

3.4.1 Inlet Source

Large inward leak of higher pressure coolant
Liquid injection for cooling
From process side of upstream equipment
Scrubber or evaporator overflow
Catchpot overflow
Excess makeup
Carry-over due to process upset
Condensation (gas migration plus build up)

3.4.2 Machine Source

Cylinder oil supply (unlikely)
Inward leak of coolant
Condensation on cylinder walls due to over cooling

3.4.3 Discharge Source

Liquid will be carried away from machine subject to piping arrangement allowing.

3.5 After Maintenance

a) Sources

Liquid left in system
a) after Hydrotest
b) after cleaning/washing
Excess level
a) inventory on filling
b) level control problems
Condensation

b) Protection

Blowing out
Heating
Draining
Pulling a vacuum on the system
Inventory control

APPENDIX 4. OPERATING AND MAINTENANCE GUIDANCE

4.1 Protective Systems Available

a) Inlet separation vessels and knock-out pots.

b) High level switches and level indicators in inlet separators.

c) Tell-tale drain points in gas piping system.

d) Vibration Monitoring (Vibration switches are of use in reducing consequental mechanical damage and possible loss of containment).

e) Machine Condition Monitoring - see below 4.3 e).

f) Operating Records.

g) Maintenance Procedures.

Mechanical (Evidence of liquid during overhaul)
Instrument/Electrical (maintenance of protective systems)

4.2 Possible Pre Start Up Checks

To be interpreted after consideration of local installation and circumstances and whether it is an immediate re-start or re-start after prolonged shutdown.

a) Check the operation of each separator drain valve and consider the possibility of chokes. This is best done with the system under positive pressure.

b) Check for liquid in the compressor inlet pipe downstream of inlet separator or knock-out pot.

c) Check cylinder cooling system is not causing condensation inside cylinders.

d) Check for liquid by opening drains on interstage gas pipes and coolers.

e) Open drain/blank on each inlet/delivery pulsation damper if condensation is possible

f) After prolonged shutdown remove the lowest valve per cylinder and check for CW leaks inwards from the cylinder liner or the cylinder head.

g) Run electrically driven cylinder lubricator and check that cut-out timer works with machine stopped.

h) In water/glycol cooled machines, check the concentration of cooling mix to guard against frost damage.

i) Check condition of insulation and operation of trace heating when fitted.

j) Bar the machine over; listen for mechanical knocking and check unusual sound or resistance to rotation.

4.3 Checks During Machine Operation

a) Check for any knocking sounds within cylinders and motion work.

b) Check the operation of each knock-out pot/separator drain valve and that flow is not excessive due to leakage

c) On refrigeration compressors, ensure that there is no excess charging of refrigerant.

d) Where top-up head tanks are provided check that the cylinder lubricator reservoir or the machine crankcase cannot be over-filled.

e) The typical temperature ratio for a particular pressure ratio for each cylinder is not reduced, ie polytropic efficiency increased - can be over 100% if liquid present.

4.4 Additional Items

a) Instrument and Electrical routine maintenance is appropriate, to include switches, alarms, thermocouples, pressure switches and trip systems.

b) Integrity of Insulation.

c) Reliability of Trace Heating.

d) Non-Return Valves to be checked for operation and overhaul on a routine basis.

e) Check condition of heat exchangers so that deterioration, eg due to corrosion, does not lead to coolant inward leakage.

4.5 General Note

Liquid ingestion is a rare occurrence so that these checks will yield zero return for most of the time.

APPENDIX 5. CASE STUDIES

5.1: Carbon dioxide Compressor

This event occurred on a large 5 stage compressor. The machine failed on start-up due to severe liquid ingestion into the second stage. The photographs (Figures 4, 5, & 6) show the damage to the second stage cylinder, which was fractured, resulting in a loss of containment. The material of the cylinder was cast iron.

The source of liquid was cooling water which leaked into the system from two fractured tubes in the intercooler between the first and second stages. This occurred over a period of a few hours while the machine was shut down.

The design concept was that the catchpot downstream of the intercooler was large enough to cope with a tube failure if it was drained once per shift.

Unfortunately at start up there was little driving force for liquid to drain (no pressure in the pot) and the layout of the drains was poor in that the operator could not see if water was draining. The pot must have been full, or nearly full, and as a result, although blow down was of the pot was carried out, it was ineffective.

No level indicator was fitted to the catchpot.

The solution adopted was to re-route the drains from the pot so that they were visible when blowing down, coupled with greater attention to draining prior to start-up.

5.2: Recycle Gas Compressor

This is a three stage machine, compressing propylene from a polymerisation reactor. In the event the machine tripped unexpectedly. On restarting it was extremely noisy and was shut down for examination. It was discovered that the wedges which lock the first stage piston rod to the crosshead had sprung out (see Figures 8 & 9). A cross section of the relevant area is shown in Figure 7.

In effect the wedges had acted as a mechanical fuse.

It was concluded that a gradual build up of liquid in the first stage inlet pulsation damper had been pulled into the machine.

The liquid was drawn in vapour form from the reactor system and was condensed by a cooler upstream of the compressor. The rate of liquid arising was low, but no allowance for it had been made in the design. No catchpot system was fitted to collect such liquid.

The solution was to fit additional catchpots with level instrumentation ahead of the first and second stages.

SOURCES

INLET	STAGE	INTERSTAGE	BYPASS
precooler leak	coolant leak	cooler leak	bypass cooler leak
system overfill	cylinder lub. oil	condensation	aftercooler leak
vapour migration	condensation	cooling injection	condensation
condensation			cooling injection
liquid carryover			
pipe holdup			from delivery via NRV
cooling injection			system overfill
jacketed pipe leak			condensation

PROTECTION

INLET, INTERSTAGE, BYPASS	STAGE
cylinder configuration horizontal	coolant antifreeze
ko pot/separator	coolant flowing
level indication/alarm/trip	coolant > condensation temperature
inlet damper free draining	barring over
insulation/trace heating	spring loaded head
gas coolant	internal relief
ko pot heater	lubrication feed rate
ko pot drain, auto/manual	cylinder lubrication pump time limit
system inventory control	minimum lubricant inventory
coolant < gas pressure	
piping layout/drains	
discharge non-return valve	
automatic isolation valve	

Fig 1 Liquid Ingestion - Sources and Protection Summary

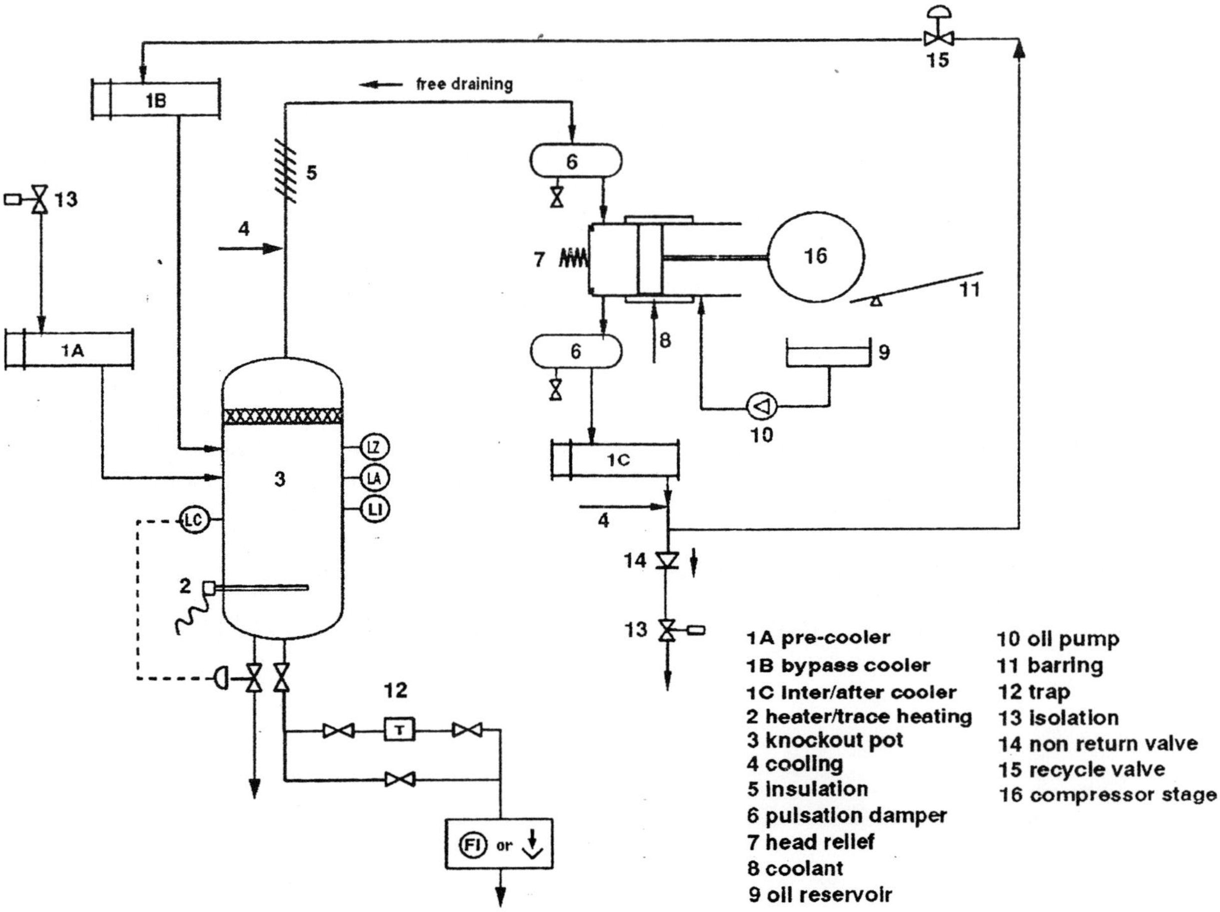

Fig 2 Liquid Ingestion - Illustrative Line Diagram

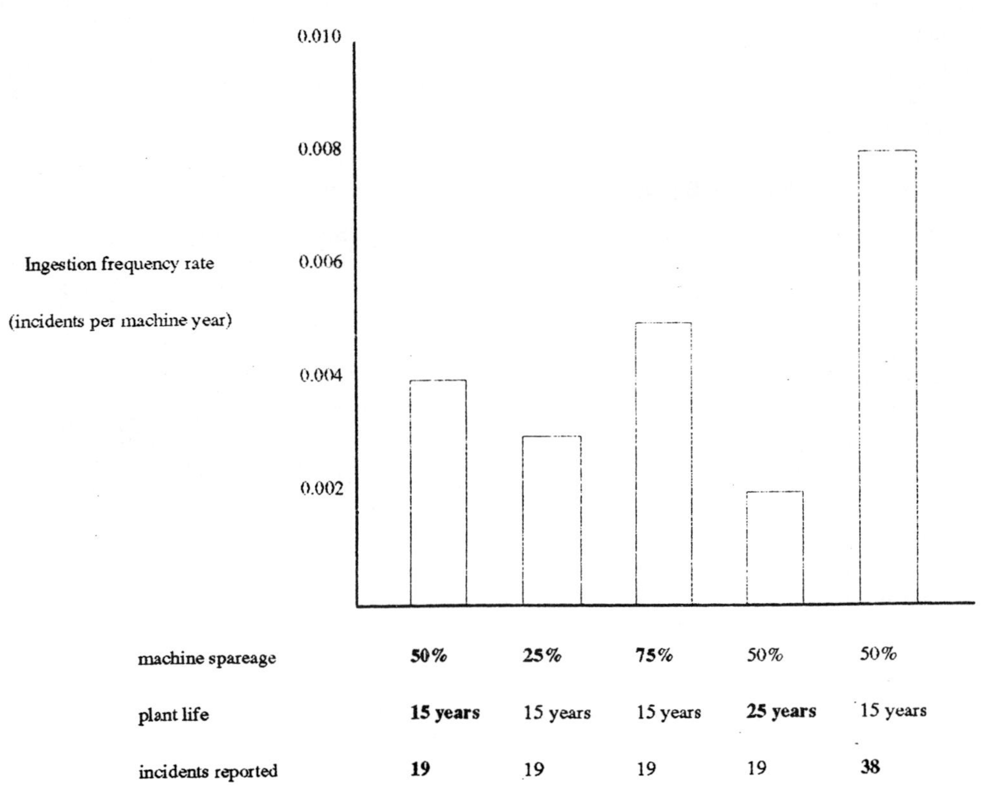

Fig 3 Variation of estimates of frequency rate with assumptions

Figure 4(a) Cylinder with damaged portion removed

Fig 4(b) side view

(a)

(b)

Fig 5 Cracks in cylinder at valve covers

Fig 6 Cracking in gas passage wall

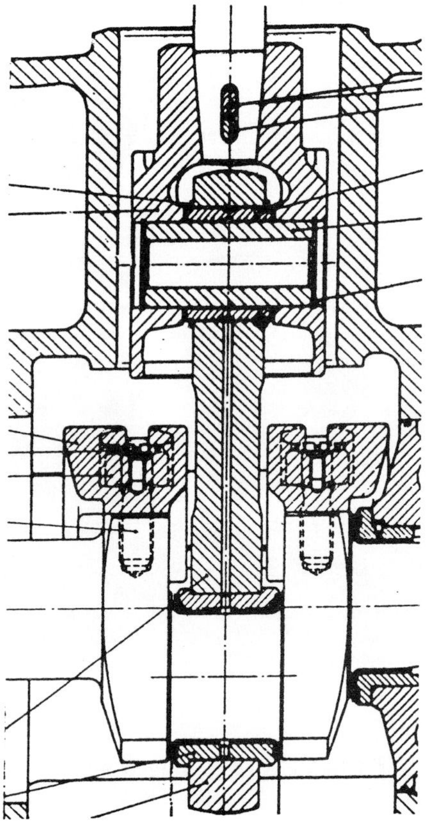

Fig 7 Cross section of motion work

Figure 8 Crosshead with wedges

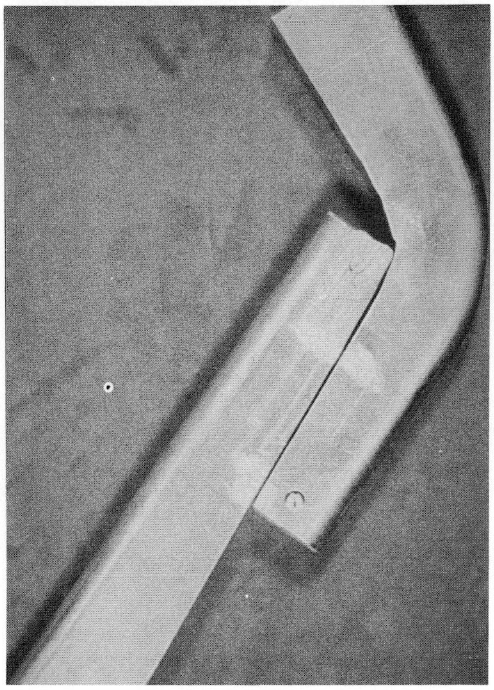

Fig 9 Wedges after removal

C449/019/93

Reciprocating compressor piston rod, a hidden design challenge

F L BOER, Ing
Thomassen International BV, Rheden, The Netherlands

SYNOPSIS

A critical component of all reciprocating compressors is the piston rod, and much attention is focussed on it by users of this category of machines.

While the traditional design is straightforward, it represents a definite challenge for a designer to improve upon it.

It is possible, by using the concept of preloaded connections, complemented by the optimization of rod geometry, to improve the performance and reliability of a piston rod in comparison with traditional designs.

This paper provides insight into the considerations employed in achieving a more advanced design using the noted approach, and relates them to the operating experience gained with this design of rod.

1.) INTRODUCTION:-

Although literature on the subject of piston rod design in reciprocating compressors does not discuss the origins of design, this most likely stems from steam engine or pump design preceding it. In retrospect it seems strange that the design has remained rather primitive for such a long period of time. The piston rod seems for unclear reasons to have escaped to a degree the attention paid by designers to other machine parts. Reference to older books on the subject of the design of piston rods showed an unexciting lack of variety: there is in fact the impression that once the very simplest approach had been introduced, it was simply copied and recopied for the sake of convenience without much further thought. This appears to have resulted in a rather vague equilibrium where on the one hand operators and owners of such machines had to accept the disadvantages in the form of fatigue fractures which occurred often enough to breed disatisfaction but too infrequently to result in a fundamental improvement in the design. It is interesting to note that the status of the reciprocating compressor piston rod has remained unchanged in the design of many compressor manufacturers, and that the fundamental advantages offered by a more modern design approach are more or less confined to European manufacturers during the last decade. It might be concluded that it would represent a definite, if somewhat low key challenge, to improve significantly on the existing of piston rod.

2.) DESIGN CONSIDERATIONS:-

Formulating design objectives:

It can be readily understood that improvements should be sought in the basic concept and geometry of the rod as materials and manufacturing processes used in production are common to all designs and are not seldom as specified in A.P.I. 618 and/or by the user. Fatigue failure is the only failure mode which occurs as a result normal machine operation, a piston rod having to withstand almost purely cyclic loading, so that any design of piston rod will have features aimed at the reduction of fatigue failure. Improvements in the design would thus fall in the following categories:-

1. Notches or stress raisers must be avoided or minimised, and combinations thereof optimised.

2. Reduction of cyclic load imposed upon the most critical areas of the rod, although in apparent contradiction with the fact that the cyclic load on the rod as a whole cannot be changed.

3. Devising a consistent and reliable fastening technique for the components to be attached to the piston rod to eliminate the variability of static stress conditions as much as possible. Statistics indicate that insufficient preload is an especially important factor in fatigue failures of large threaded fasteners.

3.) CLASSICAL PISTON ROD DESIGN:-

To have a point of reference for such evaluation the classical design of rod can be used as shown in fig 1. The design corresponds to that found in a variety of textbooks on the subject.

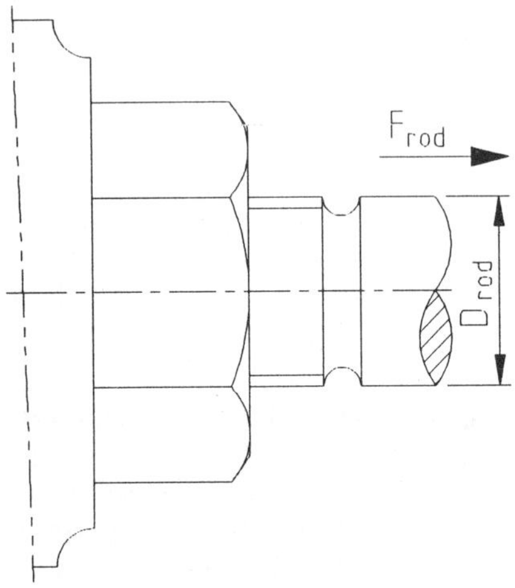

Fig 1. The classical design of recip. compressor Piston rod attachment.

The most critical notched area of the rod is the screw thread at its entry into the locknut or crosshead: in any case the thread at the crosshead side. The piston end is usually much less critical, for reasons which will still become clear. The cyclic stress amplitude which ultimately is determinant for the probability of failure of the rod during operation can be very simply defined as:-

$$S = \frac{F.K_{thr}}{A_{thr}}$$

where S = cyclic stress amplitude.
F = rod load.
K_{thr} = notch factor thread.
A_{thr} = cross sect. area thread root

For purposes of comparison it is assumed that F is of equal magnitude for tension and compression. Examination of this relationship shows that only 2 of the 4 parameters can be changed at will, being K and A. From the point of view of manufacture it is most convenient to select a screw thread diameter equal to the diameter of the rod itself, automatically imposing a "soft" limit on the maximum value of A. As commonly used screw thread profiles do not provide a basis for substantial reductions in the value of K, thread rolling and the resultant residual compressive stress in the thread gives the final, readily attainable, improvement in the fatigue resistance of the thread. The main disadvantage of this design of rod attachment lies in the fact that all of the externally applied cyclic force is transmitted through a threaded (= notched) portion of rod. Additionally a substantial bending stress in thread foot where the cyclic load is transferred to the nut/crosshead, creates at the first thread where up to 30% of the load transfer takes place, an unattractive stress peak. This becomes, understandably, the prime location for failure. Whithout going into details of the failure pathology of this design of rod, there remains the distinct impression that the fatigue failure rate of this design of piston rod is higher than desirable. It is well understood by the users of petrochemical plant compressor that each fatigue failure of a piston rod can have catastrophic results for personnel and plant, and must therefore be avoided. This then should perhaps form the motivation for attempting to improve on this design of rod.

4.) AN IMPROVED DESIGN:-

Closer examination of the mechanism which governs cyclic stress amplitudes in components forming closed preloaded systems of which the static equilibrium is distrurbed by an externally applied force, provides the basis for a more sophisticated design of rod/crosshead attachment. Whilst the construction of such a joint is physically simple, the basic impact which it has on the control of cyclic stresses in the constituent components of the joint is not widely appreciated. Fig 2. below shows the cross selection of this type of piston rod end.

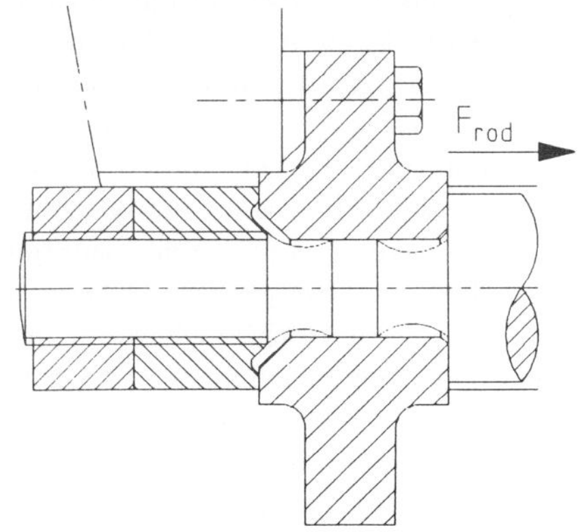

Fig 2. The preloaded pisto.ı rod end contruction.

As shown above, the rod end to be attached to the crosshead, has been substantially reduced in diameter and fitted to a coupling flange which in turn is attached to the crosshead by a suitable number of studs. Initially the loose flange is positioned on the rod end. The nut is then fitted and in this case hydraulically tensioned to 1.5x the rated rod load of the machine as required by A.P.I. 618 para.2.8.1. The coupling flange is then fitted to the crosshead whereafter the relatively small nuts used here can be tensioned with a torque wrench. The joint is, once in this condition of assembly, ready to accept its externally applied load, alternating tensile and compressive rod forces. This external force results in the redistribution of the already present static forces or preloads introduced into the joint during assembly. The amplitude of the cyclic forces thereby created in the system of preloaded components, is determined purely by the relative compliances of the components in question or in other words their relative stiffnesses with respect to one another. The first attempt to describe this analytically was apparently made by Felix Rotscher in 1927 and the current rather similar manner of describing the effect of an external load on a preloaded system is shown in fig. 3 below:-

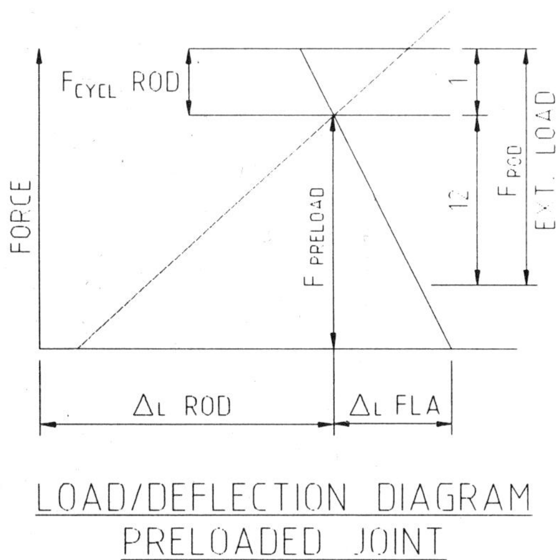

Fig 3. Load deflection diagram, preloaded joint.

The effective compliance ratio for the reduced rod end/flange, as shown, dependent on local geometry, lies typically around the value of 10 of 14. Actually, this is a combined factor which represents the ratio

$$\frac{\text{delta } L_{rod}}{\text{delta } T_{flange}} \quad \text{or} \quad \frac{\Delta L_{rod}}{\Delta L_{flange}}$$

as shown in the static load diagram fig 3., multiplied by the effective force induction height in the flange. (This latter factor is also in part responsible for the actual cyclic stress level ratio for an externally applied load). Reflection on the physical nature of this construction will lead the designer to formulate the following design features / requirements / possibilities:-

1. The higher the compliance ratio, the lower the cyclic stresses in the rod end. A maximum value is desirable, which may be achieved by the use of a minimum diameter rod end. The limiting factor determinant in this requirement, is the level of static stress which is attained during the preloading operation. Yielding or plastic deformation places the limit on the amount of waisting which the rod end may have; a loss of preload could have disasterous consequences for the rod. Choosing a high but safe prestress level with a maximum waisting would be the starting point for the design. Design objective 2. is achieved via this choice.

2. The next step is to provide sufficient bearing area at the shoulder to withstand the compressive stress resulting from pretensioning. Selecting the largest base radius here gives the lowest stress concentration or notch factor, which results in the lowest attainable cyclic stress within the tensile member of the assembly. An optimum design dictates that the cyclic stresses at the highest stress concentration, being the thread root radius, be no higher.

In other words:- $S_{thr} = S_{rad}$

i.e.

$$\frac{K_{thr}.F}{A_{thr}} = \frac{K_{rad}.F}{A_{rad}}$$

so that

$$\frac{K_{thr}}{A_{thr}} = \frac{K_{rad}}{A_{rad}}$$

K_{thr} will in practice always be greater than K_{rad}. The ratio K_{thr}/A_{thr} should equal K_{rad}/A_{thr} implying that required the cross sectional areas at respective sections should be in direct proportion to the local notch factors. From this, it becomes apparent why the screw thread diameter will in an optimised design always be somewhat larger than the shank of the waisted section. While it was just stated that the cyclic stress amplitudes at thread root and shoulder radius would be equal, the situation is more complex insofar that the equivalent cyclic stresses (i.e. the effective combination of cyclic and static stresses of different amplitudes at these cross sections.) should be equal. The discussion on the evaluation of combined stresses is set out in section 5. From following this procedure the screw thread root diameter will almost automatically be defined once the choice of thread profile (and thereby root radius) has been decided upon. Fig 2. shows, schematically, the general configuration dictated by the preceding considerations. One of the design requirements for the coupling flange is a low compliance. This can best be achieved by using a massive hub of ample dimensions. The effective cross sectional area which participates in the absorption of forces (both preload and cyclic) can be calculated by several available methods such as for examply V.D.I.2230. or F.E.M. Initially there could be some temptation to reduce the diameter of the flange boss as an increased compliance thereof effectively reduces the cyclic stress during one half of the load cycle of the external force. A moment's reflection however, reveals that this results in an equal increase in cyclic stresses in the opposing half cycle so that there is nothing to be gained by such a measure. Simply increasing the boss diameter to an extreme value is subject to the law of diminishing returns and spatial considerations determine practical limits.

Another feature of the complete design is the use of a lip nut whereby the root bending stresses in the most highly loaded threads are reduced. The use of lip nuts on conventionally designed piston rods is not unknown and the benefit afforded by it being is in the order of magnitude of 30% increase in endurance limit, a worthwhile improvement on both designs.

To achieve a maximum grade of component and machine reliability, an accurate preload control is necessary. This type of preloaded design rod has been fitted to the company's compressors over the past 15 years using a semi-hydraulic nut for the attainment of a well controlled preload. 3 years ago the introduction of a full hydraulic system eliminated the disadvantages inherent in the use of this type of nut. The scatter in the amount of preload introduced into the assembly is one of the lowest of all types of fastening methods. (see table #8 of V.D.I.2230). Another choice which had to be made was how the actual tensioning device was to be executed. One option was the integration of the hydraulic system within the nut itself. Seal failures, although unlikely, would with this design lead to irritating disassembly problems during maintenance, with extra costs involved in unnecessary downtime. Additionally, 2 dynamically loaded screw thread sections would also have been required, while the preference was for 1. Also, for multicylinder machines, the hydraulic mechanism contained in each nut is an additional complicatior and expense. It was therefore decided to use a solid nut design with a separate loose tensioning device. The same tensioning device could then be employed to tension the piston nut and, ultimately, other major machine fasteners. The extended section of rod required for the tensioning device could additionally provide a convenient feature for use in the location of the positive locking called for in A.P.I.618 para 2.8.1. The preceding considerations are to a large degree also valid for the design of the connection fo a piston on a similarly reduced section of piston rod at the other end. From this it may be understood why, with the conventional design of piston rod, failures at the crosshead end of the rod are much more prevalent. The parameters governing the cyclic stresses at the piston end are the same as the described construction at the crosshead side. (For non corrosive duties, as corrosion may on the piston end sometimes create its own particular problems). Less extreme waisting is used for the piston end as it provides the possibility for cyclic stress control for hollow, hubless piston designs. To provide a relatively simple basis for the comparison of the two design of piston rod, the following reveals the essential differences:-

Conventional

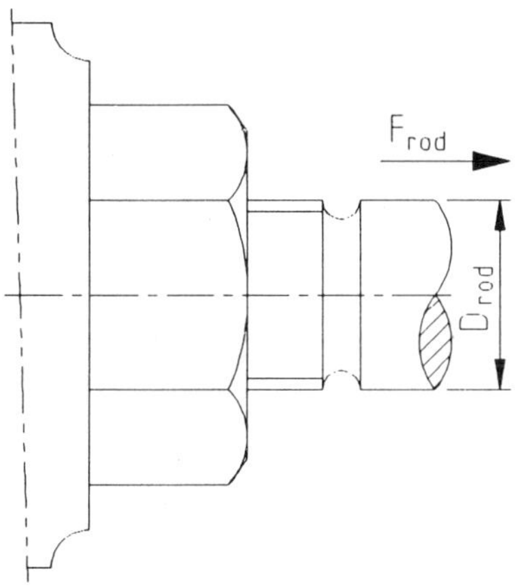

Cyclic stresses.

Symbols used:

S = cyclic stresses.
F = axial force.
A = cross sectional area.
D = diameter

$$S_{cycl} = \frac{F_{cycl}}{A_{thr}}$$

$A_{thr} \approx A_{rod}$

also $F_{cycl} = F_{rod}$

$$S_{cycl} = \frac{F_{rod}}{A_{rod}}$$

Thomassen "Hydrofit"

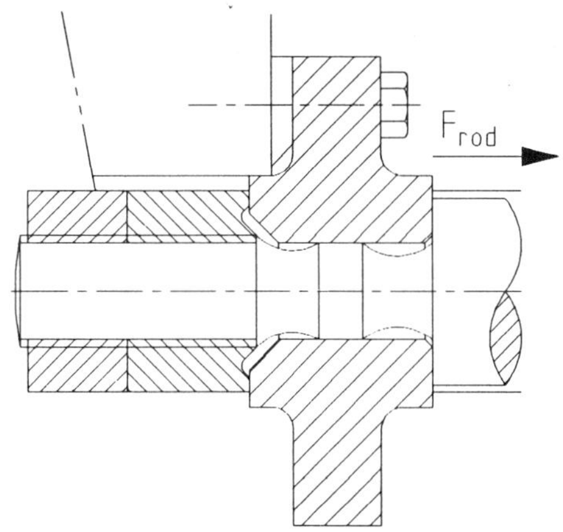

Cyclic stresses.

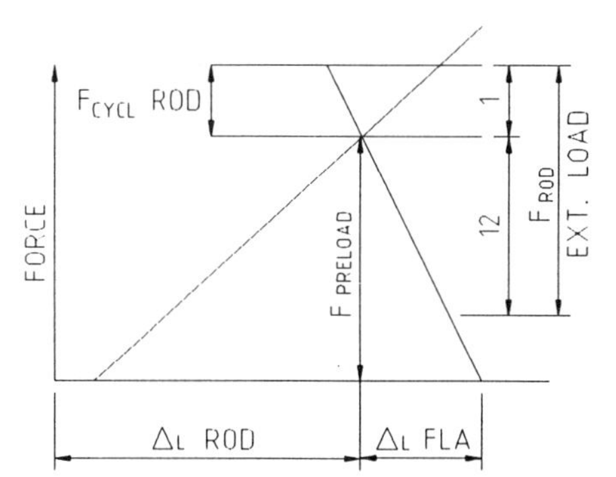

LOAD/DEFLECTION DIAGRAM
PRELOADED JOINT

$$S_{cycl} = \frac{F_{cycl}}{A_{thr}}$$

$A_{thr} \approx ¼\ A_{rod}$ $[\ d_{thr} \approx ½\ D_{rod}\ ]$

$F_{cycl} \approx {}^{1}/_{12}\ {}^{*}F_{rod}$

$$S_{cycl} = \frac{{}^{1}/_{12}\ F_{rod}}{{}^{1}/_{4}\ A_{rod}}$$

$$S_{cycl} = \frac{{}^{1}/_{3}\ F_{rod}}{A_{rod}}$$

* This is typically the compliance ratio of the waisted rod to coupling flange hub calculated in accordance with the V.D.I.·2230 standard and research by W. Thomala.

5.) STRESS EVALUATION:-

The preceding comparison demonstrates that the cyclic stresses occurring in the thread root of the piston rod could easily be reduced by a factor of circa 3 by the use of a preloading system. This is an important advantage for the reliability of the rod. Static stress should however not be forgotten, as well as other factors which are determinant for the fatigue strength of the rod. The complete procedure for a systematic assessment is shown in the form of block diagrams. The most important factor is the combination of static or prestress and the cyclic stress occurring during operation. The use of a modified Goodman approach or Smith diagrams has been popular and in use for many years. On the other hand the Gerber parabolic relationship probably offers, for piston rod materials commonly in use, a more accurate relationship. Also the Soderberg relationship could be of value as it provides a criterium to avoid yielding of the cyclically loaded component, which is a relevant parameter in case of a piston rod attachment. It may be questioned if these approcaches in unmodified form provides a good evaluation of the stress combination at hand. It is for instance, common to use an unintensified static stress component in the Smith relationship. Whilst this may give reasonable results for ductile materials, it is less suitable for brittle materials and notched components. Also, data available for the class of materials used for piston rods is somewhat contradictory in nature. Any threaded connection with a prestress value of 25%-30% of the yield stress will automatically have a static tensile stress value equal to the yield stress at the thread root as a result of the stress concentration at the thread root radius. Within the range of say 25% to 80% of the nominal cross section yield stress only the volume of material at yield stress in the vicinity of the thread root will vary and consequently, its influence on the fatigue strength of the screw thread can be treated as a size effect problem, although is the effect is not pronounced. In the V.D.I. 2230 standard for high duty bolted joints the effect of prestress on the endurance limit of bolting is ignored for "schlussvergutert" threads, but included for "schlussgewalzt" threads as a considerable residual compressive stress in the thread root is to be taken into account. Whilst residual and static stresses should not be simply combined because of stress gradients envolved, it can be anticipated that the effect of a different range of prestress should be accounted for in some manner (e.g. V.D.I 2230). To avoid problems as the result of the amount of stress relief inadvertently present in the rolled thread of the rod it was decided to use the approach outlined in A.S.M.E. code sections III & VIII div. 2 in combination with the Peterson cubic equation as is to be found in "Low cycle fatique of large diameter bolts", by A.L. Snow and B.F. Langer, in which a stress concentration factor is also applied to the static stress component, with as upper limit of the static stress amplitude, the cyclic yield stress of the material. The use of this approach for the values of nominal cross section prestress between 25% and 80% yield stress results in an equivalent cyclic stress amplitude more or less independent of prestress, comparable to the V.D.I, 2230 "schlussvergutert" recommendation. The outcome of this method of combining static and cyclic components results, typically, in an equivalent cyclic stress amplitude of 1.7 x the value of the intensified cyclic stress. A direct comparison of the cyclic stress and equivalent cyclic stress amplitudes of the two then reveals the following:-

CONVENTIONAL	HYDROFIT
$S_{cy} = 100\%$	$S_{CY} = {}^{1}/_{3} \times 100\%$
	$S_{EQ} = 1.7 \times S_{CY}$
	$S_{EQ} = 60\%$

This means that a significant reduction in the effective cyclic stress amplitude can be achieved by using the prestressed design. To achieve some idea of the impact of a reduction of this order of magnitude in the cyclic stress amplitude relatively simple statistic analysis provides th following:-

Failure probability for $S_{cy} = 100\%$ ___ 50% (P_{50})

Failure probability for $S_{cy} = 60\%$ ___ 0,02%

This represents a reduction in failure probability of 2500 x and represents fully the conventional safety factor used.

6.) COMPRESSION AND BUCKLING:-

Until now no mention has been made of rod buckling and compression. For the normally dimensioned reciprocating compressor in refinery service, this is very often not the limiting factor which estabilishes the piston rod diameter. In other words:- If the minimum dimensions for a piston rod have been determined by criteria for tensile stresses, rods will in very many cases withstand buckling loads presented them. Having an as small as possible diameter has the dual advantage of allowing a smaller stuffing box as well as providing a minimum assymetry for the double acting cylinders used. Crosshead pin load reversal can in a number of cases be influenced to an significant degree by this assymetry under part load conditions, so that a small diameter piston rod is normally considered desirable.
As the slenderness ration of the normal piston rod falls in the range of 30 to 120; even the most slender rod is therefore only just within the boundary of the Euler relationship for elastic instability of columns. Structural columns in this transition zone of slenderness ratio exhibit the largest scatter in buckling loads, attributable to residual stresses. Whilst the data is derived form standard steel sections, residual stresses in piston rods may also be present to a degree sufficient to influence the maximum buckling loads which can be withstood. On the other hand a compressive residual stress at the rod surface is advantageous from the perspective of the reduced possibility of stress corrosion cracking or sulphide stress cracking. Compressive residual stress is more or less unavoidable when using coated or induction hardened rods, and is a desirable feature. The safety factor used in the determination of the maximum allowable or rated buckling load based on statistical analysis of data on this subject, has been calculated as 3. This is nearly twice that used for tensile cyclic stresses, but results in an equal failure probability. This provides adequate allowance for residual stresses.

7.) MATERIALS, MANUFACTURE AND WORKMANSHIP:-

All the preceding considerations were based upon the premise that the materials, manufacture and general workmanship in the production of piston rods is fully comparable for rods from all sources from O.E.M. down to users manufacturing their own with far less resources at their disposal.Without an exhaustive discussion on the influence of each of these factors individually or collectively, it can be appreciated that the qualities of even the best design may be severely compromised by inferior material or manufacture. During the couse of the years we have seen piston rods made by non-O.E.M.'s of material having half of the tensile strength required by specification, having machined screw threads instead of rolled threads, and containing multiple cracks, (immediate rejection being necessary, without having been in service). These comments illustrate the importance of quality in all aspects of the piston rod to achieve a maximum of life and reliability.

8.) OPERATIONAL EXPERIENCE:-

The original execution of this type of piston rod as manufactured by this company was 15 years ago. While the design was not fully optimised in its local geometry, it was executed with a semi-hydraulic nut to achieve a constant and repeatable preload. The cyclic stress levels at the point of highest stress were comparable to the present design, so that the 12 years of experience with the previous generation of piston rods may be taken as representative for the present design. At

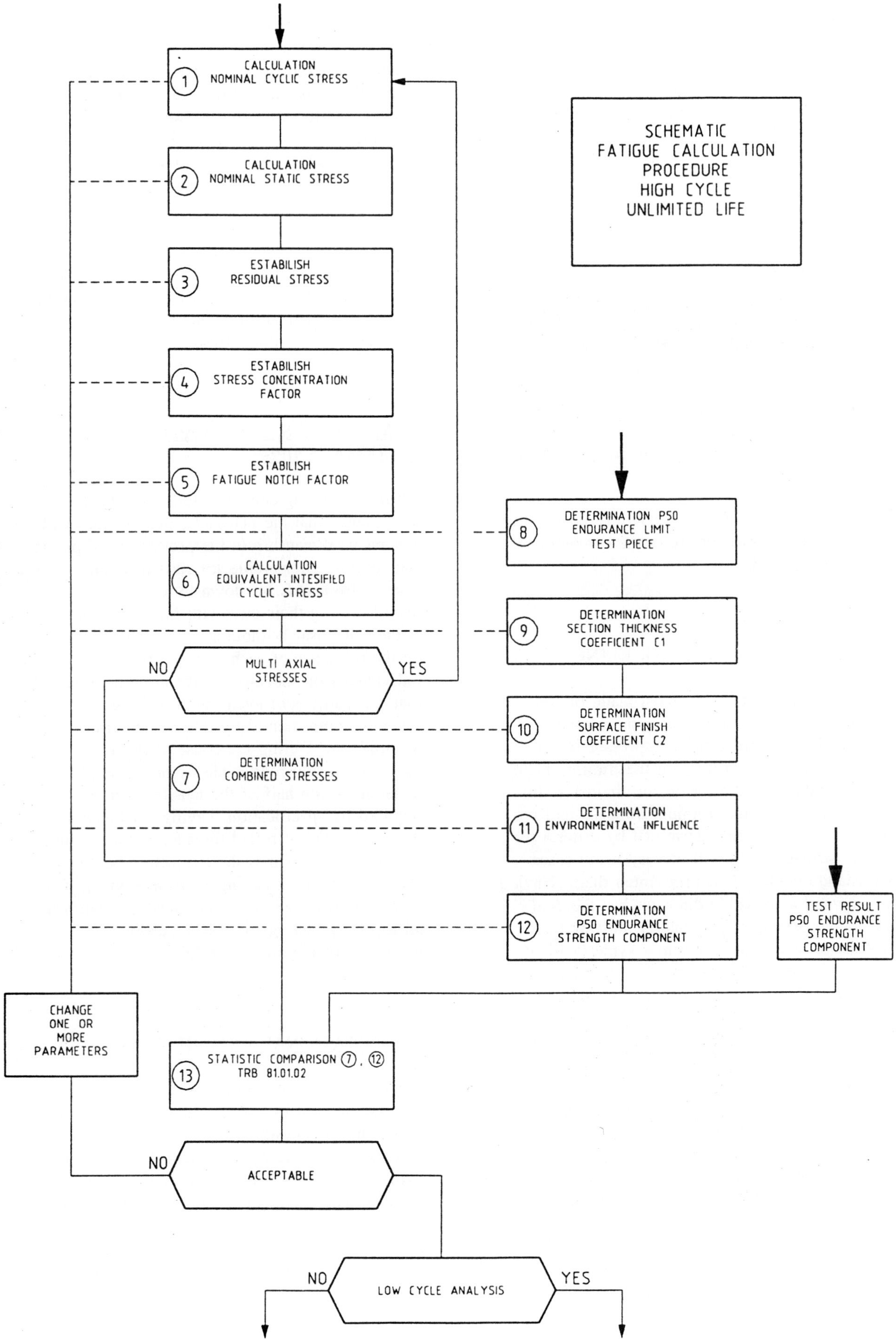
CALCULATION
1 NOMINAL CYCLIC STRESS
CALCULATION
2 NOMINAL STATIC STRESS
ESTABILISH
3 RESIDUAL STRESS
ESTABILISH
4 STRESS CONCENTRATION FACTOR
ESTABILISH
5 FATIGUE NOTCH FACTOR
CALCULATION
6 EQUIVALENT INTESIFIED CYCLIC STRESS
MULTI AXIAL STRESSES
NO
YES
DETERMINATION
7 COMBINED STRESSES
SCHEMATIC FATIGUE CALCULATION PROCEDURE HIGH CYCLE UNLIMITED LIFE
DETERMINATION P50
8 ENDURANCE LIMIT TEST PIECE
DETERMINATION
9 SECTION THICKNESS COEFFICIENT C1
DETERMINATION
10 SURFACE FINISH COEFFICIENT C2
DETERMINATION
11 ENVIRONMENTAL INFLUENCE
DETERMINATION
12 P50 ENDURANCE STRENGTH COMPONENT
TEST RESULT P50 ENDURANCE STRENGTH COMPONENT
CHANGE ONE OR MORE PARAMETERS
STATISTIC COMPARISON 7, 12
13 TRB 81.01.02
ACCEPTABLE
NO
LOW CYCLE ANALYSIS
NO
YES

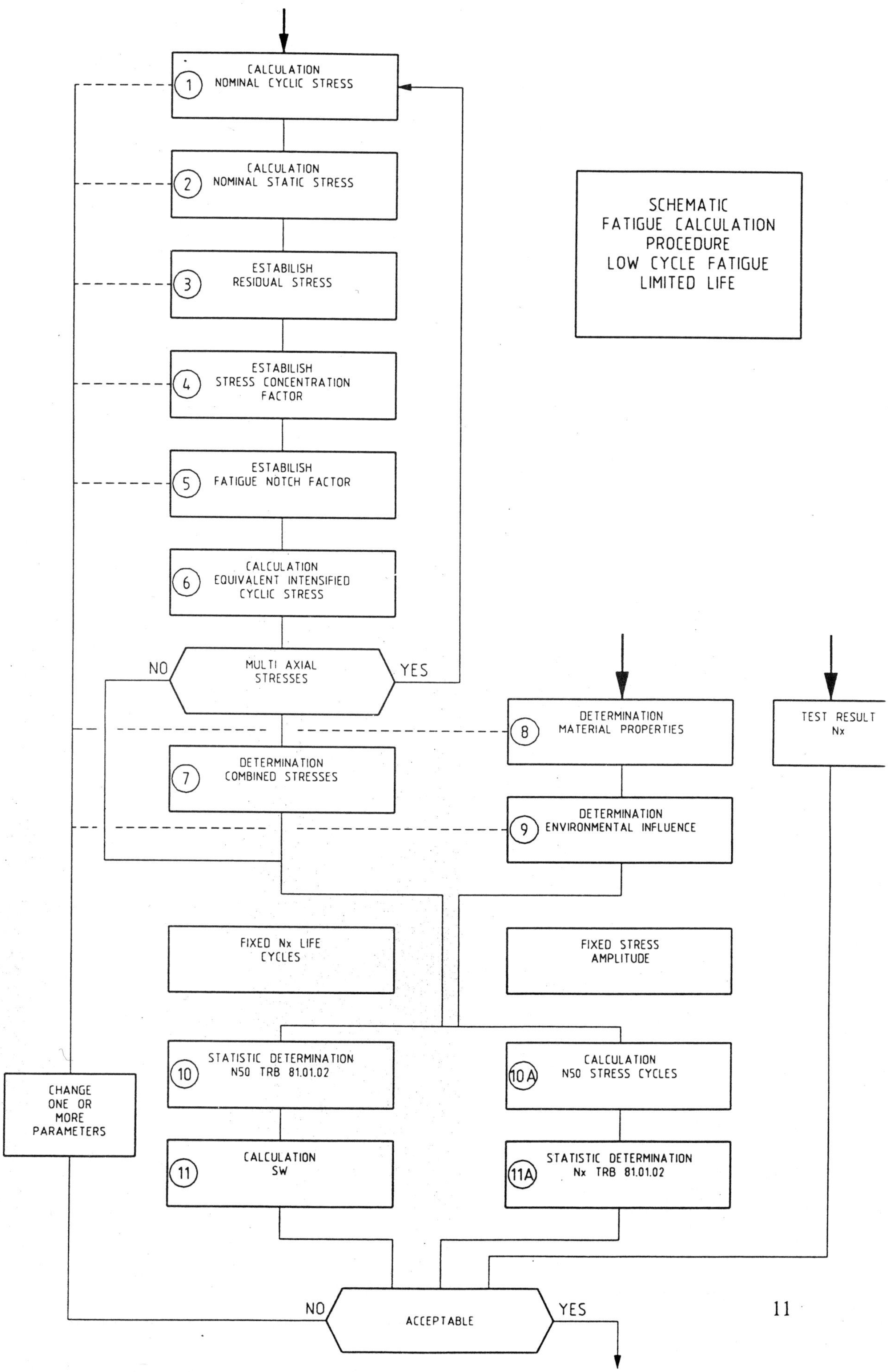
CALCULATION NOMINAL CYCLIC STRESS
1
CALCULATION NOMINAL STATIC STRESS
2
ESTABILISH RESIDUAL STRESS
3
ESTABILISH STRESS CONCENTRATION FACTOR
4
ESTABILISH FATIGUE NOTCH FACTOR
5
CALCULATION EQUIVALENT INTENSIFIED CYCLIC STRESS
6
NO
MULTI AXIAL STRESSES
YES
DETERMINATION COMBINED STRESSES
7
DETERMINATION MATERIAL PROPERTIES
8
DETERMINATION ENVIRONMENTAL INFLUENCE
9
TEST RESULT Nx
FIXED Nx LIFE CYCLES
FIXED STRESS AMPLITUDE
STATISTIC DETERMINATION N50 TRB 81.01.02
10
CALCULATION N50 STRESS CYCLES
10A
CALCULATION SW
11
STATISTIC DETERMINATION Nx TRB 81.01.02
11A
CHANGE ONE OR MORE PARAMETERS
NO
ACCEPTABLE
YES
SCHEMATIC FATIGUE CALCULATION PROCEDURE LOW CYCLE FATIGUE LIMITED LIFE

11

present there are some 1000 of these rods in operation in a wide variety of duties. From the calculations made to determine the desired safety factor for the design, the order of magnitude of the failure probability was known value. From this it would be expected that of 1000 rods in service, none should fail under normal operational loads. Examination of the nature of rod failures during this period of time can then reveal if expectations with regard to reliability are justified or not.

Reported rod failures were as follows:-

Origin of failure	Number of failures
Liquid ingestion	4
Stress corrosion cracking	1
Corrosion	1
Incorrect assembly	4
Induction hardening fault	1
Total	11

None of the known failures listed here show origins which are the result of deficient design, as no symptomatic failure mode is evident. The basic product as a combination of sound design with high quality production and N.D.T. techniques may be classified as extremely reliable in practice. Reduction of the number of failures which were traceable to incorrect assembly, can be expected with the introduction of hydraulic preloading. The magnitude of preload in the old system with the semi-hydraulic nut was governed by the thickness of a shim: errors being made possible when the rod was tensioned with a nut not snug against the coupling flange. The inadequate pretensions resulting from this were instrumental in the recorded failures. Failures associated with corrosion and stress corrosion cracking occurred on the piston end in the immediate vicinity of the shoulder at the reduced section with free gas access and not in the screw thread of the piston nut or reduced sections. It is perhaps interesting to note that the chosen materials for these failed rods were in accordance with N.A.C.E. requirements, customer specifications and known good engineering practice. Investigation revealed that in all cases the materials met specification and that the corrosivity of the process gases was unexpectedly greater than could have been anticipated upon the basis of formally specified gas compositions. Replacement rods of more corrosion resistant materials eliminated these failures. Improvement in material choice for certain types of process applications would appear to be a logical result of this experience, which indeed can now be found in the standard compressor requirements of a large process design company.

9.) CONCLUSION:-

The piston rod design as described in the preceding text has proved in the course of the years that it has been in operation to be a very reliable component. It should perhaps be classified as an example of how careful applicaton of basically sound design features can be combined to provide exceptional performance.

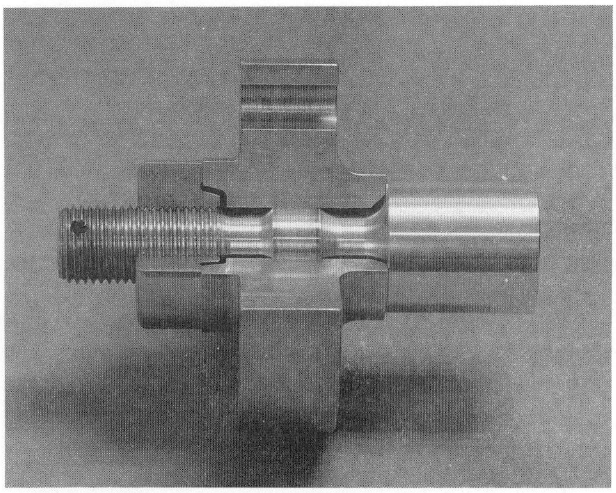

Fig.7 Assembled piston rod, coupling flange and nut.

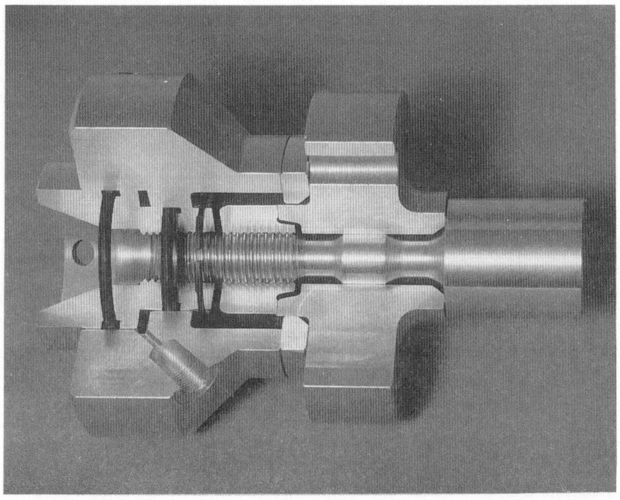

Fig. 8 Same assembly with attached tensioning device.

C449/014/93

Reciprocating compressors in a high-pressure miscible gas injection project

H SAADAWI, PhD
Abu Dhabi Company for Onshore Oil, United Arab Emirates

SYNOPSIS Two high pressure reciprocating gas compressors have been recently installed in an onshore oil field in the Arabian Gulf. Each compressor was designed to inject up to 9140 std m^3/h (8 MMscf/D) at a discharge pressure of 368 bar a (5336 psia).

This paper describes the specification, design and project implementation of the compression system. Problems encountered during the commissioning and solutions to these problems are also addressed.

1. INTRODUCTION

According to the Oil & Gas Journal (1), Bu Hasa field is the 20th largest oil field in the world in terms of total recovery. The field is located some 200 m south west of Abu Dhabi (Figure 1). A high pressure pilot gas injection project has been recently implemented in the field (2).

The purpose of the project is to evaluate the improvement in recovery efficiency that is possible from the northern part of the field by high pressure miscible gas injection.

The project was intended to investigate high pressure miscible gas injection of two alternate types of gas, namely rich gas and lean gas into the reservoir. The injection pattern is an inverted 5 spot with one injection well and four producing wells. Rich gas is supplied from the existing gas/oil separation facilities in Bu Hasa field. The lean gas is the residue gas, from a nearby NGL recovery plant. Table 1 gives the analysis of the both lean and rich gas.

The surface facilities were designed to inject rich gas during the first three years of the investigation at injection flow rates between 4570 and 9140 std m^3/h (4 and 8 MMscf/D). The original plan was that upon completion of the rich gas investigation, the project will be resumed using lean gas at injection rates between 6800 and 16 000 std m^3/h (6 and 14 MMscf/D). This phase was to be conducted for a further three years.

2. PROCESS DESCRIPTION AND COMPRESSOR SPECIFICATIONS

The process flow schematic is shown in Figure 2. Saturated gas from the production separator flows through a low pressure pipeline into the gas compression plant. The gas first enters a slug catcher and then a filter separator to remove any solids before a glycol dehydration package. After water has been removed from the gas, it is compressed in two parallel compressor trains. Each train is made of a four stage reciprocating compressor with inter-stage and after stage cooling. The process flow scheme of the compressor package is shown in Fig. 3. After compression, the gas, in the dense phase, passes through a metering station and is then transferred via a buried high pressure pipeline to the wellhead area where it is filtered before injection.

During the optimization phase of the project, both centrifugal and reciprocating compressors were considered. However, at the relatively low volume flows and high pressure required on this project, centrifugal compressors become inefficient and uneconomical. For this reason, reciprocating compressors are used. Also, both mechanical and electrical drivers were considered. The availability of surplus power, generated locally in the field, made an electric motor an attractive option.

The compressor trains were tendered as one package. The scope of supply included the compressors, drivers, coolers, knock out drums and inter-connecting piping. The motor drivers were manufactured in the U.K and compressors in the United States. The package was assembled at the vendors shop in the United States and a mechanical test using air with discharge valves opened was carried out.

Each reciprocating compressor is driven by an 11 kV, 50 Hz induction motor rated at 1450 kw. It is (Ex)N classified for zone 2 and has a temperature class of T3. The compressor is a 4 stage, 3 cylinder machine with a rated speed of 368 rpm. The use of slow speed machines for high pressure applications is now well established in the industry (3), (4). The slow speed allows better removal of the compression heat from the cylinder.

Compressor valves are of the channel type. The distance piece is a two compartment API type 1C.

Each compressor is sized to handle up to 9140 std m^3/h (8 MMscf/D). The suction pressure is 15.6 bar a (227 psia) and the discharge pressure is 368 bar a (5231 psia).

Capacity control is achieved by a 5 step control scheme which uses a combination of four pockets and one unloader. Two head end pockets and one crank end pocket are in the first stage cylinder. A head end pocket is on the second stage cylinder. An outer end port (plug) unloader is also used in the first stage cylinder. A first stage bypass is used to smooth the flow between the steps. This is in addition to an overall by pass around the compressor.

3. SPECIAL DESIGN CONSIDERATIONS

Due to the nature of the gas injection process and the environment, special features of the design have to be considered.

3.1 Injection pressure

In selecting the injection pressure for miscible gas injection projects, two limitation are considered, the minimum miscibility pressure and the fracture gradient.

The minimum miscibility pressure is defined as the minimum pressure required to achieve a single homogeneous phase between the injected gas and the reservoir fluid.

The surface injection pressure should be high enough to ensure that the bottom hole pressure is above the minimum miscibility pressure. It should be emphasized that this holds true for any flow rates lower than design flows the condition of meeting the minimum miscibility pressure is still valid. i.e. the system resistance curve is a flat one. Designers of surface facilities should have full understanding of this point.

The injection pressure at the wellhead should not exceed the fracturepressure, otherwise the formation can be damaged. This is normally protected by a relief valve. This should not be confused with other relief valves that may be required to protect the piping system.

The compressor discharge pressure is specified according to the injection pressure after taking into consideration the pressure losses in the high pressure pipeline and well tubing. For Bu Hasa, the key pressure values are as follows:

Minimum miscibility pressure = 324 bar a (4700 psia)
Wellhead injection pressure = 352 bar a (5100 psia)
Compressor discharge pressure = 368 bar a (5336 psia)

3.2 Discharge temperature and number of stages

In order to maintain an effective lubrication, particularly at the last stage, the maximum discharge temperature was specified at 125°C, well below the maximum given in API 618 (5).

At the proposal stage, a three stage machine which meets the specified discharge temperature at the design conditions was considered. However, the discharge temperature was found to be excessive at part load operation.

Part load operation of multistage reciprocating compressors is best understood by examining the theoretical formula for the volumetric efficiency

$$VE = 100 - C\,(R^{1/k} - 1)$$

Where VE is the volumetric efficiency
C is the clearance %
R is the compressor ratio
k is the ratio of specific heats.

At part load operation, the actual intake gas volume into the compressor cylinder decreases. This results in a decrease in the volumetric efficiency. For high pressure cylinders, the use of clearance pockets is impractical. Therefore, for multistage reciprocating compressors, the compression ratio of the last cylinder increases at part load operation. Hence, the discharge temperature at part load conditions is higher than that under full load conditions.

The adiabtic discharge temperature is also higher for the lean gas case than for the reach gas case. This is because the lean gas has a higher ratio of specific heats than the rich gas.

In other words, the maximum discharge temperature is experienced under the following operating conditions.

- part load operation.
- maximum ambient condition
- lean gas case.

To avoid temperature increase during part load the capacity control system must maintain the interstage pressure at the full load values, requiring complicated arrangement such as additional clearance pockets, also in the second stage plus a recycle line in the 3rd stage (clearance pockets at this high pressure are impractical).

For these reasons, a four stage compressor was specified instead of a three stage machine.

3.3 Rod loadings

The load carrying capacity of a reciprocating compressor involves two considerations: horsepower and frame load. The driver, crank and bearings determine the horsepower limitation.

Frame load is known by a variety of names including rod load and pin load. Frame load limitations are established to prevent stresses in the piston rods, connecting rods, bolting and other frame parts from exceeding safe values.

The actual rod loading is the sum of two components; the first is the inertia force and the second is the force created by pressure differential across the piston (Figure 4).

Because of the high pressure involved, a conservative approach was used. The specification stipulated that the maximum actual rot load should be less than 85% of the manufacturer's maximum allowable continuous rod load for the compressor.

As noted in section 3.2, the interstage pressure of a multistage reciprocating compressor 'floats' at part load operations. The actual maximum rod loading was calculated taking account of part load operation by considering, for each stage of compression, the relevant relief valve set pressure and the minimum suction pressure.

3.4 Cylinder lube oil

When selecting the type of cylinder lube oil for the Bu Hasa Pilot gas compressor, comprehensive investigations where conducted which took into account published technical literature, vendors recommendations and feedback from other operators.

Two potential types of problems which can result from the reciprocating compressor lube oil coming into contact with the injection gas stream were considered.

The first potential problem is related to the maintenance of the reciprocating compressor. When hydrocarbon gases dissolve in the mineral oils, they can cause degradation of viscosity and flash points resulting in scoring and high rates of wear of piston rings and pressure packing. The problem is aggravated at high pressures since the solubility of gases increase with pressure. Another problem with compressors operating at pressures above the critical point is that at this supereritical pressure the dense phase hydrocarbon will tend to dissolve or absorb the mineral oil from the cylinder wall to an enormous effect. Some operators reported that the lube oil consumption has increased to ten times the rate recommended by the compressor vendor when using mineral oil (6).

As a result of these problems, there has been an increasing tendency in the past few years among vendors and operators to use synthetic oils for cylinder lubrication of high pressure injection compressors. This is because of the insolubility of synthetic oils in hydrocarbons and because they do not suffer the serious thinning that occurs with mineral oils when in contact with hydrocarbon gases.

The second potential problem concerns the reservoir. The carry over of lube oil aerosols and other chemicals such as corrosion inhibitors and glycol can form polymer-like material. This material can plug the pores of the formation and cause decline in the well injectivity. Because Bu Hasa North reservoir is a tight formation, particular attention was given to the type of lube oil to be used and the compatibility between this oil together with the chemicals in the gas stream and between the reservoir fluids. To investigate this problem, a service company was commissioned to perform core tests in order to select the most appropriate (in terms of least formation damage) type of compressor lube oil, synthetic or mineral (7). A coreplug from the formation was put in a test cell and subjected to gas flooding with entrained chemical aerosols and retained for seven days. The permeability of the coreplug was measured after the test. The chemical blend consisted of compressor lube oil, corrosion inhibitor and triethylene glycol. Reservoir oil was also added. Two tests were conducted, one using a synthetic oil and the other using a mineral oil. The core tests showed marginal preference for synthetic oil.

Based on these analyses, it was decided not to use mineral oil and instead to use synthetic oil for the lubrication of the cylinders of the compressors.

3.5 Potential liquid carry over

Reciprocating compressors are positive displacement machines. Liquid carry over can destroy the lubrication and cause mechanical damage to the machine. As mentioned in section 2, the gas flows from the first stage suction separator via a 0.4 km pipeline into the compression plant. A conservative approach was adopted by designing both a slug catcher and a filter separator upstream of the compressors (Figure 2). The slug catcher was designed to handle any upset in the upstream production facilities or condensation in the pipeline. When sizing the vessel, the slug volume was taken as 4 seconds of flow with the normal feed velocity and 100% liquid filling of the low pressure pipeline. Since the facilities were commissioned there have been no incidents or problems of carry over of liquids into the compressors.

4. COMMISSIONING PROBLEMS

This section outlines some of the problems encountered during the commissioning of the compressor package.

4.1 Level sight glass

One of the most serious problems was the failure of the level sight glass of the fourth stage scrubber. The sight glass was shattered at 5000 psi. Fortunately, the operator was in the control room at the time and reacted immediately by pressing the emergency shutdown button. After a lengthy investigation of the problem with the compressor vendor and the sub-vendor for the sight glass it was suspected that the damaged sight glass was rated for a lower pressure rating than the operating pressure.

4.2 Congested compressor scrubber skid

The compressor scrubber skid was extremely congested. This presented a maintenance problem and made access difficult. It is also made personnel escape in a reasonable time unlikely in the event of an incident.

Project specifications called for double block and bleed valving arrangement for maintenance purpose for piping class 900 and above. This was incorrectly interpreted as adopting this valving arrangement "across the board" for these piping classes. One of the problems was that a comprehensive maintenance philosophy was not developed during the engineering phase of the project.

Furthermore, the specifications of the compressor packages when tendered did not give minimum dimensions for the skid. The dimensions of the package, access platforms and walkways were the subject of a dispute between the Operator, Engineer and the compressor vendor.

To solve the problem, the following steps were taken to improve accessibility:

- A maintenance philosophy was developed to identify which instruments required servicing while the compressor was running. For example, it was agreed that the machine would be depressurized prior to undertaking maintenance work on certain instruments, hence single valve isolation could be used instead of double block and bleed (Figure 5).
- A post commissioning HAZOP was conducted and the additional valves which were not used were removed.
- Access platforms and walkways were constructed on site.

4.3 Piston rod and packing

During the early days of commissioning, a blockage was suspected in the circulation of the packing oil. Investigations revealed that the cause of the blockage was a mismatch of the horizontally drilled passages in the piston packing rings of the third and fourth stage. Later during commissioning a failure was experienced with the piston rod packing of the third/fourth stage cylinder. When the piston rod was stripped, scores and chipping of the coating were discovered.

The third and fourth stage integral piston and rod assembly are made of AISI 420 stainless with tungsten carbide coating on the rod in the packing travel area. The bond between the coating and base metal was suspect.

The failed rings and piston rod were sent to the laboratory for examination. The laboratory investigations showed no evidence of inadequate bond. The conclusion by the compressor vendor was that the most likely cause of failure was an upset in lubrication.

4.4 Vibrations

When the compressor packages were first commissioned, excessive vibrations were observed. Vibrations were monitored and the situation improved somewhat after removing the extra valves mentioned in section 4.2. However, the vibrations remained excessive. It was found that when the vibrations were measured in the cool of the morning (25°C) and in the evening heat (46°C). The heat produced a 50% decrease in vibrations. It was evident that changes in the gas temperature affected the acoustic response of the system. One of the difficulties was the lack of vibration acceptance criteria. The compressor vendor sent a specialist to site to carry out a vibration survey. The acceptance criteria was established at the following basic check points:

- cylinders (on cylinder head)
- frame (on lubricator end)
- scrubber (above scrubber base)
- skid (on the main outside runners)

The vibration survey confirmed that the problem was resonance with the piping and natural frequencies on the scrubbers.

Although the vibrations problem was a complex one, it was overcome with simple but correct solutions. Essentially, pipe supports were removed, shimmed and adjusted until all vibrations were brought below the acceptable levels.

4.5 Failure of Valve 0-Rings

Failure was experienced in the O-ring of the valve assemblies on the third/fourth stage cylinder. Each valve cover in this cylinder has an O-ring which was made from viton. Upon examination of the O-rings, they were found to be cracked and blistered. It was concluded that the failure is due to the phenomenon known as explosive decompression which affect elastomeric materials when subjected to a high pressure gaseous environment.

During operation, the gas permeates or "soaks" into the O-ring material until the pressure inside the O-ring is equal to that outside the O-ring (in the gas passage). When the unit is shut down and depressurized, or "blown down" the pressure in the gas passage falls off to atmosphere. When that happens, the O-ring material cannot bleed off the gas trapped within itself and the gas literally blows the O-ring apart as it escapes.

The O-rings were replaced with new ones made of an upgraded material which has so far have performed satisfactorily.

4.6 Miscellaneous problems

Two other major problems were experienced during the commissioning. Although they were not related to the compressor package, they caused considerable delay in the commissioning activities.

- During the depressurization of the discharge piping downstream of the compressor, the temperature dropped to minus 80°C on the ball valve in the carbon steel piping upstream the stainless steel flare header. The problem was the wrong installation of depressurizing system which caused damage to the ball valve. This valve was replaced and the restriction orifices and globe valve were installed downstream of the ball valve as shown in Figure 6.
- The flare is of the horizontal type. The flare wall refractory brick work collapsed during the initial commissioning. Cracks were noticed from the first day and the bricks at the centre of the wall began moving out. The cause was identified as a lack of expansion joints. Also, the flame licked back over the wall when the wind was in the wrong direction causing damage to the thermocouple junction box.
- The wall was rebuilt and its height was extended by one meter. A protective tunnel was installed over the flare lines in the vicinity of the flare wall. The junction box was also re-located.

5. CONCLUSIONS

This paper described the design criteria, specification and selection of a high pressure reciprocating compressor for Bu Hasa Miscible Gas Injection Project in Abu Dhabi. Problems encountered during commissioning and solutions to these problems were also outlined. After overcoming the commissioning problems, the gas compression facilities have been operating satisfactorily. At the time of writing this paper, the cumulative running time for the two units was 25 000 hours. Based on our experience with the project, the following points should be noted when specifying and selecting high pressure reciprocating compressors for miscible gas injection projects:

- As part of the Basic Engineering, the operation and maintenance philosophy for the compressor package should be developed, reviewed and endorsed by the Operator, before enquiries are issued. Other philosophies that should be developed during the Basic Engineering phase of the project, are shown in Figure 7.
- For high-pressure gas injection projects introducing formal reviews at key stages in the project. These could be concept, basic design, detailed design, construction, precommissioning and post commissioning.
- A quality assurance/control programme should be employed throughout the life of the project. Particular attention should be given to high pressure components in the system.
- Consideration should be given to constructing a plastic piping model to check accessibility of components. The model should be reviewed at different stages during the design. Operations and maintenance staff should participate in these reviews.

- Despite careful considerations during the design, the odds are that vibrations will be experienced in the compression system during the commissioning. The Operator should be prepared by having trained staff, appropriate vibration measuring instruments and agreeing with the vendor on criteria for vibration acceptance.
- Part load operation of the reciprocating compressor should be carefully reviewed to ensure that the increase in the compression ratio in the last stage

will not result in a higher temperature or a higher frame loading than specified.

- Tests should be carried out to establish the compatibility of the lubricant used for the compressor cylinders with the reservoir fluids and any chemical additives used in the process (e.g chemical inhibitors or glycols) to establish plugging tendencies. For the Bu Hasa gas injection project, a synthetic oil was used.
- The importance of after sale service by compressor vendors to users in remote locations of the world can hardly be overstated.

REFERENCES

(1) "Giant fields : Size distributions of the world's largest known oil, tar accumulations", Oil & Gas Journal, vol. 84, No. 8 Feb 24, 1986.

(2) Saadawi, H., "Design of Surface Facilities for a High Pressure Miscible Gas Injection Pilot Scheme", SPE Paper No. 17953 presented at the SPE Middle East Conference, Manama, Bahrain, 11-14 March, 1989.

(3) Raubenheimer, D.S.T. "Offshore Gas Compression: An Operators Experiences", First European Congress on Fluid Machinery, I.Mech.E. Conference Publications 1981-3.

(4) Traversari, A., Petrini, P. and Agostini, M, "Operational Experience of reciprocating compressors for offshore platforms", Second European Congress on Fluid Machinery, I.Mech.E. Conference Publications 1984-2.

(5) API Standard 618, "Reciprocating Compressors for General Refinery Services", third edition, 1986, American Petroleum Institute.

(6) Hatton, D.R., "Lubrication of high pressure gas compressors", Shell International Petroleum Co.

(7) "Bu Hasa and Bab Fields Gas Injection Projects: core test work for compressor lube oil selection", Serck Baker Oil Services, July 1987.

ACKNOWLEDGEMENT

The author wishes to thank the Management of Abu Dhabi Company for Onshore Oil Operations (ADCO) for permission to publish this paper.

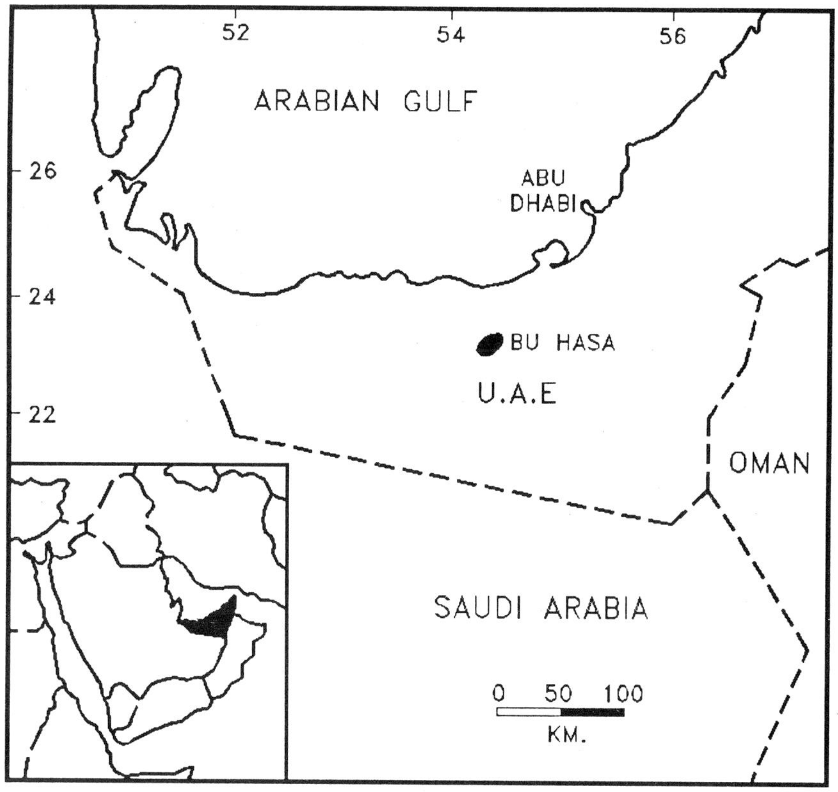

Fig. 1 Location map

COMPONENT	RICH GAS mol %	LEAN GAS mol %
N^2	0.42	0.55
C_1	69.18	81.84
C_2	10.65	8.94
C_3	7.32	2.72
C_4	3.99	0.52
C_5	1.68	0.08
C_6	0.67	0.04
C_7	0.64	TRACE
H_2S	0.02	0.02
CO_2	5.43	5.29
Water Content	Saturated	70 Kg/10^4 Std. m^3
Avg. Mole Weight	24.8	19.9

Table 1 Gas analysis

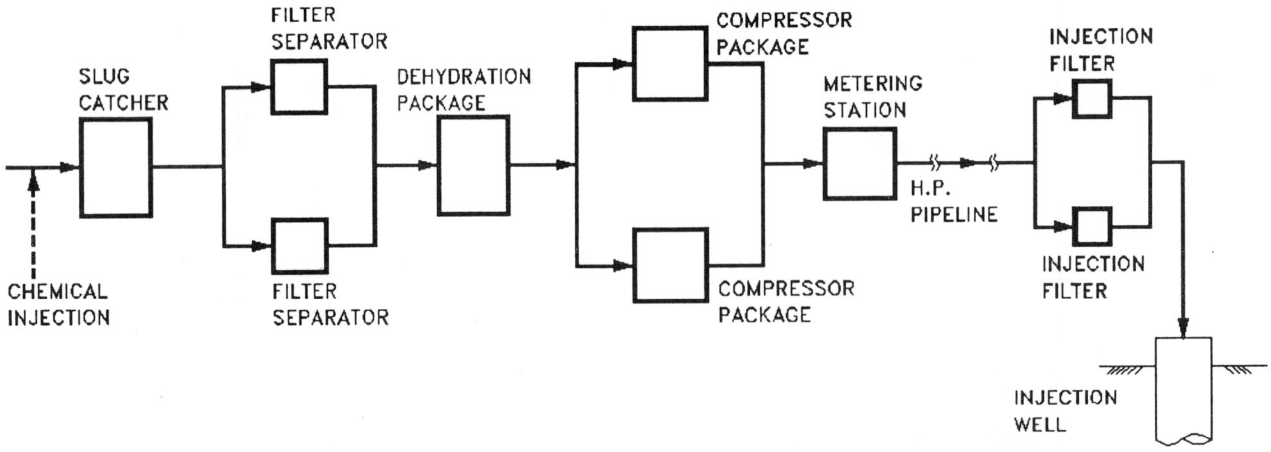

Fig. 2 Block diagram of the surface facilities

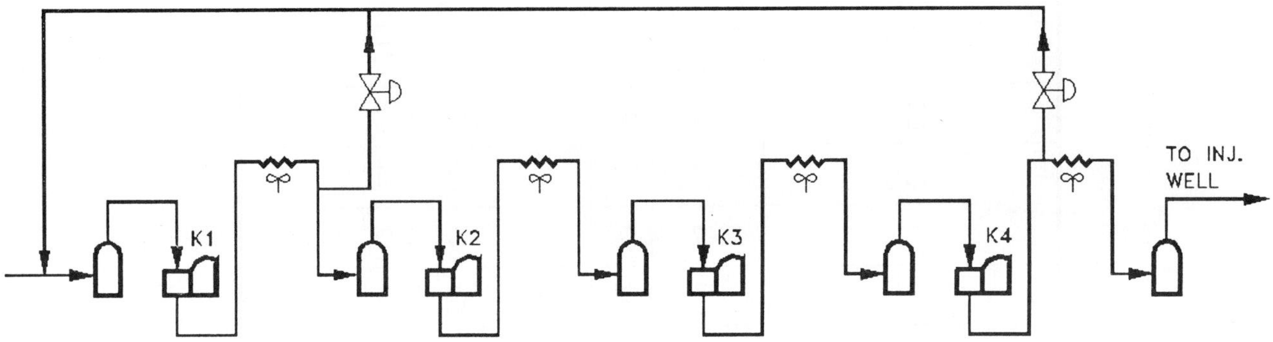

Fig. 3 Process flow schematic of the compressor package

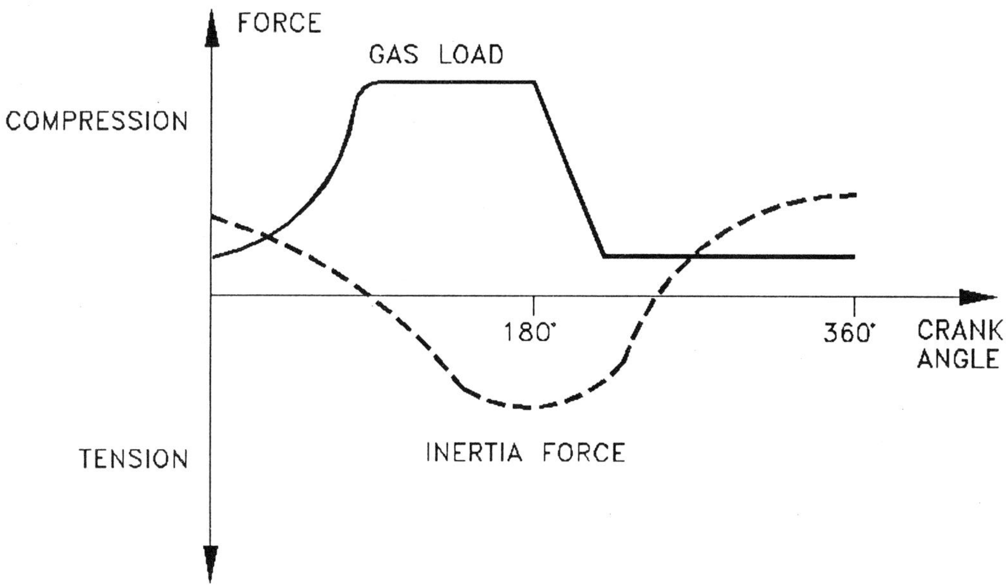

Fig. 4 Gas inertia forces acting on a typical single cylinder

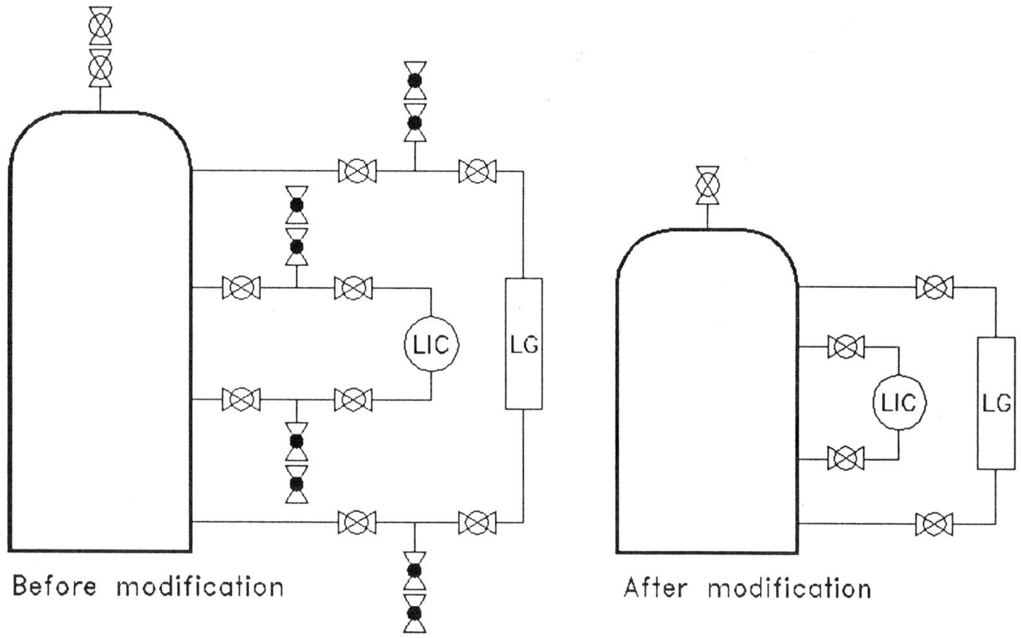

Fig. 5 Valving arrangement for high pressure scrubber

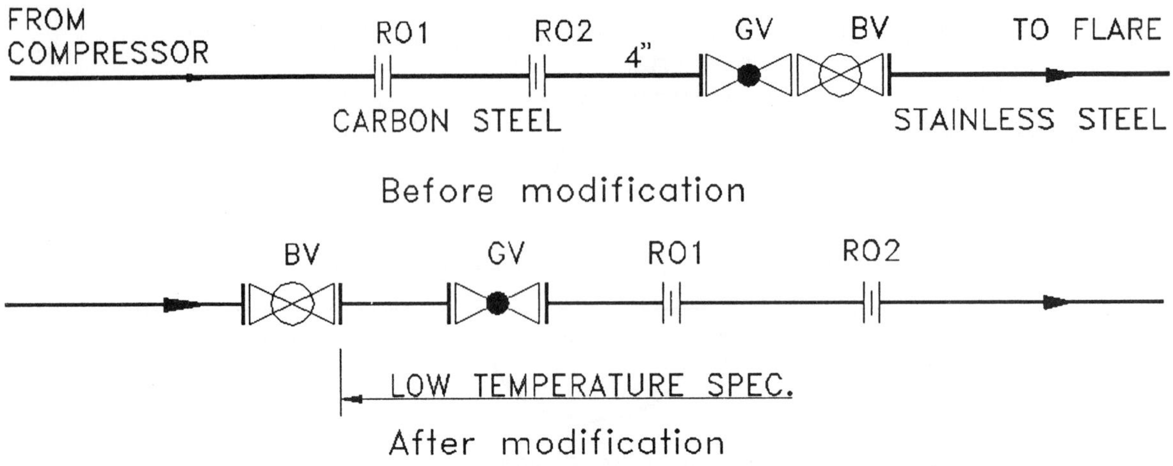

Fig. 6 Depressurizing line

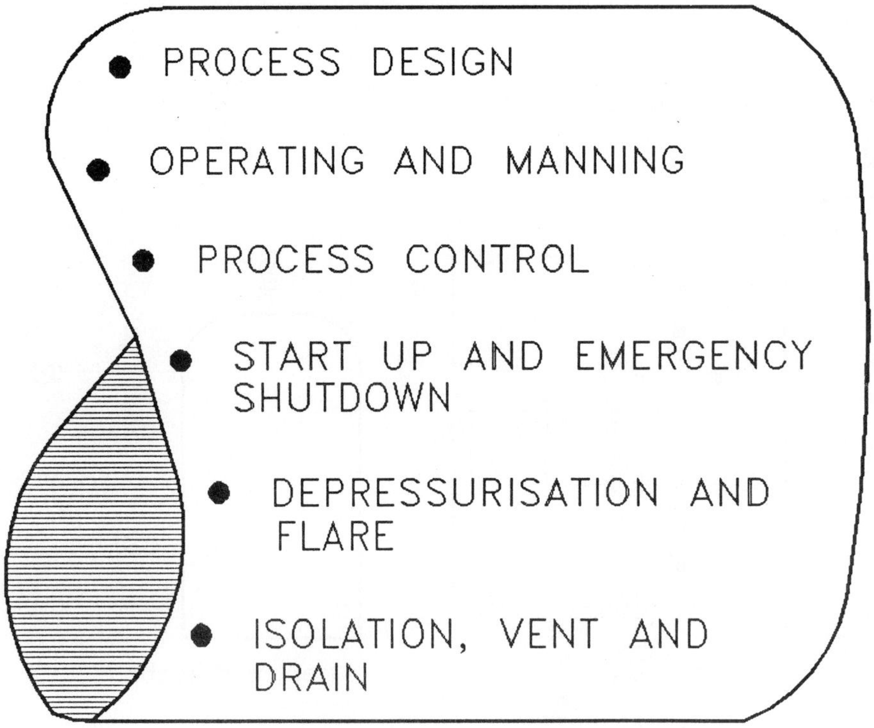

Fig. 7 Basic engineering design philosophies

C449/002/93

Wet gas compression – state of technology after successful first field trial

P NOACK, Ing
Nederlandse Aardolie Maatschappij, Assen, The Netherlands
C S CHURCHFIELD, BSc
Shell Internationale Petroleum Maatschappij, The Netherlands
O H JACOBSEN, MSc
Framo Engineering AS, Nesttun, Norway

SYNOPSIS The development of Wet Gas Compressor Concepts was initiated some years ago by Framo Engineering AS in co-operation with the Shell Group to achieve direct boosting of the untreated production from crude oil/gas reservoirs.

This paper gives an overview of the Wet Gas Compressor Development Programme describing operating principles, performance and mechanical aspects. In addition, the results and experience gathered during the first successful field trial on a productive gas gathering oil field station in the Netherlands are discussed.

1.0 INTRODUCTION

The field trial carried out at "De Lier", the Netherlands, in mid 1991 was the first ever-field trial of this type of technology. The trial objectives were achieved and demonstrated the feasibility of pressure boosting untreated high gas fraction multiphase well streams.

This trial was the culmination of several years of development carried out by the sponsors A/S Norske Shell, Shell Internationale Petroleum Maatschappij BV (SIPM), Nederlandse Aardolie Maatschappij (NAM) and Framo Engineering AS.

2.0 WET COMPRESSOR DESIGN

2.1 Operating Principle

The CR400 wet gas compressor is based on a dynamic principle to enable a compact design coping with the large volumetric throughputs typically required for the predicted duty. The main challenges are thus:

- To provide the required pressure boosting by using aerodynamic vane profiles in a mixture of gas and liquids, containing typically 2% liquids by volume.

- To survive and operate during gas and/or liquid slugs generated by the upstream production facilities.

- To be capable of long term operation whilst being exposed to sand production combined with an aggressive mixture of liquids and gas.

- To adapt to varying and highly transient operating conditions dependant on the actual production rates.

2.2 Design Principles and Assumptions

The aerodynamic design of wet gas compressor profiles requires an approach different to designing profiles for conventional machines. As a first approximation it may be assumed that the multiphase head can be calculated from single phase velocity triangles by using Euler's equation, and further that the differential pressure can be derived by using the mixture density as for a conventional machine.

In addition, several aspects particular to multiphase flow, need to be considered in the layout of the vane profiles and the procedures for performance estimation. Performance on pure gas forms the reference condition. The multiphase mixture is then treated as a homogeneous mixture to account for its effect in the equations. Deviations from this state are treated separately.

The phase slip ratio is defined as the ratio of local gas to liquid velocity. This ratio is unity for homogeneous flow, which is required for the procedure outlined above to be valid.

Accounting for phase slip in the energy equations, either at inlet or exit of a compressor stage, results in a degraded head for phase slip different from unity. A homogeneous mixture of liquids and gas is thus required to provide optimum operating conditions for each stage within the machine. This is achieved at suction by having a flow mixer at the inlet of the machine. Interstage mixing to maintain homogenity is enhanced by arranging contra-rotating impeller stages rather than conventional stator vanes. The contra-rotating stages not only assist with maintaining the homogeneous mixture, but also add energy and thereby additional head to the fluid.

There are, in addition to gravity, two major "body" forces acting on the fluid elements within a rotating compressor stage. These are the centrifugal and the Coriolis forces. They are proportional to density, and therefore produce different magnitude of forces on gas and liquid elements. These forces promote phase separation.

Phase separation within a compressor stage will lead to different flow patterns for liquid and gas, resulting in different velocity triangles and work input to the liquid and the gas. Incidence angles will also be different, all of which contribute to the negative effect of phase slip.

It is a known fact that liquid droplets can retard the speed at which acoustic waves propagate through a mixture by a factor of ten. This may produce multiphase shock waves and choking which severely reduces the booster performance.

Several different profiles were designed and tested for use in CR400. The selected type has a moderate lift coefficient of 0.25. A small blade angle is used to provide a broad range of acceptable performance to cater for flow rates different from the nominal.

2.3 Mechanical Aspects

The general layout of the wet gas compressor is shown in fig. 1. It consists of two direct drive electric motors, one driving the inner rotor and the other the outer rotor.

The original specifications are:

Capacity	:	265 m^3/h
Pressure boosting	:	15 bar
Suction pressure	:	10 bar
Gas Volume Fraction		
(GVF) - nominal	:	98%
- transient	:	0 - 100%
Speed (each rotor)	:	2 - 6000 RPM
Power rating	:	2 x 200 kW

The mechanical design strategy was as follows:

The complete rotor assembly was designed to operate subcritical to ensure stability independant of gas volume fractions and actual operating speed. This has been achieved by using hollow compressor rotors supported by anti-friction bearings at each end.

The unit was designed without the need for a balancing drum. Such a device would be exposed to wear due to sand production, and its performance under multiphase flow could be unstable. The thrust bearings are thus capable of carrying the unbalanced loads.

Minimising wear was a significant consideration, both with regard to component configuration and material selection. The compressor blades are unshrouded. Stellite overlay was used on the mating surfaces. Passage ways are flush without extrusions or cavities which would otherwise lead to high wear rates due to sand impingement or accumulation. The mechanical seals are of the internally pressurized type with minimum interaction with the flow path.

The complete rotating assembly is encapsulated within a housing designed to the same pressure rating as that of the production system. The motors are submerged in a barrier fluid pressurized above the production pressure. This approach minimised the possibility of environmental discharge due to a failure, and also limits the instrumentation needed to protect the machine from overpressure. Both features considerably enhance safety.

A Variable Speed Drive (VSD) system was selected, both to provide flexibility with regard to actual operating conditions, and also to provide regulation in case of liquid slugging.

2.4 Factory Testing

Factory testing consisted of component qualification in addition to an extensive full scale loop test of the complete skid mounted unit, including all subsystems.

The functionality of each of the critical components was tested under a variety of load conditions. Electric motor testing included standard bench tests to confirm the performance data, and additional specific tests to confirm the integrity and interaction of the motor/VSD combination.

Mechanical seals were tested in a specially constructed test rig to verify the design and materials. Operating conditions were thoroughly investigated and reproduced in the test where possible. Various blade profiles were tested hydraulically in an impeller test rig with regard to multiphase pressure generation, efficiency and surge limits.

The loop test was performed with nitrogen and diesel as the test medium at the test facilities of Frank Mohn Fusa AS, Norway. The flow diagram is shown in figure 2.

Testing included the complete system with regard to performance at various speeds, gas volume fractions, suction pressures and capacities. Functionality of the system under transient operation was also verified.

Thorough investigation of vibration levels, axial thrust, various operating temperatures and subsystems performance were included.

3.0 MULTIPHASE FIELD TRIAL

3.1 Site Requirement

The test was carried out in NAM's "De Lier" production facilities near Schiedam, Rotterdam, the Netherlands.

The gas volume fractions of the producing gas lift wells whilst at manifold pressure of 5 barg are approximately equal to or higher than the 98% GVF design point of the CR400 wet compressor. The site also provided ready access to the involved engineers and could be supported by Framo's facilities in Rotterdam.

3.2 Objectives

- To operate a wet compressor on well fluids.
- To correlate loop test performance with field recorded performance.
- To demonstrate the potential for boosting high GVF well streams.
- To establish the integrity of the concept.
- To gain operational experience.
- To gauge the potential for an unmanned operation.

3.3 Test Installation

The wet gas booster package was installed at the "De Lier" site and connected to the existing production facilities.

Wells flowing into the production station could be open to both the normal production manifold and to the booster. Flow from the booster discharge was returned to the main process via a pressure control valve.
The booster was fully shutdown protected by instrumentation within the booster skid (and associated control panel) as well as being integrated into the station ESD logic.

The piping arrangement for the performance test was designed to give maximum production flexibility (see fig. 3). Any of the flow line streams into De Lier could be routed to the booster suction via the existing test manifold, thus achieving maximum flexibility in terms of flow rates and gas/liquid ratios.

A water pump with capacity control was installed to increase the fluid percentage, if required, by pumping formation water into the suction lines loading to the booster. This formation water was stored in tank T-301.

To monitor the amount of fluid and gas passing through the CR400, a skid mounted mobile test separator was placed downstream of the booster. The test separator was equipped with liquid and gas flowmeters to measure the flow rates entering the CR400. The liquid and gas streams from the separator were combined directly after the test separator and returned to the main process.

3.4 Test Programme

The field trial consisted of the following two tests:

- Performance test
  This test was the verification of the performance data from the loop test in Norway with nitrogen and diesel as medium, versus these actual field conditions as derived from the various, directly connected well streams.

- Integrity test
  Accumulate 200 hours operation as a limited endurance test. During this period frequent flow measurements were made to document the test, and to check any deterioration in performance or integrity of the pump.

4.0 OPERATING EXPERIENCE

4.1 Test History

The CR400 wet compressor skid and control container arrived at the De Lier site July 1991. At this stage the complete system had already been commissioned at the factory. It was therefore only required to hook up to the process suction and discharge lines, the mains power supply to the control cabin and the power and signal lines between the skid and the control cabin.

Thereafter the functionality of all subsystems was tested and reference performance verified on inert gas.

On 17th July 1991, the system was operated on unprocessed well fluids for the first time. The average volumetric liquid content was expected to be less than 2%. However, liquid slugs generated by the upstream pipework passed through the unit with a frequency of approximately 4-5 per hour. These were larger than expected, and in some cases the mixing unit at suction was flooded. During this phase of testing, the regulation system of the variable speed drive was adjusted to minimise stresses imposed on the assembly. Operational experience with various wells was accumulated prior to a detailed performance investigation. For a period of one week, the unit was connected to a test separator. At the same time, transient measurements of aerodynamic, mechanical and electrical performance were collected. In this way, the overall performance and interaction with the production system could be established and compared against performance in the multiphase test rig in Norway.

Thereafter the unit was run, primarily taking suction from the station's inlet manifold until an accumulated run time of 200 hours was achieved.

4.2 Performance Correlations

The reference multiphase performance of the CR400 wet compressor had been established under stable operating conditions working with a mixture of diesel and nitrogen. Gas and liquid rates were measured prior to mixing such that accurate rate measurements could be obtained.

Infield the multiphase discharge was directed via the test separator for the same purpose. The gas rate was measured continuously by using an orifice plate, whereas the accumulated liquid rate was logged by using a rotating positive displacement meter. The differential pressure across the orifice was unstable due to the slugging nature of the production. Relatively long operation was therefore required to establish a representative average measurement. Performance data were accumulated over a minimum of three hours to establish each single data point.

The test confirmed that accurate booster performance data from a live production system is very difficult to obtain. However, the data collected is considered to give a good indication of the performance.

A typical comparison of reference performance against field performance is shown in fig. 4. It shows that the correlation between the two is good. This demonstrates that multiphase loop testing with diesel and nitrogen is an adequate tool to establish booster performance. The fact that the field performance appeared better than the data collected in the test loop is most likely due to inaccuracy of the field flow measurements, and not attributed the booster.

In De Lier, the CR400 was able to almost halve the upstream flow line pressure when individual wells were selected, whilst maintaining the discharge at the process set pressure. A change of this magnitude would have a positive effect on gross production rates, especially if the booster was located close to the well.

4.3 Liquid Slugging

Liquid slugging was known to exist in the production lines at De Lier. How the booster would handle this situation was therefore envisaged to be one of the critical items. Slug lengths were estimated prior to installation, and a suction flow homogenizer included to both optimize the pressure generating capability and also to avoid severe transients.

A major finding from the trial is the extremely transient nature of the flow. Despite operating in average 99% gas environment at 5500 RPM, the machine ingested large liquid slugs greatly in excess of the mixer capacity whose volume was based on calculated slug volume. The test conditions were therefore more severe than expected. However, although the mixer was flooded, it enabled a much more gradual transition from nominal operation to slugging. Speed reductions of 40-50% was experienced for prolonged periods, and the power consumption fluctuated from 150 kW to 400 kW. A typical response to a slug is shown in figure 5. It shows that the transients downstream of the mixer are gradual and not instantaneous.

During continuous transient operation, there was no measurable increase in motor winding temperature, demonstrating the suitability of the oil-submerged high speed electric motors for this duty. The booster vibration levels were not dramatically increased during a slug. However, the skid and booster configuration had a natural frequency around 4000 RPM which was excited during run-down, run-up and transient operation.

Testing of the VSD system provided valuable operating experience. Whilst steady state VSD operations are well documented, transient operations are not. Early in the trial there was a number of shutdowns due to over-current which occured when a large liquid slug flooded the mixer and the booster. Later in the trial, with improved programming of the VSD, shutdowns were minimised.

Pressure levels were surprisingly stable as the liquid slugs passed through the machine. This is inherent in the dynamic booster principle. Although the potential pressure generating capability of the booster was much higher with the large liquid contents, high pressures were not generated since the downstream piping back-pressure did not change instantaneously as a result of the liquid slug (fig. 5).

The performance of the mechanical seals through monitoring the use of seal/lube oil showed that under certain slugging conditions, the seals would open allowing seal oil to be lost at a high rate.

4.4 Findings after Disassembly

The complete system was disconnected and returned to Norway for disassembly and inspection upon completion of the field trial. A preliminary strip revealed that the general condition of the booster after this test was good.

Internal surfaces were covered by a layer of heavy crude containing some sand and solid particles. The crude could readily be washed off leaving some black colouration towards the discharge side where operating temperatures had been elevated. Additionally, crystallized salt was adhered to the surfaces surrounding the contra-rotating mechanical seal on the suction side. However, indication of rubbing, erosion or corrosion could not be identified.

Thereafter a detailed compressor strip was carried out and components inspected in detail. This confirmed the initial impressions. The rotor assemblies had maintained their dimensional specifications as measured prior to the test. The contra-rotating mechanical seals being one of the more novel components and also highly loaded, were in reasonable condition after dismantling. It was, however, evident that more development work is necessary to provide a seal system which is capable of long term operation at high speeds in a multiphase environment.

5.0 CONCLUSIONS

The test verified that the contra-rotating wet gas compressor principle is a technically feasible method to add pressure to an untreated high gas fraction multiphase well stream. Furthermore, it is a suitable means of reducing the back-pressure on the reservoir, potentially increasing the production and the recovery.

The experience from De Lier therefore forms an important step in the commercialization programme for this principle.

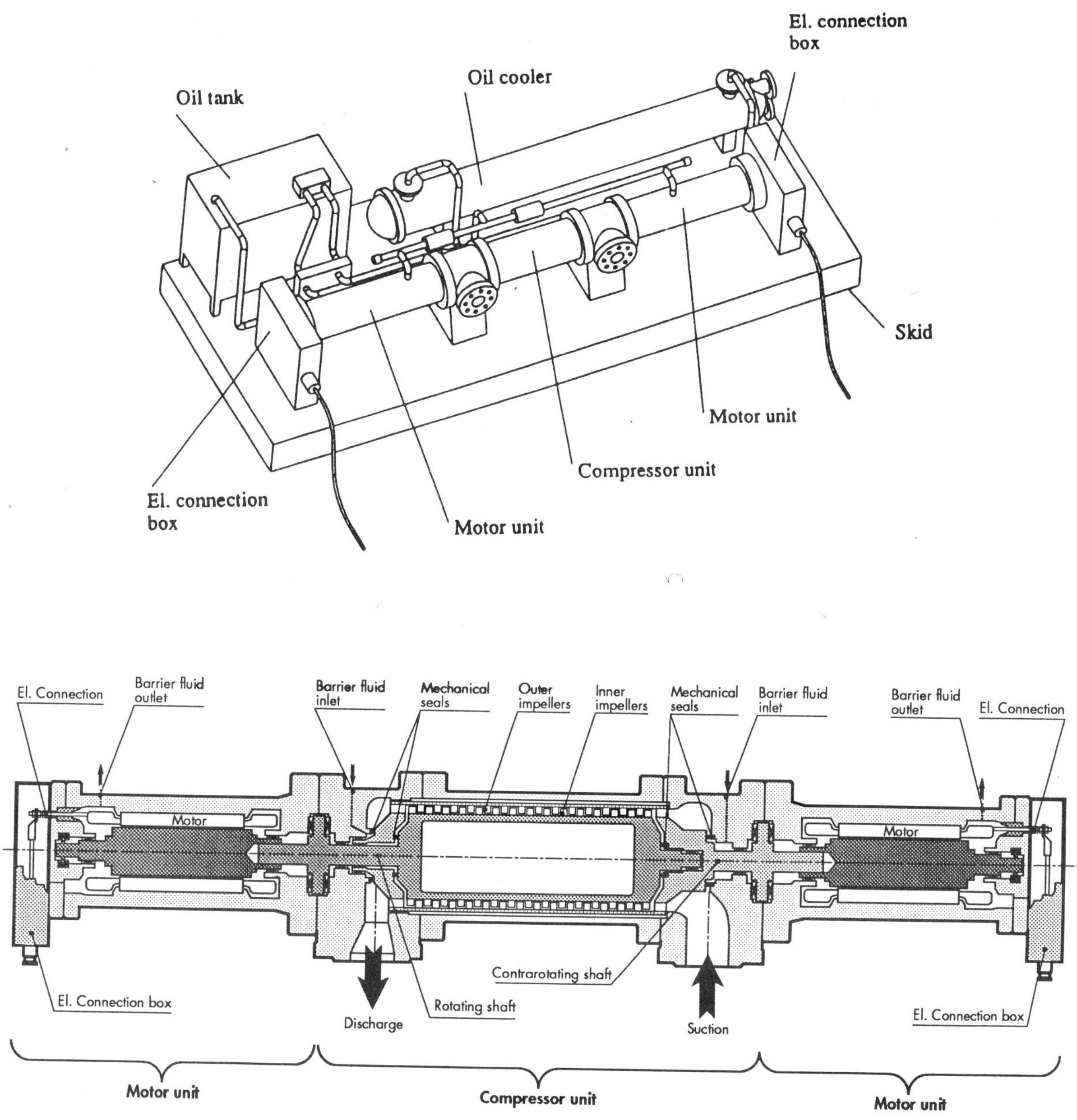

Figure 1: Layout of the CR400 Wet Compressor

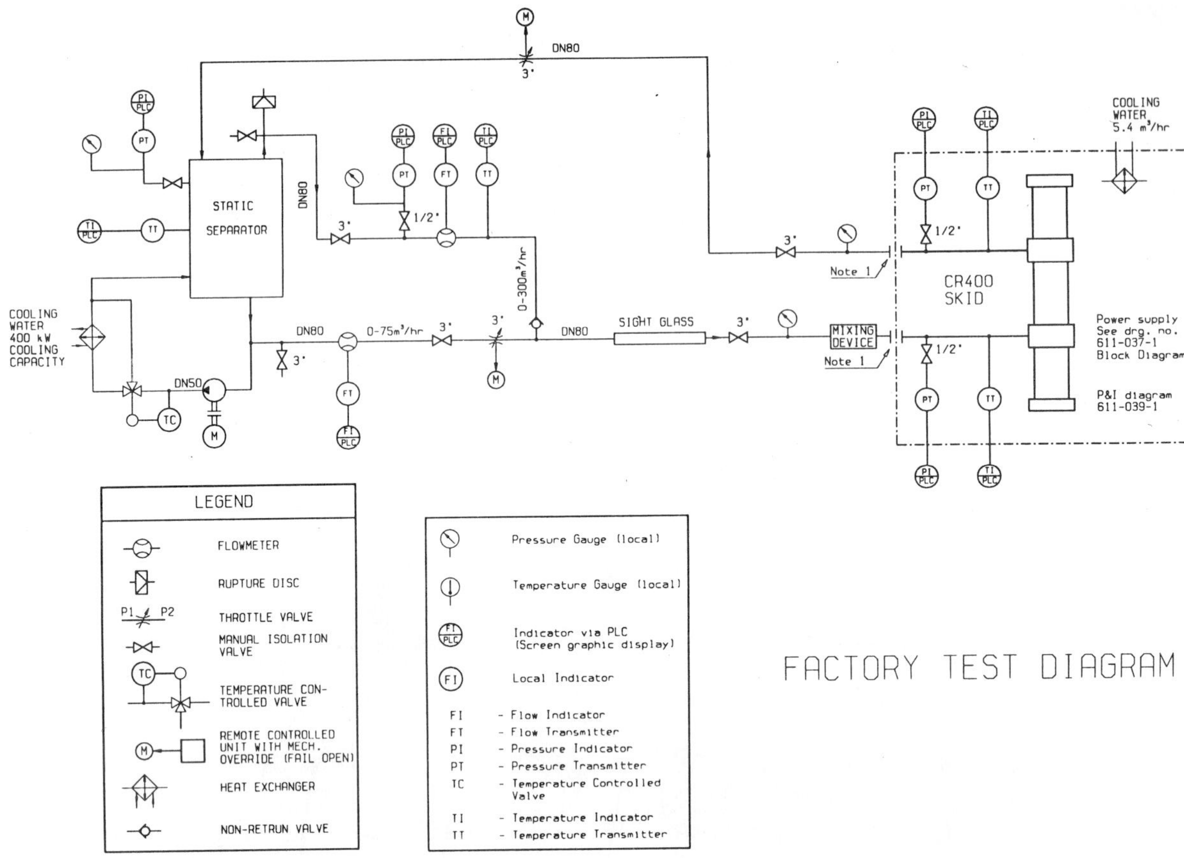

Figure 2: Test Loop Flow Diagram

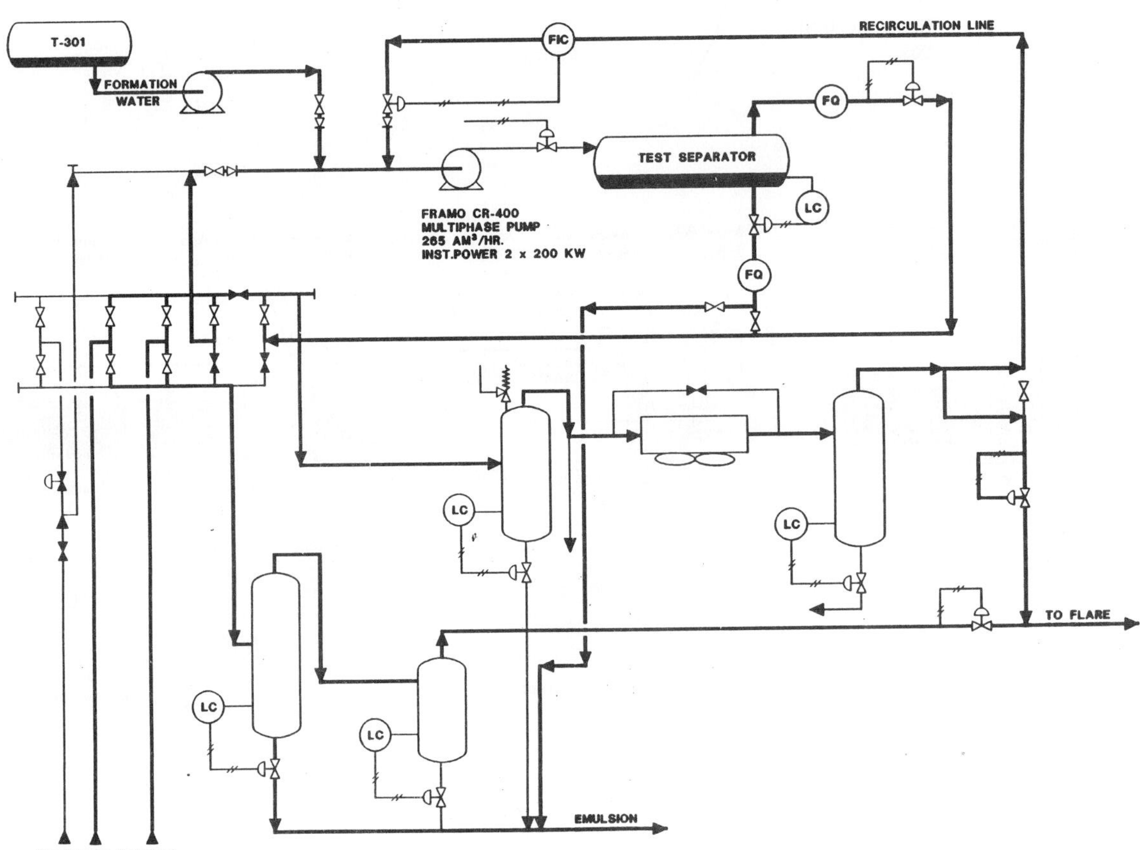

Figure 3: De Lier / CR400 Process Flow Scheme

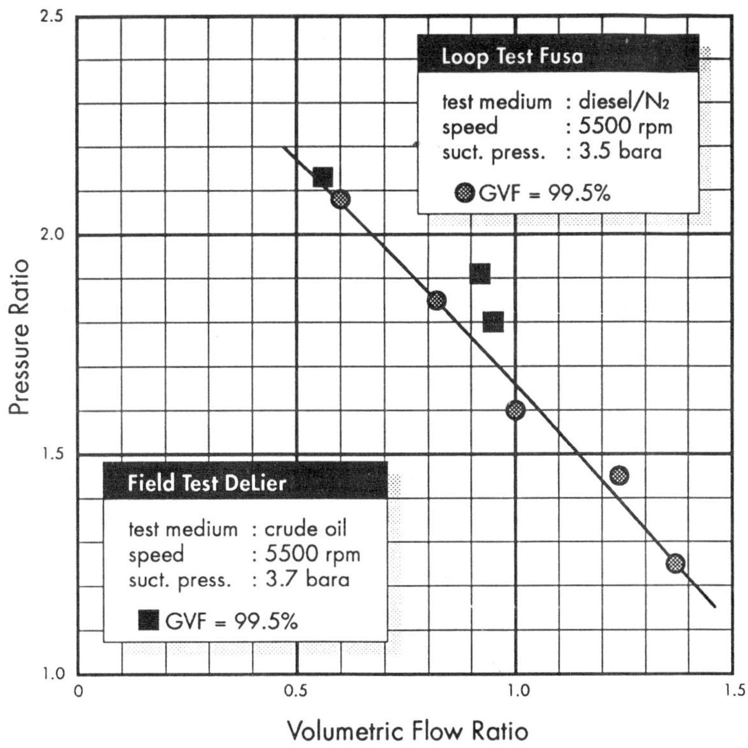

Figure 4: Comparison of Reference Performance against Field Performance with Gas Volume Fraction GVF = 99.5% at Suction

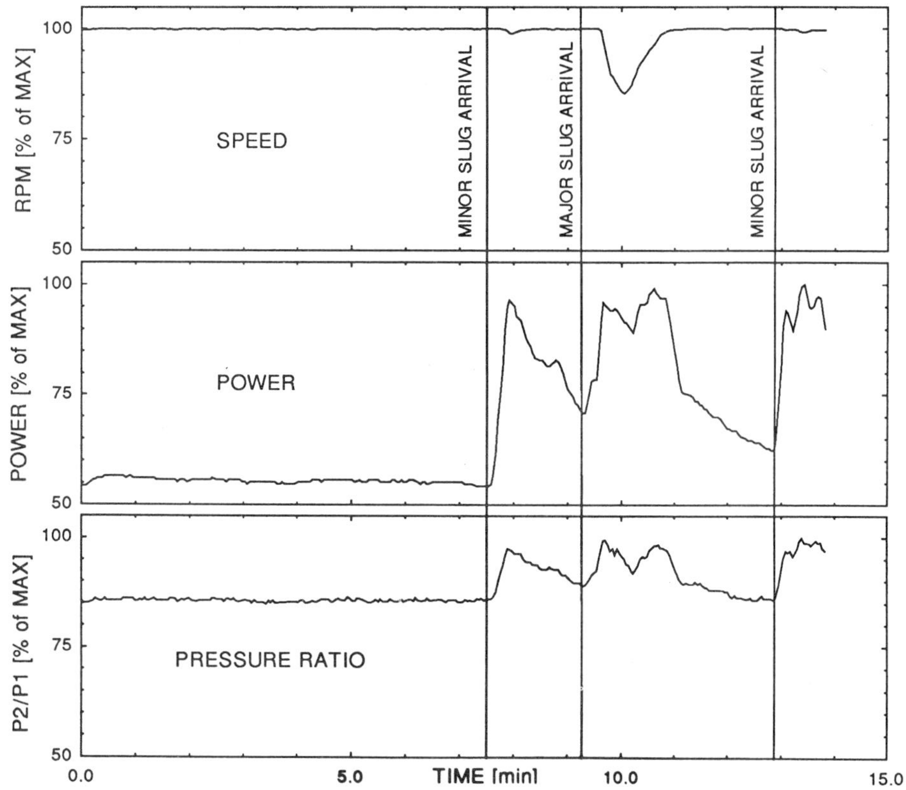

Figure 5: Transient Response to Slugging

C449/043/93

Rotodynamic pump development for liquid/gas mixtures

A M SHARP, BSc and N D BARNES, BSc, PhD
BHR Group Limited, Cranfield, Bedford

SUMMARY The development of pumps for handling liquid/gas mixtures is currently the subject of a number of programmes world-wide. Conventional rotodynamic pumps are known to be severely limited in these applications due to loss of head and efficiency at even modest gas void fractions (GVF's) and ultimately de-priming. Consequently, work on multi-phase pumps has concentrated on the adaption of existing positive displacement pumps and the development of new machines with combined liquid pumping and gas compression capabilities. These machines are inevitably expensive owing to the extensive, and on-going, development of novel technologies necessary to their successful operation. In addition, they will be costly to install subsea, control and maintain, perhaps too costly for many marginal satellite fields.

Some operators are considering a return to more conventional technology. Inevitably this includes a re-evaluation of the use of simpler rotodynamic pumps subsea. It is therefore important to know the precise limitations of these pumps and to explore ways in which these might be overcome.

This paper reports work undertaken at BHR Group Limited to establish the limits of multiphase performance for four rotodynamic pumps and points the way to further development which can extend their performance significantly.

1. BACKGROUND

It is well established that rotodynamic pump performance deteriorates with increasing gas content (see, for example, reference (1)). The deterioration in head is marked by a series of discontinuities relating to inlet flow regime changes. At very low gas content the flow is essentially homogeneous, with little slip, and head degradation is minimal or may even improve slightly. At higher gas content performance drops off rapidly due to gas accumulation and stratification in the impeller channels. At some point the gas accumulation will extend from the blade suction surface to the pressure surface and the pump will deprime. Clearly for acceptable multi-phase pumping performance the point of rapid head degradation must be pushed to as high a gas content as possible by careful impeller design, control of the inlet flow regime. The ability to control head degradation is well defined in the aerospace industry where the pumped medium is a single component fluid, and a gas content in excess of 50% can be achieved.

Applications for pumping of multi-phase fluids have increased in recent years. The original impetus for research into gas/liquid pumping arose in the U.S.A. in the pressurised water reactor (PWR) programme. There remains a specific need to consider the performance of reactor circulating pumps under loss of coolant accident (LOCA) conditions. In such accidents there is a danger of the reactor core overheating and of 'meltdown' occurring. It is therefore essential that the coolant circulating pumps continue to perform well when the coolant temperature and hence vapour content rise, or when air is entrained into the system. In the oil and gas industry there is a need to develop marginal and deep water reserves at minimum costs. There is significant interest in the development of subsea production systems that use multi-phase pumps to transport the well fluids to remote offshore or onshore processing facilities. In the aerospace industry pumps are used for suction fuel supply systems for gas turbine engines. These include small, high speed rotodynamic pumps used to supply aviation fuel to the engine. As a result of altitude and temperature changes experienced in aircraft, the vapour content in aviation fuel is continuously changing. These pumps are therefore required to handle varying gas content in a critical area where loss of performance in civil and particularly military aircraft could prove to be disastrous. Within the processing industries many applications of multi-phase pumping exist. Sewage in waste treatment plants often contains high levels of gases. Fermentation processes are another example where gas content is high.

In 1985 BHRA conducted a series of tests on a small centrifugal pump to determine its basic gas handling capability (2). These tests showed that gas quantities as low as 7% by volume were capable of depriming the pump but that the bubble size, the inlet whirl, and the homogeneity of the mixture all affected the de-prime point. Enough evidence was extracted from the results to suggest that much higher gas/liquid ratios could be achieved and as a result further research was undertaken. The results from part of this work form the basis of this paper. The work was supported by a consortium of industrial companies (see Table 1) and the UK Department of Trade and Industry.

2. OBJECTIVES OF TEST WORK

(1) To determine the basic gas-handling capacity of typical examples of modern rotodynamic pumps. This includes the effects of pump type and installation attitude.

(2) To evaluate the benefits of easily embodied pump modifications and additions.

(3) To suggest ways of overcoming gas-handling limitations by the use of pumping systems.

3. TEST APPARATUS

In order to meet the objectives a programme of experimental testing of four standard rotodynamic pumps operating under two-phase air/water operation was undertaken. To perform these tests a closed loop water system was built with an open loop air injection system.

Each of the test pumps is an end suction, radial discharge machine, which allows it to be mounted either horizontally at ground level, or vertically at the upper level.

3.1 Experimental Test Loop

The facility is shown diagrammatically in Figure 1. The main loop is manufactured from 152mm diameter galvanised steel pipe. Water is supplied from a large holding vessel which allows the air and water to separate fully. The separator has a volume of 10m^3 which gives a residence time of 1.5 minutes at the maximum system flow rate of 400 m^3/hr. The separator can be vented to relieve pressure build-up, and can be pressurised or evacuated to modify the suction pressure at the pump inlet.

Air is introduced into the system through eight nozzles mounted radially into a 152mm diameter pipe section which is positioned just upstream of the pump. The air is supplied through a manifold allowing individual selection of the eight nozzles, and the nozzles can be adjusted radially. Each nozzle has eight discharge holes. Perspex pipe sections have been manufactured and can be fitted to allow visual inspection of the flow just upstream and just downstream of the pump. The facility is rated to a maximum working pressure of 10 bar.

3.2 Control and Instrumentation

The pumps are mounted to a base plate and coupled through a flange and tyre coupling to a 112KW variable speed motor. Control of the motor speed is provided by a Thyristor Drive providing for a range of 0 - 3000 rpm.

Fluid flow control is achieved using a valve positioned just ahead of the separator. Air flow is controlled by a diaphragm valve just upstream of the air supply manifold and by a pressure regulator. Flow measurement is through ½" and 1" Turbine Meters with the air made to pass through either of these or through a 1" bypass line. Pressure and temperature are monitored just downstream of the air turbine meters. Water flow is monitored using an electromagnetic flowmeter mounted in the straight pipework upstream of the air injection system.

Pump suction and discharge pressures are measured using bourdon tube gauges and using pressure transducers. Water temperature is monitored using a thermocouple inserted into the fluid upstream of the injection system.

Motor power consumption is determined by monitoring the motor supply voltage and current. Pump power consumption is then calculated from the motor power consumption and the motor efficiency assuming no losses occur in the motor/pump coupling. Pump speed is measured using a tachometer fitted to the motor.

3.3 Data Acquisition

Meters for all the parameters except water temperature and motor supply voltage are interfaced to a computer for automated data acquisition. All automated readings are supported by manually collected data. 100 samples are taken for each reading of each parameter at a rate of 2 samples/second, and the data subsequently reduced to an average value and a standard deviation value.

3.4 Dimensionless Parameters

For ease of comparison, flowrate and head rise were non-dimensionalized as follows:

- Non-dimensional head coefficient

$$= gH/r_t^2\ \omega^2$$

- Non-dimensional flow coefficient

$$= Q/2\pi r_t^2\ b_t\ \omega$$

where:	g	=	gravitational acceleration
	H	=	head rise
	Q	=	flowrate
	r_t	=	tip radius
	b_t	=	tip width

ω = angular speed

4. TEST PROGRAMME

The effect of pump type has been investigated by conducting tests on four pumps of varying specific speed as outlined in Table 2. All four pumps were tested at their design speed (see Table 2), except for Pump C which was also run at a higher speed of 1800 rpm. Each pump has been tested in both a horizontal and a vertical orientation, as outlined in Table 3. The inlet flow regime is generally different for the two orientations described above. Furthermore a number of easily adopted modifications have been made that will also affect the inlet flow regime, and these have been tested as outlined in Table 3.

The options outlined above have been assessed by conducting a number of water only and two-phase air/water tests. These tests also serve to assess the effect of void fraction and of flow coefficient on the performance of the pumps.

4.1 Water Only Performance Tests

Water only performance tests were carried out by varying the flow rate through the pumps. For the pumps in their standard configuration, mounted in the horizontal orientation and running at their design speeds, the data were compared with the manufacturers data. For all options the water only tests provide the 0% gas volume fraction against which the effect of increased air content can be assessed.

4.2 Air/Water Performance Tests

These were performed for all options at fixed air contents of 2%, 4%, 6%, 8% and 10% by volume.

4.3 Fixed Valve Tests

Performance tests were conducted where specific duty points were taken at 0% air, the necessary valve setting was made, and when the air content was increased in steps up to a point where the pump had deprimed or was no longer developing any head rise. The particular duty point adopted was for best efficiency (BEP).

5. RESULTS

5.1 Fixed Valve Tests

The results for the fixed valve tests (Figure 2) serve to illustrate the dramatic loss of performance experienced by all four pumps. Without exception, effective pumping has ceased once values of 10% GVF have been reached. At lower GVF's the performance is very sensitive to small changes. These results are purely indicative: comparison between different pump types is not valid.

5.2 Pump A

The plots for head coefficient and efficiency versus flow coefficient for pump A are given in Figure 3. These curves are typical of data generated for the other three pumps. Pump A appears to handle very small amounts of gas (<2%) with no deterioration in performance, except at low flow coefficients. However, higher GVF's result in a dramatic drop in both head and efficiency. For example at a flow coefficient of 0.08, the efficiency at 0% GVF is 57% but only 17% at 10% GVF. Similar sensitivities to GVF were experienced by all four test pumps. The data for all Pump A tests are summarised in Figure 4. The data for the horizontal alignment with the pump running at design speed are used as a baseline for comparison of the effectiveness of other modifications namely: vertical alignment, inducer fitted (horizontal alignment) and inlet guide vanes fitted (vertical alignment). For comparison purposes the results presented are for a flow coefficient of 0.08.

The results for vertical alignment suggest a modest improvement in head performance for higher values of GVF , although efficiencies are consistently better. The fitting of an inducer is particularly effective at GVF's up to about

8% but the results suggest that, thereafter, there is no benefit to be gained. The results for the inlet guide vanes are inconclusive. The drop-off in performance is slightly more gradual, suggesting some flow control improvement.

5.3 Pump B

Pump B was tested in horizontal and vertical alignment. The head coefficient and efficiency plots for horizontal alignment are given in Figure 5. The results are summarised in Figure 6 for a flow coefficient of 0.08.

This pump coped well with low GVF's, less than about 5% but suffered significantly at higher GVF's. However, vertical alignment provided a significant gain in both head and efficiency across the range of GVF.

5.4 Pump C

Performance curves at design speed (1450 rpm) and a higher speed (1800 rpm) are shown in Figures 7 and 8 respectively. Figure 9 shows the results for Pump C with variants: horizontal and vertical alignment plus inlet guide vanes and higher speed (both in horizontal alignment). Flow coefficient is 0.08.

Like Pump B baseline performance is good up to GVF's of 5% and very poor thereafter. However, unlike Pumps A and B, vertical alignment has a detrimental effect on performance. Similarly, inlet guide vanes offer no advantage. Some improvement with higher speed (25% above design speed) is apparent.

5.5 Pump D

This pump had somewhat different characteristics from the other three, as is evident in Figure 10. Although running efficiencies were generally good, a significant amount of surging was evident. This accounts for the erratic performance curves.

A summary of data at a flow coefficient of 0.16 is presented in Figure 11. Baseline results offer the best performance in terms of both head and efficiency. Both the inlet guide vanes and vertical alignment cause deterioration.

6. DISCUSSION

6.1 Pump Type

It is generally considered that axial flow machines offer the greatest potential for gas handling. Neumann and Lualdi (3) suggested that two-phase performance increases with specific speed and therefore recommended the use of axial flow pumps. However Kosmowski (4) suggested that a centrifugal pump minimises the effect of separation losses and should be redesigned to include vanes on the casing wall. For example, within the aerospace industry centrifugal pumps are favoured. This is primarily because suitable head rises can be achieved with the smallest possible impeller to ensure that the outlet fluid is once more single phase. Nevertheless mixed flow impellers of high solidity have also been successfully applied.

The results obtained here are not conclusive. There is evidence that the axial flow device (Pump D) maintains a higher level of performance (measured in terms of % reduction in head and reduction in efficiency) to a higher level of gas content than the other pumps tested, (with the possible exception of Pump C). However there is also evidence that Pump D surges badly over some of its operating range. Conversely, Pump A appeared to operate in a very stable manner, albeit at much reduced performance levels.

6.2 Pump Speed

Tests were performed at the design speed for all pumps and at a higher speed for Pump C. An increase in speed would normally be expected to improve gas handling ability as it compensates for the reduced mixture density and may reduce bubble size, although this may well be contrasted by a move away from optimum blade incidence angles. For this reason many current developments are based on variable speed operation with a speed control coupled to upstream monitoring. Pump C, tested above its design speed, showed a modest improvement in gas handling ability.

6.3 Pump Orientation

All pumps were tested both horizontally and vertically. There were some difficulties encountered when testing vertically because of the higher pump position, and hence lower suction pressure, which could not be compensated for until the separator was given a full pressurising capability some way through the project. As a result some performance curves were curtailed by cavitation particularly those for Pumps C and D. Nevertheless there was clear evidence to support the use of vertically mounted pumps. This is partly because of gas buoyancy helping to strip gas away from zones within the impeller where it might tend to accumulate, and partly because the inlet flow was always homogeneous, in marked contrast to horizontal operation.

6.4 Inlet Flow Regime

Within these tests the pumps experienced both stratified flow and homogeneous flow at inlet, but were not tested under transient inlet conditions such as plug and slug flow. The improvements to be gained from vertical operation lead to the conclusion that the inlet flow should be homogenised prior to pump inlet rather than relying on the pump itself to perform this exercise. Inlet Guide Vanes may ensure that unwanted pre-rotation is removed, but for gas handling they offer no appreciable benefit. An inducer has the potential to help or to hinder. By forcing pre-rotation on the incoming fluid a gas core may be generated which can be directed towards the hub (or back shroud) of the pump. An inducer was fitted to Pump A and made a quantifiable improvement in performance between 2% and 8% air content, but offered no further improvement in performance at higher percentages. A carefully designed inducer, matched to a pump with low loading in the hub region is likely to enhance performance.

6.5 Void Fraction

All 4 pumps tested should be regarded as having stopped pumping air/water mixtures by 40% air content by volume. A small air content (<2%) is not normally significantly detrimental, and may even enhance performance slightly. Thereafter performance drops rapidly with air content up to 10-15%, and then degrades more slowly up to 30-40%. It should be remembered that gas handling with hydrocarbons will be better than with air/water mixtures.

7. CONCLUSIONS

The main conclusions of the work were as follows:

(i) All four pumps had ceased to be effective at GVF's in excess of 25%.

(ii) Efficiency and head generation at GVF's in excess of 10% were very poor.

(iii) Vertical alignment of two of the pumps was of marked benefit.

(iv) Inlet guide vanes did not enhance gas handling ability.

(v) The fitting of an inducer was of benefit in a range of GVF's from 2 to 8%.

(vi) The axial flow pump appeared to

offer the best efficiency but experienced some severe surging. Pump A ran with the greatest stability but at the expense of reduced efficiency.

8. THE WAY FORWARD

The work reported in this paper has shown that GVF's of less than about 8% can be tolerated by conventional rotodynamic units, particularly if run with some of the modifications reported. Many operators, however, are interested in much higher GVF's at suction for which these pumps, on their own, are clearly unsuited.

An attractive approach is to envelope the pump in a simple system which reduces the effective GVF at pump suction. Two basic concepts are available: (i) to by-pass most of the gas around the pump and co-mingle the fluids at pump discharge (Figure 12) and (ii) to re-cycle some of the high pressure discharge liquid and re-inject it into the suction line (Figure 13). Indeed, an optimum system might involve a composite of both concepts.

Work on such systems is currently on-going as a direct follow-on to the work reported in this paper and results have been very encouraging.

ACKNOWLEDGEMENT

The authors are grateful to BHR Group Limited for permission to publish this paper.

REFERENCES

(1) A Review on the Pumping of Gas/Liquid Mixtures,
S J Hughes, BHRA Report No RR 2340, Dec 1984.

(2) Feasibility Study on Gas/Liquid Flows in a Centrifugal Pump,
N L J Silburn and A Goulas, BHRA Report No RR2551, Jan. 1986.

(3) Pumping of Highly Aerated Liquids,
B Neumann and R Lualdi, Proc. NEL Fluid Mechanics Silver Jubilee Conference, Nov 1979.

(4) The Design of Centrifugal Pumps for the Delivery of Liquids with High Gas Content,
I Kosmowski, Pumps - the heart of the matter, Eighth Technical Conference of the BPMA, March 1983.

Andritz AG
British Petroleum International Limited
BW/IP International BV
Elf UK plc
Enterprise Oil plc
Esso Exploration & Production UK Ltd
Frank Mohn UK Limited
Ingersoll Rand Sales Company Limited
Mather and Platt Limited (now part of Weir Pumps)
Nuovo Pignone SpA
Sulzer Brothers Limited
The Department of Trade and Industry
Total Oil Marine plc
Worthington and Simpson

Table 1 Industrial sponsors

	PUMP TYPE	DESIGN SPEED (rpm)	NON-DIMENSIONAL SPECIFIC SPEED	FLOW AT BEP[m^3/hr]
Pump A	Radial flow diffuser	2200	0.4	135
Pump B	Mixed flow diffuser	3000	0.98	233
Pump C	Mixed flow diffuser	1450	1.6	350
Pump D	Axial flow diffuser	3000	2.5	240

Table 2 Summary of test pump types

OPTION	ORIENTATION	PUMP A	PUMP B	PUMP C	PUMP D
Standard	Horizontal	X	X	X	X
	Vertical	X	X	X	X
Inducer	Horizontal	X			
	Vertical	X			
Inlet Guide Vane	Horizontal			X	X
	Vertical	X			X

Table 3 Summary of test conditions for each pump

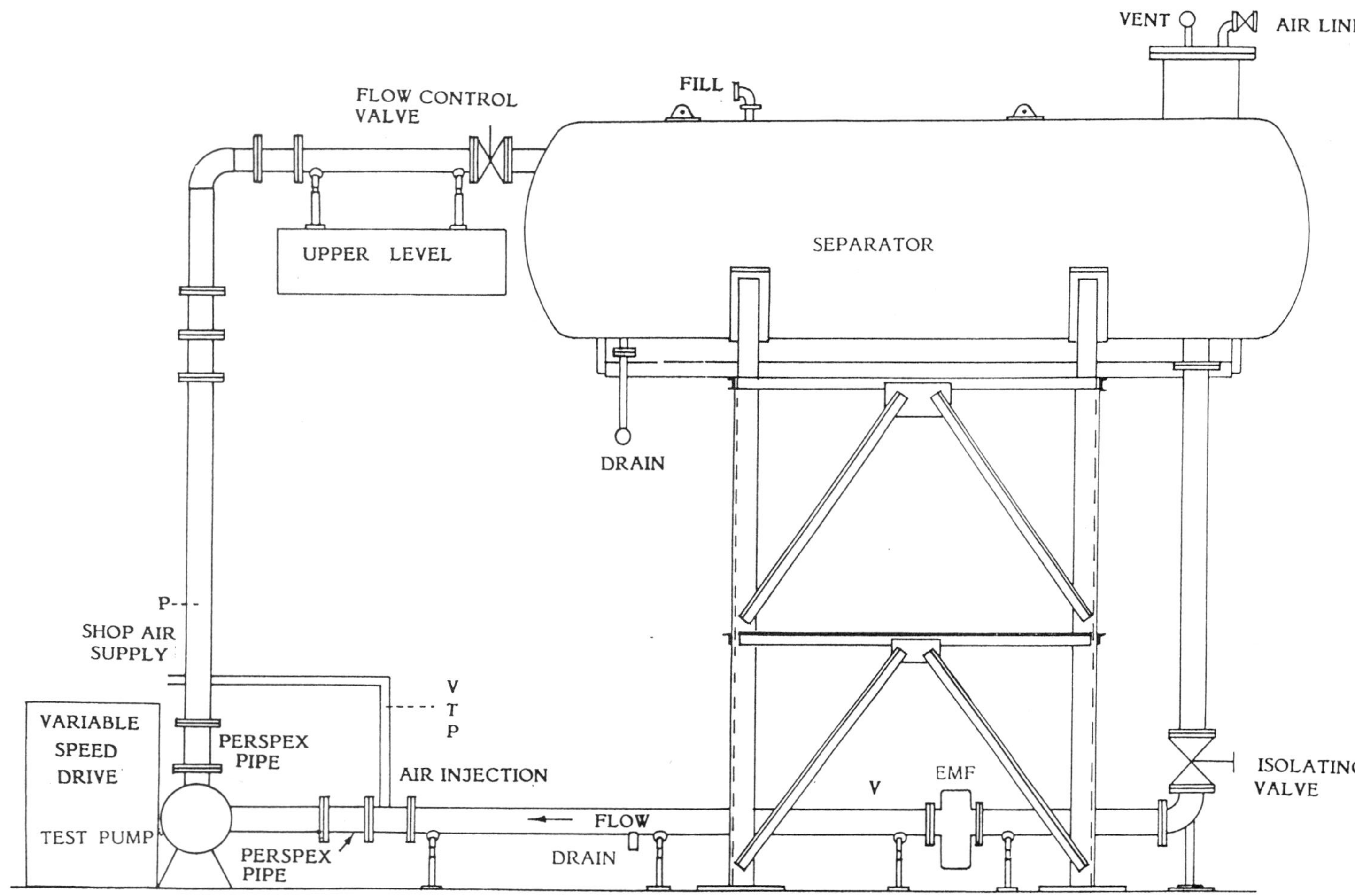

Figure 1 Schematic of liquid/gas pump test loop

Design Speed (2200 rpm)
Horizontal alignment

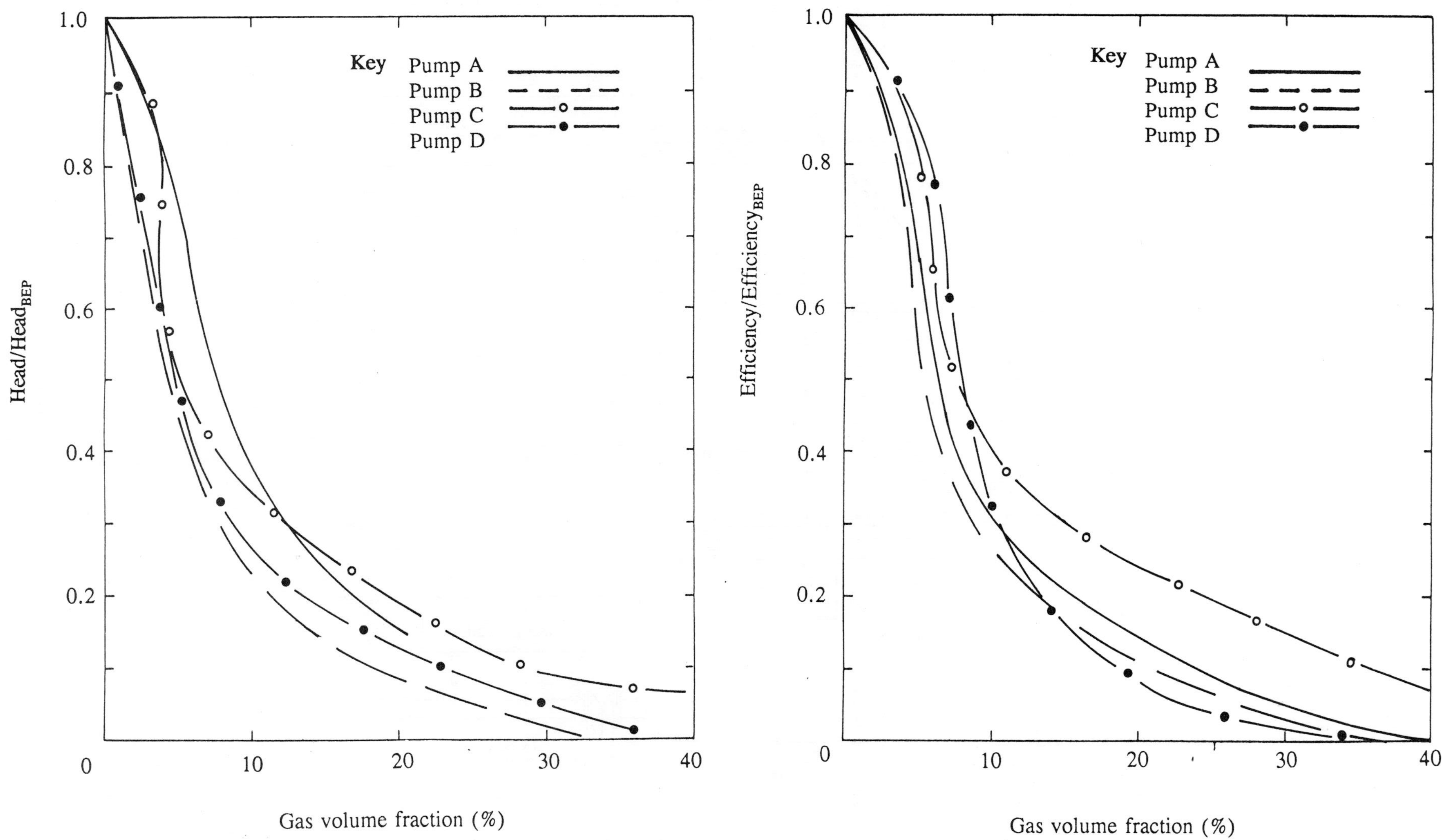

Figure 2 Results for fixed valve tests

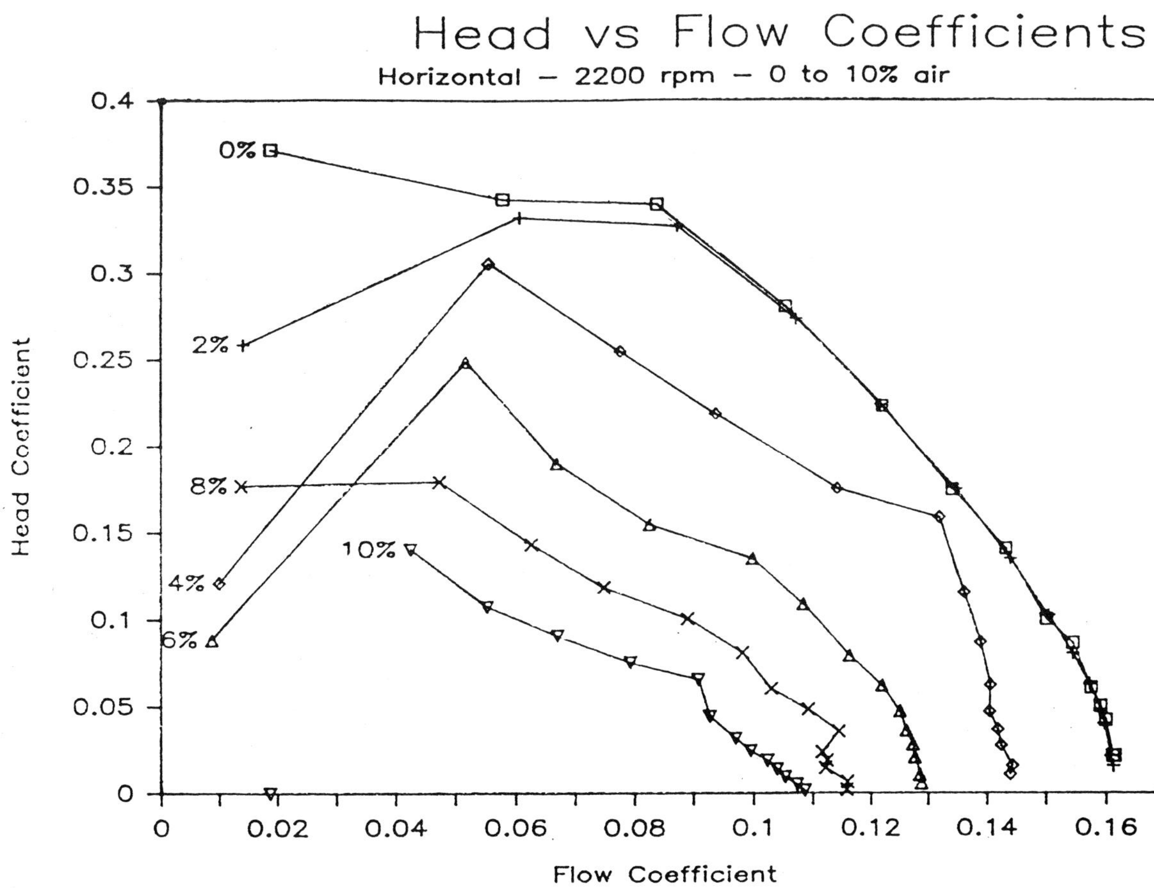

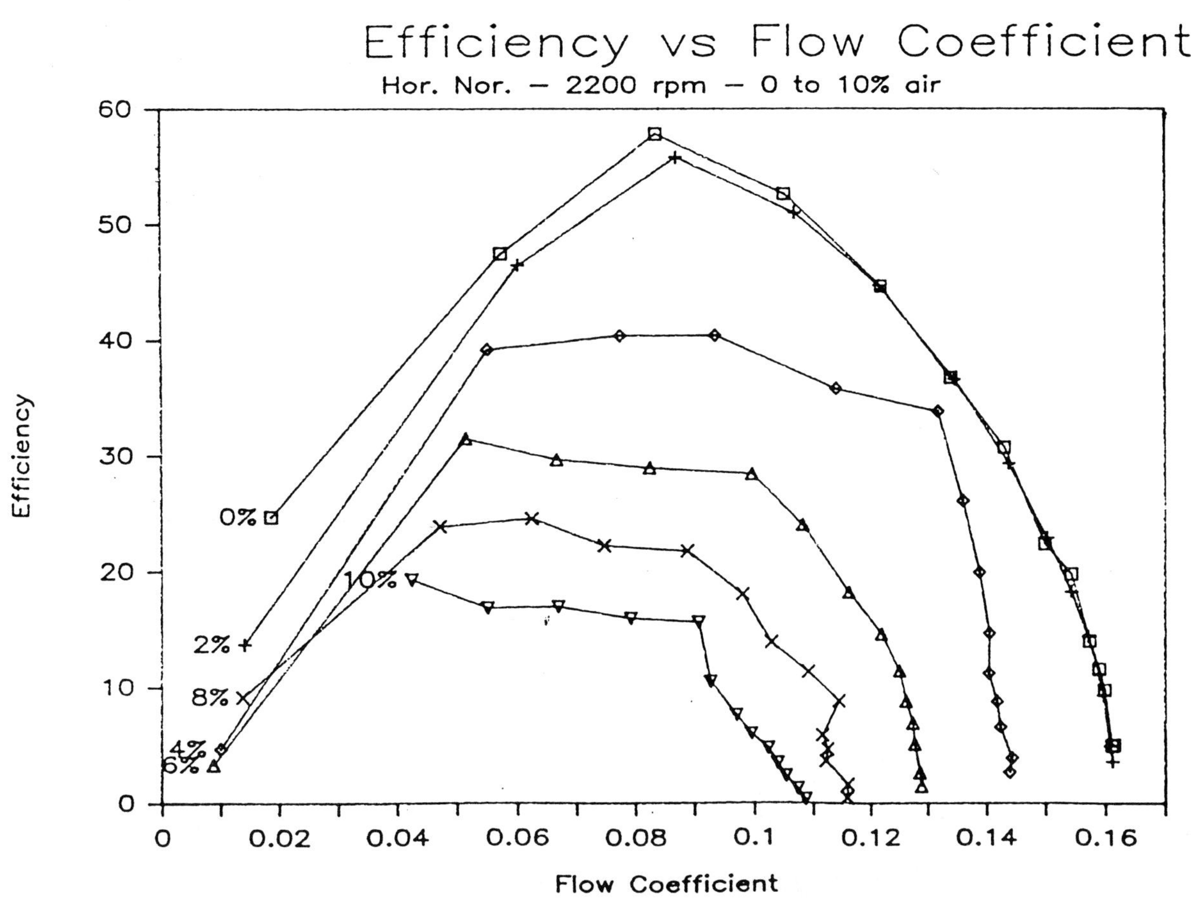

Figure 3 Performance curves - Pump A

Pump A
Design speed
Flow coefficient = 0.08

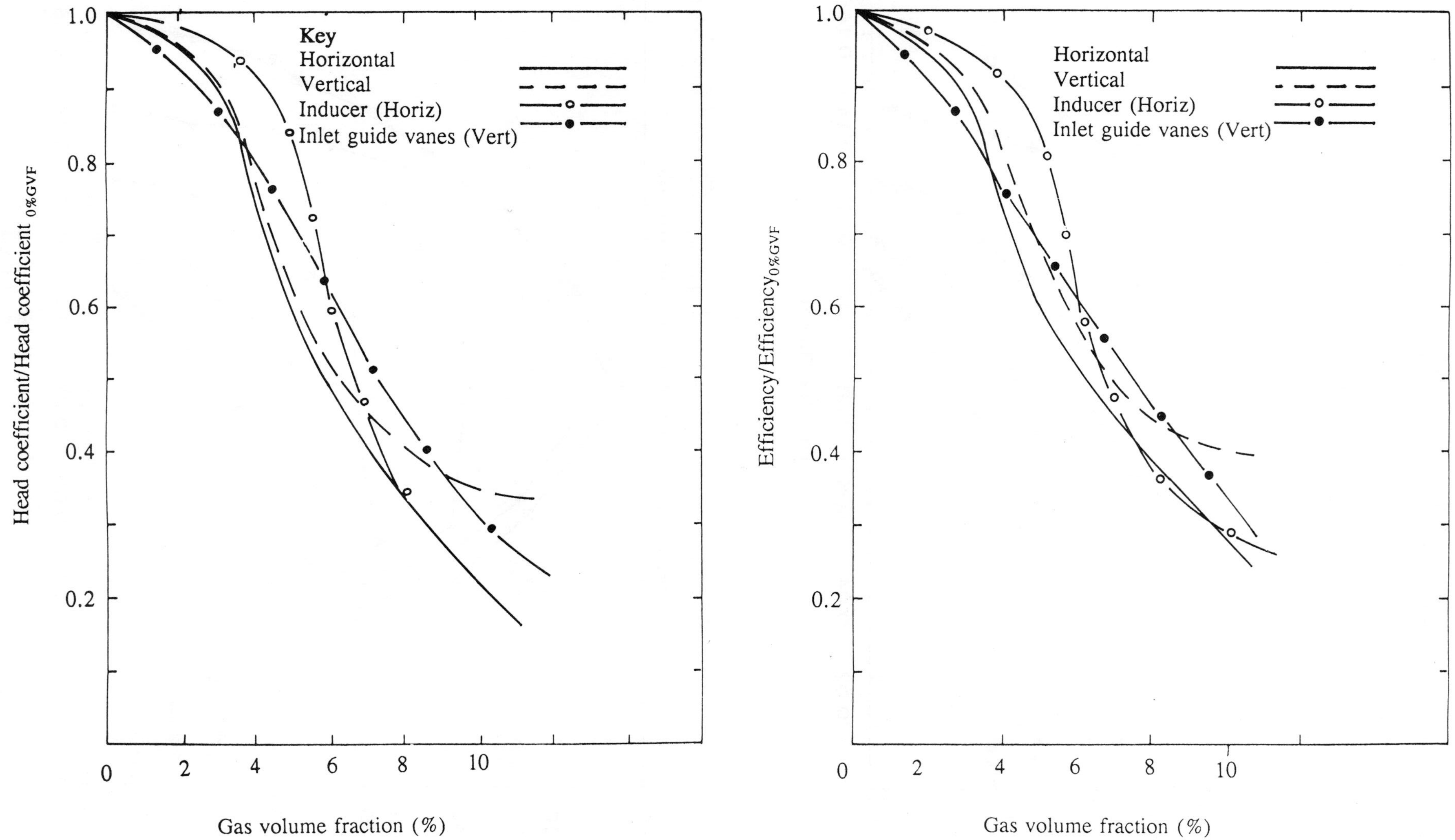

Figure 4 Results summary - Pump A

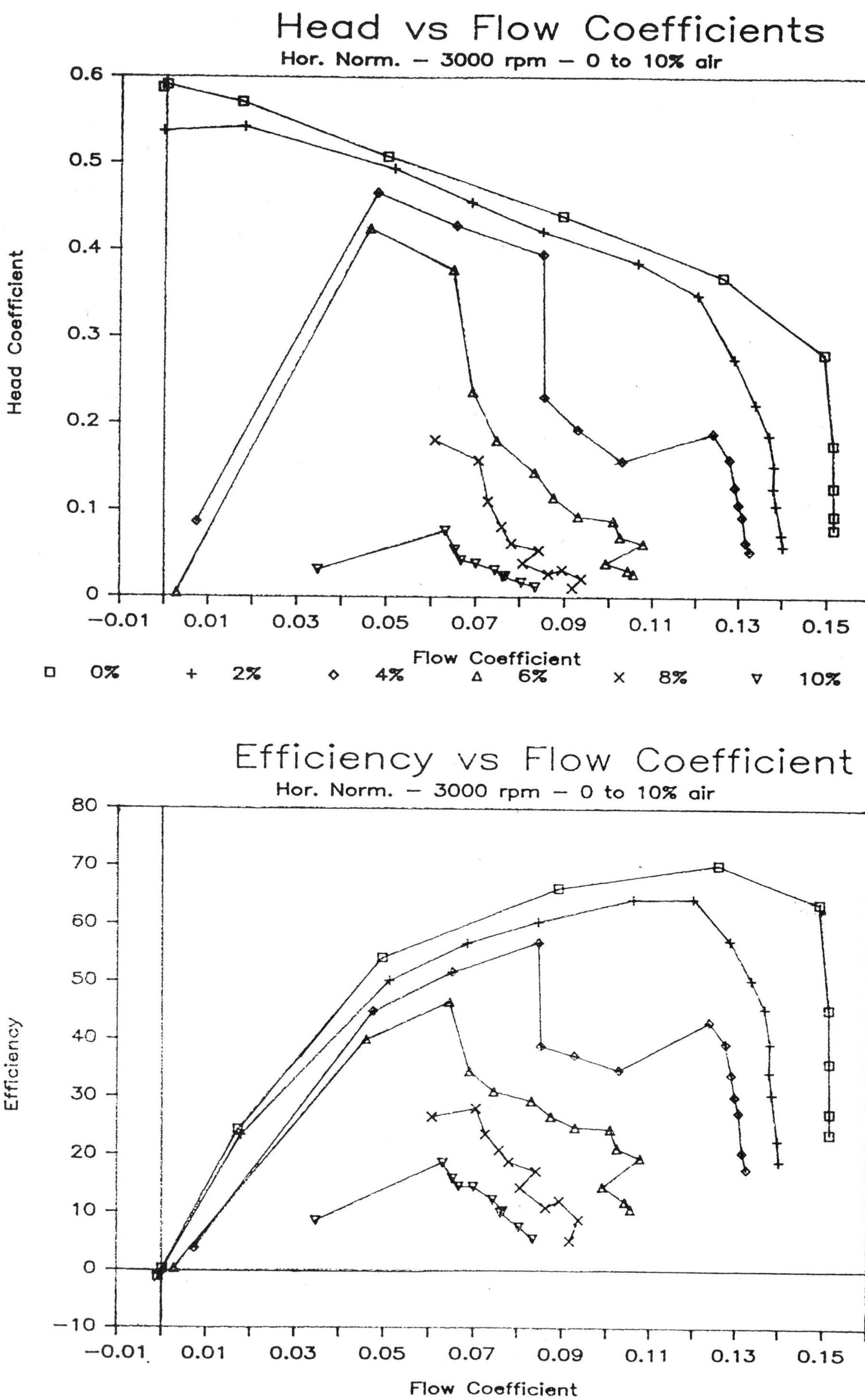

Figure 5 Performance curves - Pump B

Pump B
Design speed (3000 rpm)
Flow coefficient = 0.08

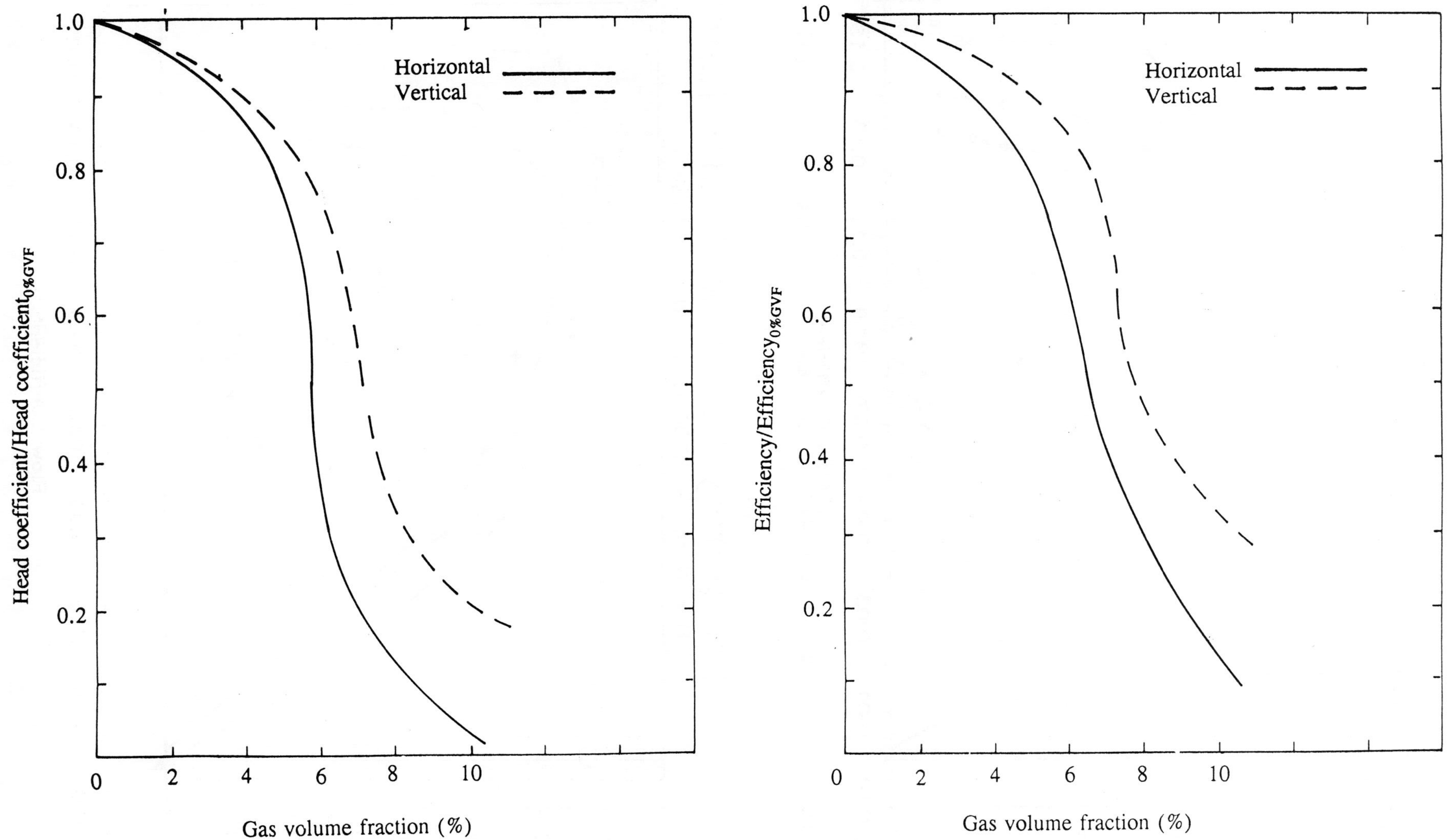

Figure 6 Results summary - Pump B

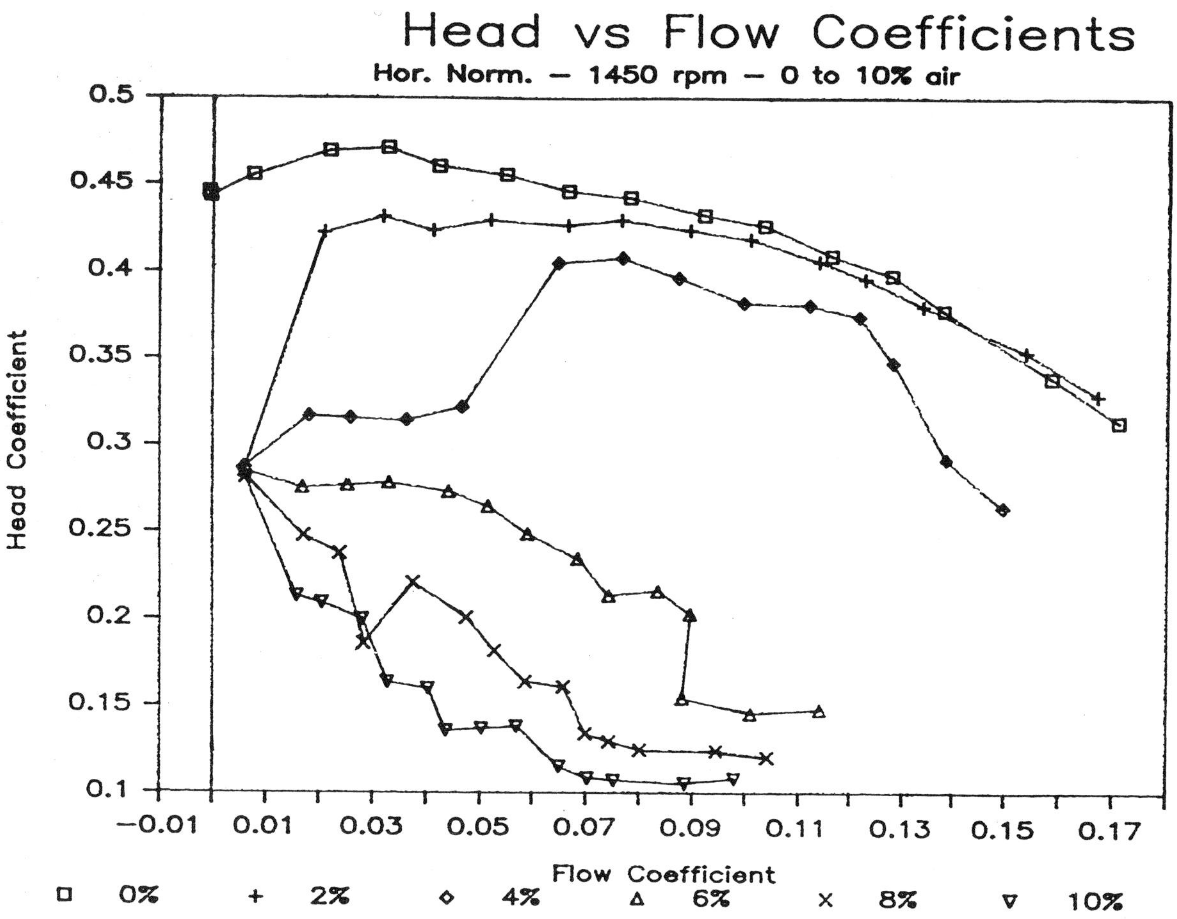

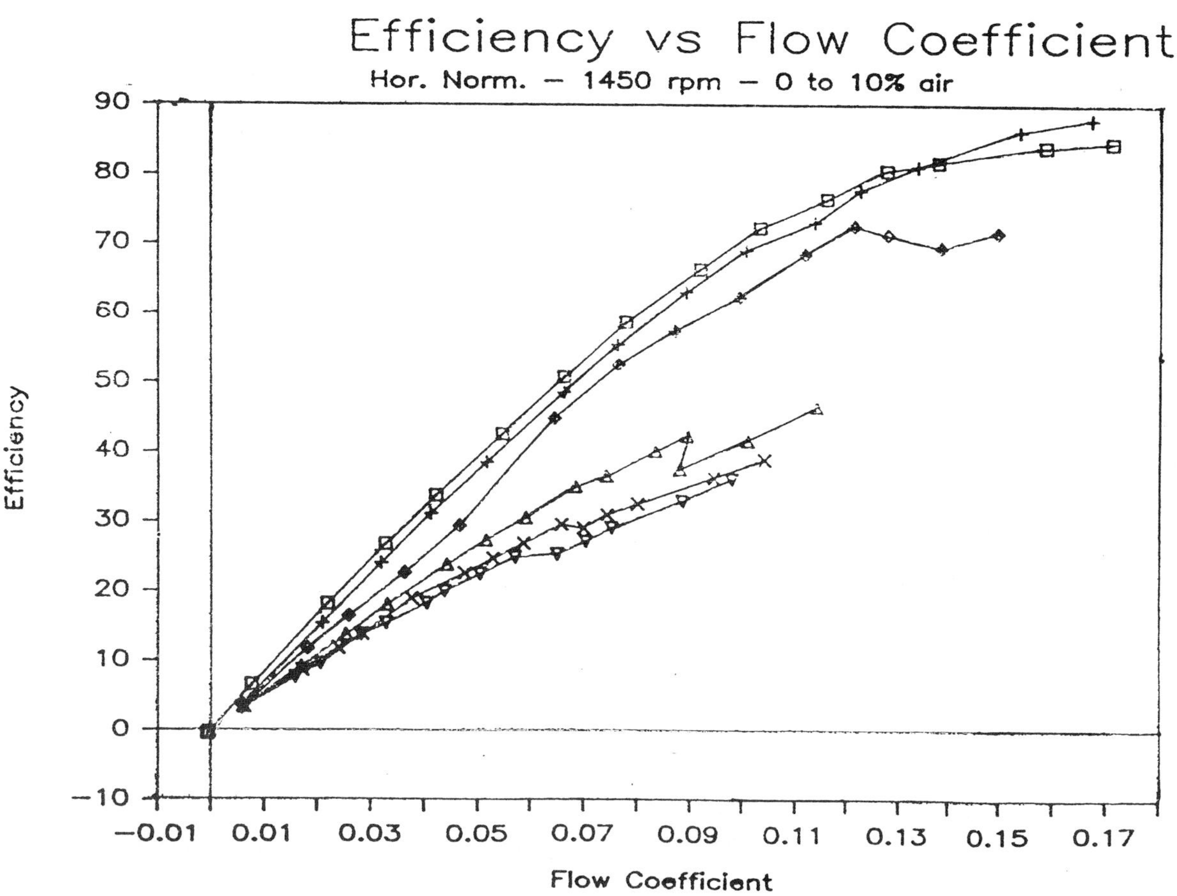

Figure 7 Performance curves - Pump C, design speed

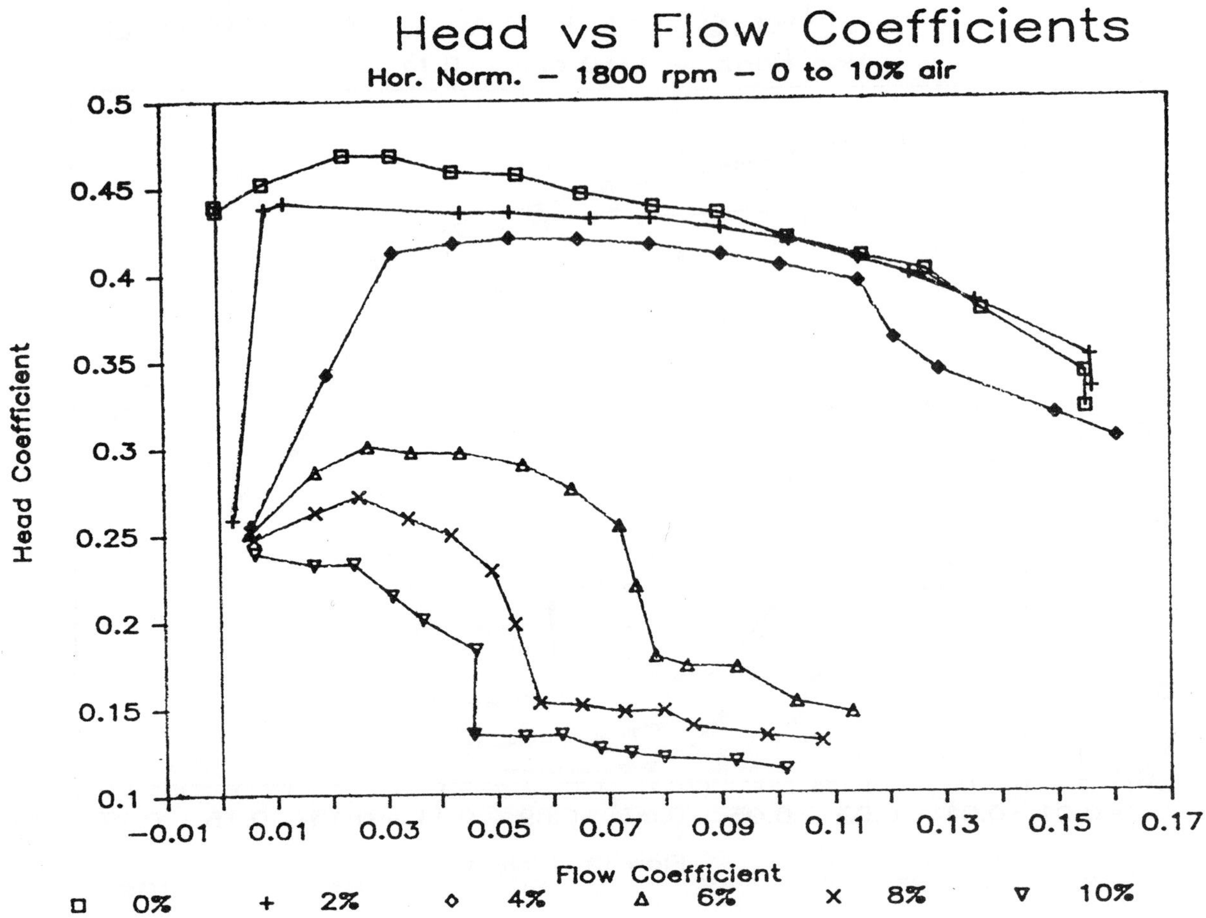

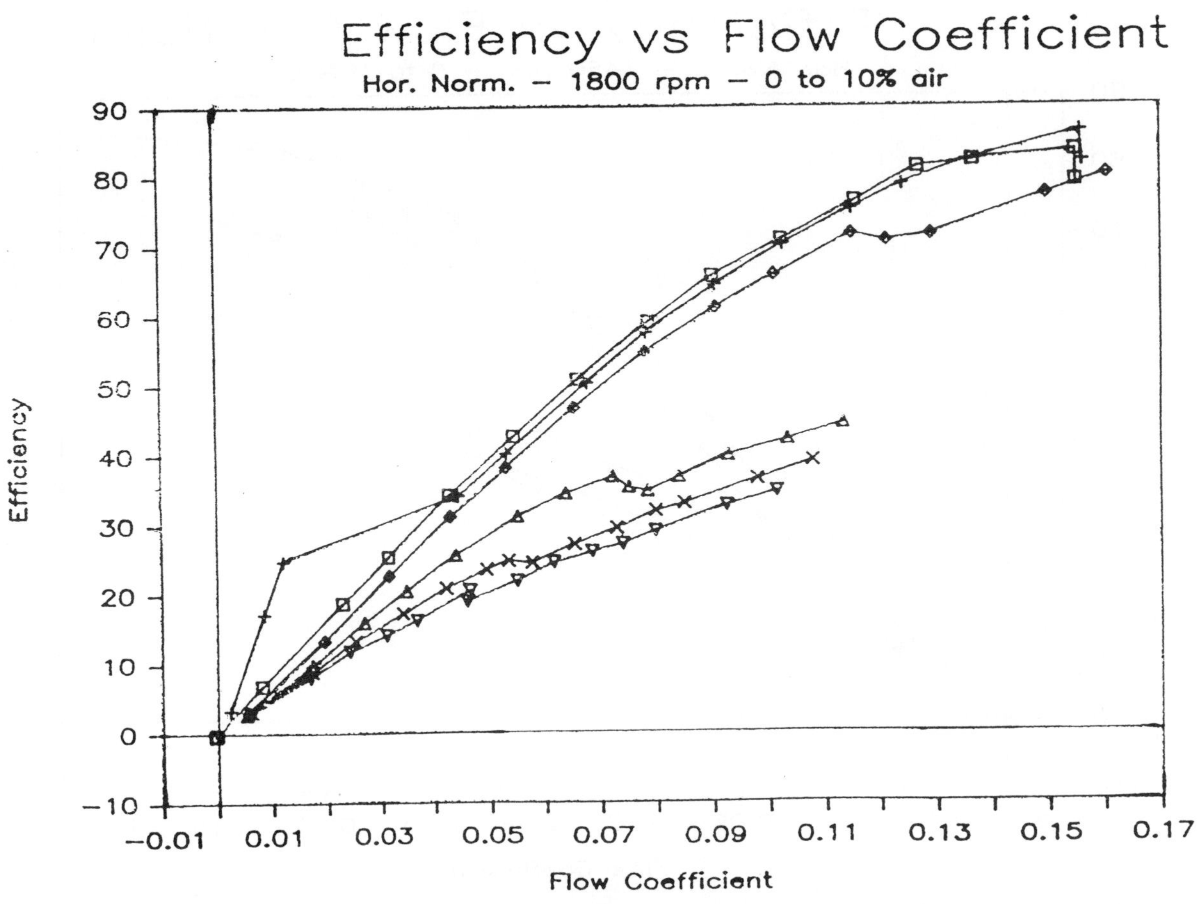

Figure 8 Performance curves - Pump C, 1800 rpm

Pump C

Design speed (1450 rpm)/1800 rpm
Flow coefficient = 0.08

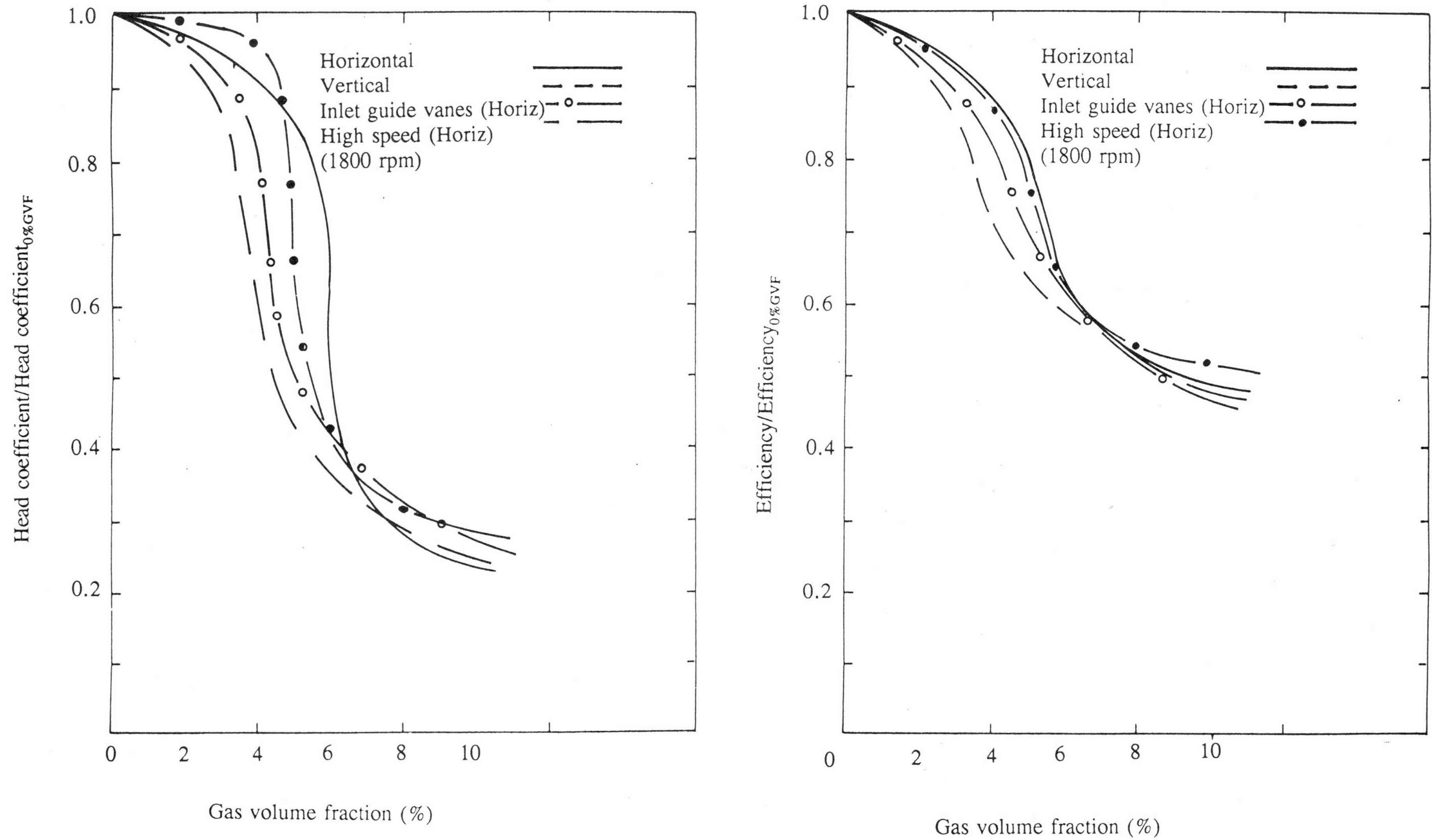

Figure 9 Results summary - Pump C

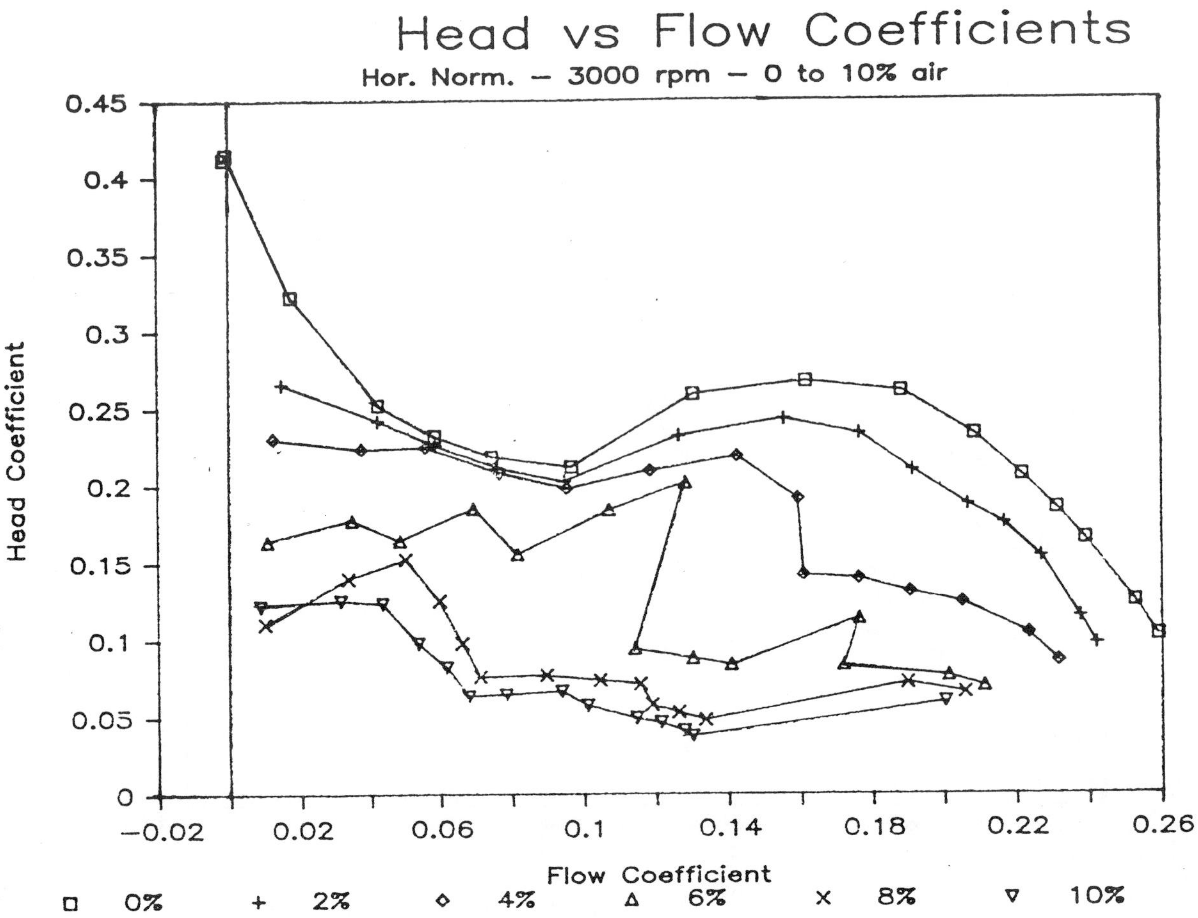

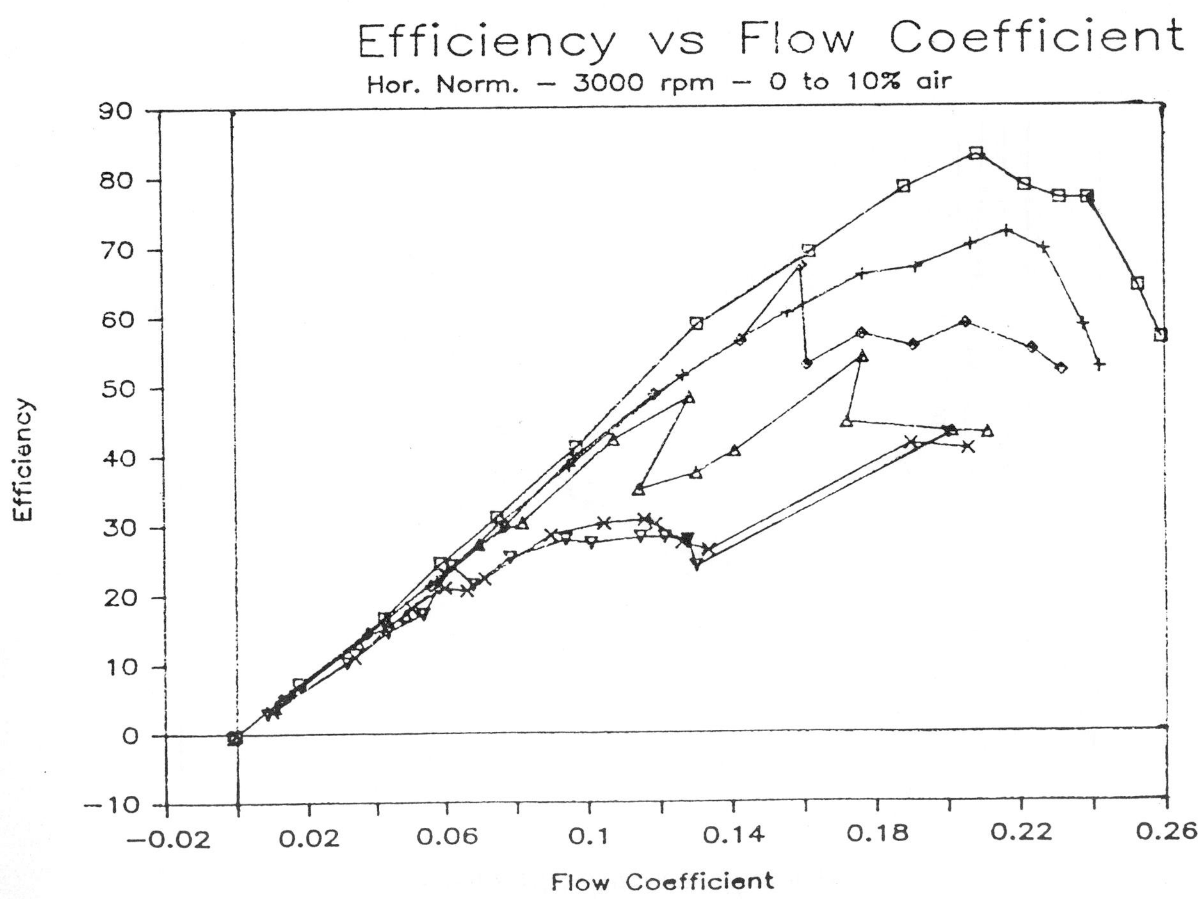

Figure 10 Performance curves - Pump D

Pump D

Design speed (3000 rpm)
Flow coefficient = 0.16

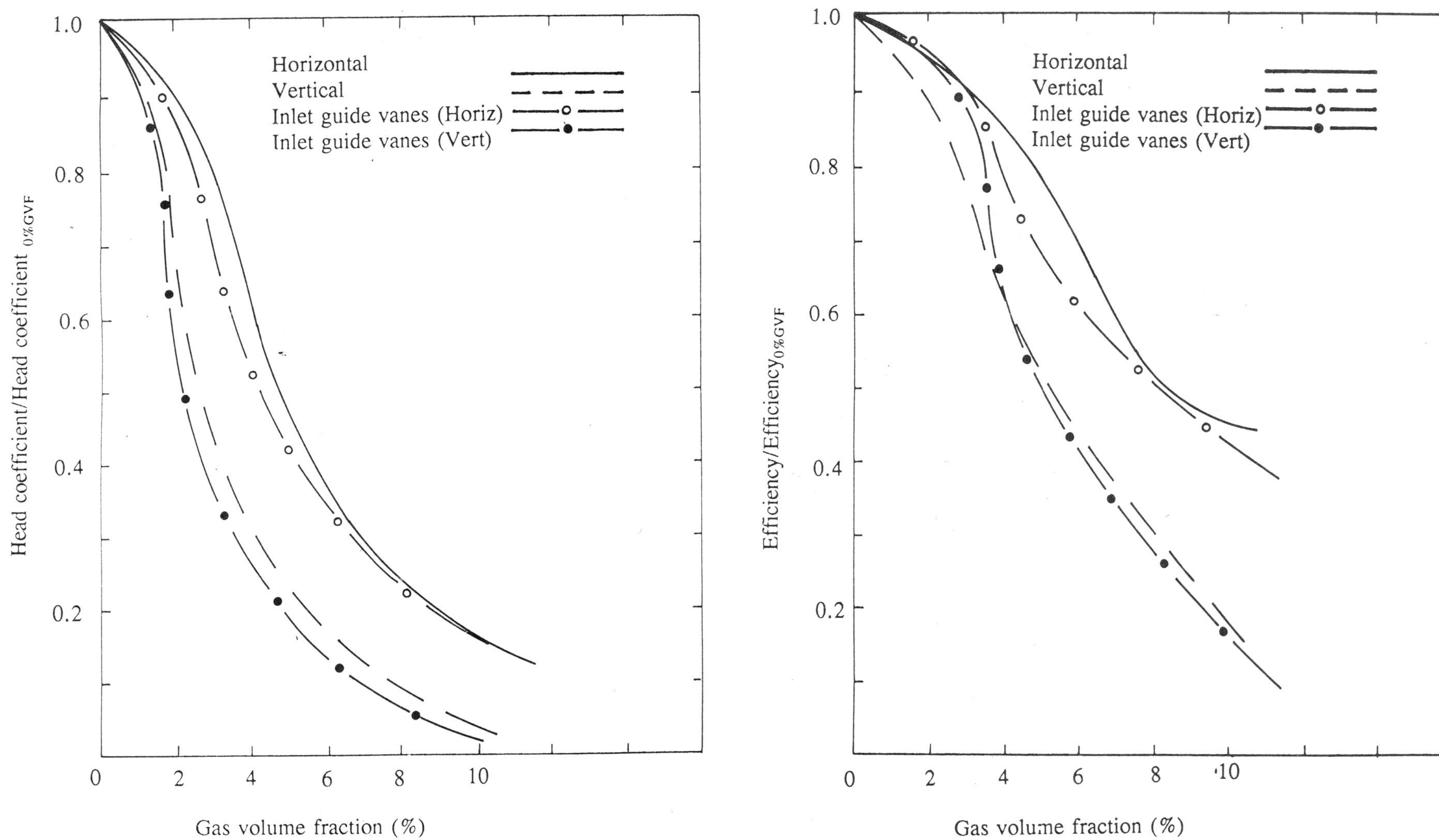

Figure 11   Results summary - Pump D

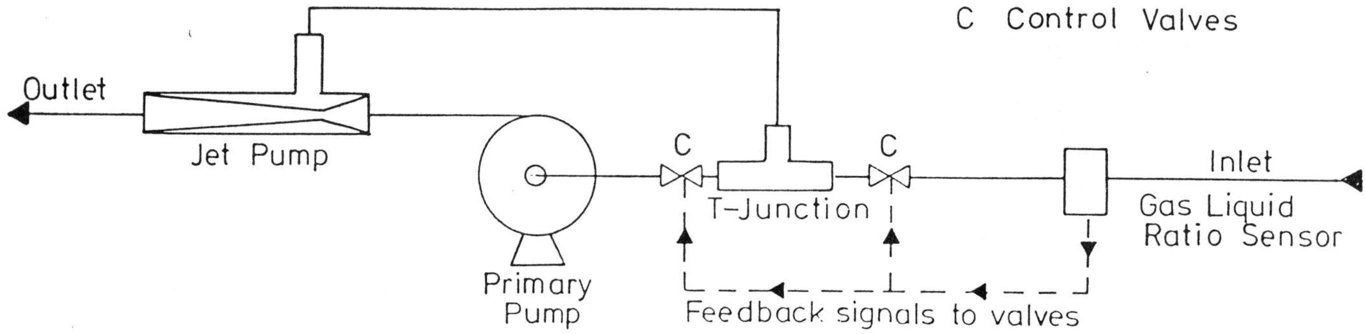

Figure 12 Gas-bypass system

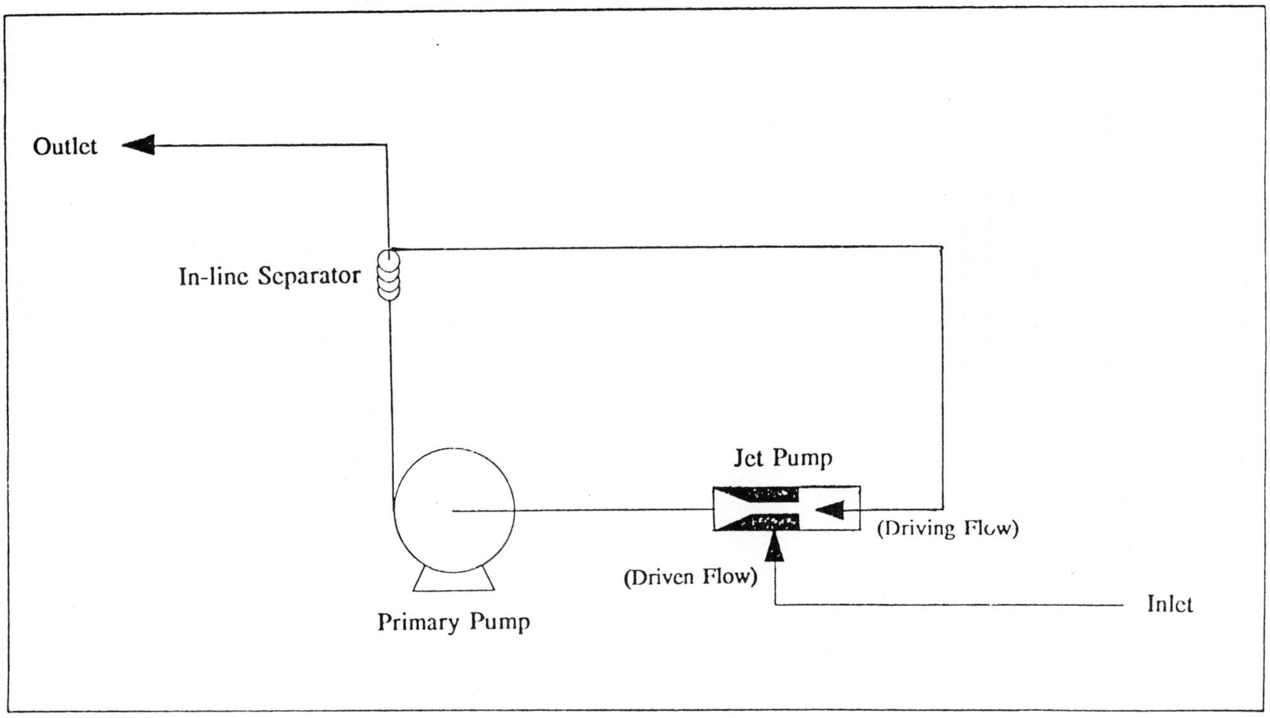

Figure 13 Liquid re-cycle system

C449/038/93

Initial experience with a new high-speed, high-pressure, oil-free motor compressor unit (HOFIM)

A C MARRIOTT
Sulzer Escher Wyss Limited, Zurich, Switzerland
D C GILON, BSc
ACEC Energie, Marchienne au Pont, Belgium

SYNOPSIS: Design details and test programme are presented for a new motor compressor unit with the acronym HOFIM (High speed Oil Free Intelligent Motor compressor). The unit features separate motor and compressor directly coupled, with both machines supported on magnetic bearings. The prototype application is in a natural gas storage facility, for which the key parameters are:

motor nominal speed	20000 rpm
motor nominal power	2000 kW
compressor inlet pressure	5000 kPa
compressor discharge pressure	15500 kPa
compressor speed range	70 to 102%
compressor flow	38000 NM3/h

The motor is an asynchronous squirrel cage induction machine, driven by a solid state variable frequency drive (VFD). The compressor is a six stage barrel machine of fully modular design. Dry gas seals are used to minimise internal and external leakage. An active balance system controls residual thrust of the whole unit to a level compatible with the capacity of the axial magnetic bearing. The unit was developed jointly by four companies in conjunction with the european BRITE programme.

1 INTRODUCTION

This paper presents the main design features and details of the test programme of a motor-compressor unit which combines several relatively new technologies. The background to this new machine was a unit developed jointly by four companies between 1987 and 1990, a concept called MOPICO, which stands for MOtor PIpeline COmpressor (see figure 1). Mopico comprises a high speed induction motor with an overhanging impeller attached at each shaft end and supported on magnetic bearings. The motor and compressor impellers are installed in the same pressurised casing, thus making the system totally free of leakage and thus emission free, since there are no rotating seals to atmosphere. Other advantages gained by this concept are completely oil-free operation, practically no wearing parts, short installation time, low maintenance costs and a high degree of operating flexibility. This arrangement however is dedicated to natural gas transportation duty and thus is inherently limited to low pressure ratios and clean, non-corrosive gases. This limitation of the range of possible MOPICO applications was the main reason for the partner companies to develop the HOFIM concept, which is not limited in pressure ratio and is applicable for many types of gas, since the motor is cooled independantly and not by the process gas.

The paper will first describe the design and development of the high speed motor, pointing out the major differences from the existing technologies. Then the selected application for the demonstration unit and the resulting compressor design will be presented. Finally the test and demonstration programmes will be described. Additional information on the variable frequency drive and the use of magnetic bearings in MOPICO may be found respectively in [1] and [2]. MOPICO itself is described extensively in [3] and [4].

2 DESIGN AND DEVELOPMENT OF THE MOTOR

This chapter will be divided into four parts. First of all a general overview of the design principles of the high speed motor will be given. These principles were applied already in the MOPICO concept and also in other high speed induction generators manufactured by ACEC in the early 80's. Then follows information on the maximum power speed capability curve which may be achieved with this state of the art technology and the choice of the HOFIM target performances will be discussed. Another section will describe the innovative aspects of HOFIM motor design and the last section will present the R & D activities related to this new design.

2.1 Basic design principles for high speed machines

The first and most important choice to be made when designing a high speed motor is the rotor construction. Based on the experience gained from previous prototypes or industrial applications, the conclusion was that for such high speed (8000 to 20000 rpm), high power (1 to 15 MW) machines the best choice is the "classical" squirrel cage laminated induction rotor. This solution leads to the best compromise between good mechanical and electrical properties. Other solutions (solid rotor induction motor or homopolar synchronous motors) offer better mechanical properties but with poor electrical performance. On the other hand wounded synchronous rotors are not compatible with such high speeds and powers.

The main differences between standard and high speed motor designs are related to the high frequency supply, high and variable speed operation and cooling systems. Operation on magnetic bearings may also have an influence on the motor design.

The high frequency supply leads to additional losses in both stator and rotor. In the stator the conductors are divided into individual strips (figure 2) and cross-connected in the end winding area in such a way that each one occupies all the possible locations in the slots (figure 3). This helps to reduce the circulating currents between individual strips in parallel. In the rotor the shape of the rotor bars must be compatible with high frequency harmonics. The round shape, the most simple from the mechanical point of view, does not lead to excessive harmonic losses.

Another factor is to protect the massive parts,such as the casing or the magnetic circuit end plates, against leakage fluxes. This is done by using non-magnetic or copper shielding.

The high speed operation must be compatible with the stress levels in motor components such as laminations and end-rings. This stress level increases with the rotor external diameter, but may be reduced by selecting a smaller inner shaft diameter. A compromise can then be found between a big rotor diameter and a small shaft diameter, but which will limit the admissable rotor length range from the critical speed point of view. The rotor volume, and consequently the available power in relationship with the speed, is a direct consequence of this compromise.

Supplying a motor with a variable frequency drive means also that the motor has to tolerate harmonic losses related to non-sinusoidal wave form from the supply as well as oscillating torques due to non-sinusoidal currents, which must be analyzed through a torsional model of the rotor.

Cooling is also one of the major problems of a high speed motor design. By increasing the speed the volume of the rotor of an electrical machine may be reduced for a specified power, as volume is roughly proportional to torque. Unfortunately, the losses in the motor are not reduced in the same proportion. As a consequence by increasing the speed the ratio of losses to dissipating area increases and this requires more and more powerful cooling systems. In the MOPICO concept, where the motor is running in a pressurized enclosure, the cooling is very efficient, at design conditions the heat transfer coefficient by convection may be as much as 17 times higher than with atmospheric air. But, as explained in the introduction, this is not valid for the HOFIM motor, where we have to use an independent cooling, which will be described below. Attention must also be paid to the efficiency of the selected cooling system in variable speed operation and especially in processes where high torques are required at low percentages of nominal speed.

The selection of magnetic bearings for a high speed electrical motor is very important for the rotor design, we must consider that an electrical machine rotor with different materials (shaft, laminations, copper bars, end rings) may present higher thermal unbalance than a compressor rotor for instance. To avoid any overloading of the bearings due to these possible thermal unbalances, the motor rotor is designed to operate below the first shaft bending mode. Thus the stiffness of the rotor must be very high, and this is obviously difficult to achieve with a laminated rotor, especially with two overhung impellers such as the MOPICO.

2.2 Power-speed capability of existing high speed machines

The above considerations explain the mechanism of the power limitation of a high speed motor in relationship with the speed. The speed determines the maximum rotor external diameter in terms of maximum stresses in the laminations and the end rings, and the length of the rotor is then limited by the requirement to operate below the first shaft bending mode. In this way the motor volume is determined and consequently the torque and the power.

Based on the tests and field experience, the limit for a MOPICO design is at 8 MW at 10000 RPM; this maximum power being roughly proportional to the cube root of speed. This limit however applies for a machine with the three main features of MOPICO design : overhung impellers, magnetic bearings and high pressure process gas cooling.

In the HOFIM, the motor is driving a separate compressor and is independently cooled, so two of these three features are changed. The first modification leads to a higher power-speed capability, because removing the overhung impellers increases

significantly the critical speed of the motor; it is then possible to increase the length of the motor. Unfortunately, the second modification reduces the power-speed capability, because an independent cooling system is less efficient than the high pressure process gas cooling.
In addition to these two basic differences, it was decided in the HOFIM development to increase the power-speed capability of our motors by developing a new rotor technology and by using new materials for the rotor components.
The selection of a prototype machine rated at 2 MW and 20000 rpm was made in order to build a HOFIM of a size suitable for proving the industrial feasibility of the technology but remaining within a reasonable range of speed and power, and thus financial outlay.

2.3 HOFIM motor design

The HOFIM motor includes all the features mentioned above in the basic design principles for high speed motors with two major exceptions, the rotor assembly technology and the cooling system.
In the preceding machines, as explained before, the rotor construction is nearly standard with a rigid shaft on which the laminations and the end rings are shrink-fitted and copper alloy round bars to form the squirrel cage with the end rings.
Increasing the power-speed capability of such a motor demands to increase the external rotor diameter. The corresponding rise in the peripheral speed and in the stresses in the laminations is compensated by the use of new materials for the laminations and the replacement of the central shaft by multiple straighteners located at a diameter slightly lower than the bars. The elimination of the central shrinkfit on the shaft reduces significantly the stresses in the lamination and the end-rings.
Concerning the main materials of the rotor, lamination and end rings, a difficult compromise has to be made between mechanical properties and magnetical or electrical properties. Materials with superior mechanical properties may be found but they offer high magnetic losses for the laminations or high ohmic losses for the end rings. Furthermore these mechanical properties are often related to the size of the component and to the operating or assembly temperature.
The cooling solution finally selected is an internal closed loop system with pressurized helium. This gas has been chosen for its good heat transfer properties combined with a low specific weight, thus reducing the windage losses. In addition this inert gas simplifies the safety aspects. The closed loop is constituted by the motor, a water to helium heat exchanger and two pipes between the motor and the heat exchanger. A high speed fan directly mounted on the shaft circulates the helium in this closed loop.
The internal pressure (22 psig) has been chosen as the best compromise between a good cooling efficiency and a low level of windage losses. At these conditions the use of helium instead of air corresponds to a gain of 1.75 % of efficiency. To minimize the helium leakage a dry gas seal is installed on the shaft coupling-side. The motor is classified as an Exp motor (pressurized motor), prior to any energization a purge using helium is mandatory. During motor operation the internal pressure is regulated by a pressurization system, which will adapt the pressure level in relationship with the operating point and the cooling requirement.

2.4 Research and development programme for the HOFIM motor

Before finalizing the design, several studies were necessary. The critical speed was calculated with a finite element computer code, the problem here being to estimate the exact contribution to the stiffness of the lamination pressed by the straighteners. The solution was found by testing the rotor on conventioal lubricated bearings and by measuring the first bending mode in rotation, this allowing us to calibrate the finite element model.The calculation of the stresses in the laminations and in the end-ring was also made with a finite element code.
The problem of the shielding of the massive metallic parts of the motor was studied with a boundary element program. The motor casing, the stator core end plates and the magnetic bearing cartridges were analyzed in this way.
From the cooling point of view we investigated first a system with a water cooled stator and a rotor with a rotating heat pipe. An experimental study was made on the associated corrosion problem and a theoretical thermal model was developed for both stator and rotor. Although this solution was found to be feasible, it did not lead to a satisfactory optimum and further investigation was decided. Consequently, we finally decided to use the helium cooling system already described above.
The high speed fan was manufactured and separately tested with atmospheric air. Flows and pressure rise as well as efficiency were measured up to 20000 rpm.
Finally, in order to maximise the margin to the first bending mode, the distance between the magnetic bearings had to be minimized. This means that the magnetic bearings must be inserted below the stator end windings. A special shape is therefore necessary for the end windings (Figure 4). A dedicated CAD 3D design programme has been written to study this special type of end winding.

3 SELECTION OF THE COMPRESSOR APPLICATION

From the outset of the HOFIM project it was recognised that short term operation on a test bed, although providing documented evidence of the achievement or non-achievement of the performance goals, would not provide proof of the suitability of the unit for industrial operation. A demonstration period following the test bed trials was thus considered to be essential. However, the size restriction placed upon the prototype machine (for the reasons given in the section on power:speed capability) initially caused some difficulty in the identification of a suitable application. In order to provide a meaningful demonstration effect the application chosen should preferably have the following characteristics:

- typical compressed medium (no exotic gases)
- industrial operating environment (conditions, personnel)
- continuous periods of operation
- operation over a wide range of conditions

A number of possible candidate applications were identified, some of which are:

- nitrogen recycle in air separation duty
- oxygen compression
- gas injection / gas lift
- gas storage

In each of these applications the design performance parameters of 2000 kW and 20000 rpm can well be utilised. After due consideration of such aspects as customer reference requirements, range of duty, etc., the application finally chosen for the prototype and demonstration machine was that of gas storage.
High pressure storage of pipeline quality natural gas is becoming increasingly important as a means of coping with seasonal fluctuations of gas consumption as well as optimising gas transport systems. Storage can take place in suitable porous geological formations or in leached out salt dome caverns. It is the latter type of storage duty that has been selected for the HOFIM demonstrator.

4 DESIGN OF THE COMPRESSOR

An application for the first HOFIM unit was found in an existing gas storage facility in the northern part of Germany. In this facility gas is taken from the pipeline during the low consumption period in summer and stored in salt dome caverns at a nominal pressure of 155 bar (2250 psig). Pipeline pressures vary between 40 and 60 bar with a nominal condition of 50 bar. At the nominal pressure conditions and nominal motor rating, the thermodynamic design of the compressor resulted in a six stage unit with impeller diameters of 212 mm and 224 mm. Discharge flow from the compressor is 38000 NM3/h at nominal conditions, corresponding to an inlet flow of about 0,2 M3/s.

4.1 Stage design

The non-dimensional flow coefficients for the six impellers range from 0,018 at the inlet stage to 0,010 at the final stage, where non-dimensional flow coefficient phi is,

$$\Phi = \frac{V}{(U \times D^2)}$$

where, U = impeller tip speed
D = impeller tip diameter
V = impeller inlet volume flow

Impellers with such low flow coefficients have inherently low efficiency due to the relatively high losses caused by friction in the blade channels, leakage across labyrinth seals and windage of the hub and shroud discs. Thus careful optimisation of impeller geometry is necessary in order to achieve acceptable efficiency levels. For applications such as that described in this paper a family of impellers has been designed and tested [5]. A typical geometry of an impeller from this family is shown in figure 5. The figure shows a meridional section through a typical low flow coefficient stage,where,

1 = inlet guide vanes
2 = impeller
3 = vaned diffuser
4 = return channel
5 = shroud disc
6 = hub disc
7 = labyrinth seals
8 = de-swirl vanes

This type of impeller features relatively wide flow channels for a given flow coefficient, a factor which allows more accurate manufacture of small impellers, such as those used for the prototype machine, and also reduces the tendency to fouling. The latter point, while not being significant for the gas storage application, will be important for future applications with wet and dirty gas. Further information on the development of this type of impeller can be found in [5].

4.2 Integration of magnetic bearings and dry gas seals into the design

The absence of lubricating oil and the requirement for maximum possible efficiency dictate the use of dry gas seals to minimise the loss of gas to atmosphere at the drive end of the compressor. At the non-drive end it would be possible to install the magnetic bearings in the suction pressure environment existing in this region, however this leads to losses in power due to

- windage loss of the radial and particularly the axial bearing
- parasitic loss of the additional cooling gas required to remove the heat generated by windage

In order to reduce these losses it was decided to install the bearings outside the pressurised zone. At the drive end this is relatively easy, however at the non-drive end a low pressure zone must be created. This is done by means of a further dry gas seal, the leakage from which acts as cooling gas for the magnetic bearings. In the low pressure thus created the power loss due to windage of the bearings becomes almost negligible. The final arrangement of the rotor/bearing/seal system can be seen in figure 6. In arriving at this arrangement a number of factors in addition to the windage considerations had to taken into account,

- position of the radial magnetic bearings and their displacement sensors relative to the mode shapes of the rotor. In contrast to the HOFIM motor and also the MOPICO concept, the compression machine of HOFIM operates above the first bending critical speed. Thus the task of avoiding destabilizing effects due to unfavorable relative locations of nodes and bearings and sensors becomes much more difficult (see also ref.[2]).
- position and capacity of the radial and axial auxiliary landing bearings. Again in contrast to Mopico [2] the relative size of the radial and axial loads during a landing are quite different, in the case of HOFIM the axial loads are much higher. For this reason the axial landing support function has been separated.
- heat removal path from the bearings and seals. Seal flow alone may be insufficient to maintain acceptable temperatures at the non-drive end, for this reason provision has been made to apply water cooling in the casing, if required.
- accessibility to seals, magnetic bearings and auxiliary bearings for maintenance. As far as possible a modular design concept has been chosen and the axial landing bearing- the component with the highest probability of wear - is placed in a location which allows simple and rapid replacement.

4.3 <u>Absorbing axial forces generated in the unit</u>

The motor itself generates relatively small axial forces due to the residual magnetic thrust of the rotor and the thrust generated by the helium circulating fan. These forces are small in comparison to the axial forces generated in the compressor and thus it is logical to absorb all axial forces by means of one bearing installed in the compressor and to connect the two machines by means of an axially rigid coupling. This type of coupling has the additional advantage of considerable lateral flexibility at low overhanging weight, an important factor in developing a dynamically stable motor-compressor rotor system [6]. The axial thrust generated by a high pressure, in-line compressor is however very much higher than that of a back-to-back arrangement, such as the MOPICO concept. It is not feasible to install an axial magnetic bearing which can absorb all of these forces since the restraints of rotor dynamics and material properties do not allow this. Thus some other method must be found to compensate that part of the axial thrust which cannot be absorbed by the magnetic bearing. Normally this function is carried out by means of a balancing piston, however for compressors of the size and pressure level of HOFIM the power losses associated with this device are prohibitive. The arrangement chosen for the HOFIM compressor is an active thrust compensation system in which a digital control valve regulates an intermediate pressure at some point on the shaft sufficient to balance the residual thrust. The dynamic characteristics of this valve must be matched to those of the axial magnetic bearing and to the behaviour of the whole system. With this system the magnetic bearing controls the high frequency variations of thrust, and the control valve those thrust changes of high amplitude (such as surge cycles) as well as the "baseline" thrust. The requirements placed on this control valve are very large dynamic range and very small dead band. The current in the axial magnetic bearing is used to generate the input signal to the control valve.

5 OVERALL DESIGN OF THE UNIT

Figure 7 shows a general view of the complete unit installed on a simple baseframe. This figure gives some idea of the degree of compactness achieved by the use of high speed, direct drive. The overall length of the unit as shown is 4300 mm , including the compressor cartridge insertion rails, the maximum height being 2100 mm from floor level. The total weight of the unit as shown is about 9 tons.

6 TEST PROGRAMME

Prior to installation in the gas storage facility the prototype HOFIM was fully tested on the test bed in Zürich.
In the first part of the test programme the compressor was tested alone using the standard test bed drive motor and two gear boxes. In this first phase the

initial tuning of the compressor magnetic bearings was carried out, both stationary and at various speeds up to about 17000 rpm, the speed limit of the test bed drive system. Performance testing was also carried out using CO_2 gas and at reduced speed. Functional testing of the compressor dry gas seal systems completed this phase.

In the second testing phase the motor and compressor were installed on their common base frame and coupled together. The site transformer and variable speed drive were connected to the motor-compressor unit, thus making up a full scale string test.

The first tests carried out were to determine pressurisation and purging times for the motor when operating in a closed loop circuit with helium as cooling medium. Then further tuning of the four magnetic bearing rotor system was necessary in order to take account of the mutual dynamic influence of the two machines. This, and the previous phase of compressor bearing tuning were of considerable importance since the magnetic bearing system of both machines incorporates a new digital control system. The above involved operating the system over two speed ranges, one below the normal operating range and at the normal range itself. The thermal characteristics of the motor were verified by operating with air and helium as cooling media and at different motor internal pressures. The active thrust compensation system was tested over a range of static and dynamic conditions, including surging of the compressor. The test programme was completed with a full speed, full pressure, full load run using a nitrogen:helium mixture in the compressor loop.

The test programme was completed by the end of January 1993 and results analysed to date indicate that all performance targets of the various system components have been met or exceeded.

7 DEMONSTRATION PROGRAMME

On completion of the test programme the whole HOFIM system was shipped to a site in the north of Germany where it will be installed in an existing gas storage facility. The system will be connected to one of the leached out caverns of the facility and will be required to operate over one complete filling cycle during the period from April to September 1993. During this time the compressor will operate over a wide range of pressure and speed conditions covering most of the operating envelope of the machine, see figure 8. During the demonstration phase an intelligent monitoring system will be used to collect data on the performance of the components of the system. Following successful completion of the demonstration programme the HOFIM system will be released for full industrial operation at the site.

8 CONCLUSION

The HOFIM concept represents a further step in the application of high speed, oil-free technology in the field of turbomachinery. It is to be expected that in the future this concept will find increasing application in certain segments of the turbocompressor market where the features described are of advantage, for example in the offshore area. The experience gained in the test and demonstration phases of the prototype HOFIM system will provide valuable data for further development.

9 ACKNOWLEDGEMENTS

The authors would like to thank their respective companies for the permission to publish this paper and the many colleagues involved in the development of the HOFIM concept, also the partner companies involved in the development, S2M (magnetic bearings), Hill Graham Controls Ltd., (variable frequency drive) and Samtech (non-linear analysis). Finally our thanks to the European Economic Community for partial sponsoring of this project in the framework of the Brite-Euram Program.

10 REFERENCES

1. Graham, D.H., Lattimer, R.H., and Steel, J.G., "Capacitor Compensated Variable Speed Induction Motor Drives", IEE Fourth International conference on Electrical Machines and Drives, pp 339-343.
2. Schmied, J.,"Rotordynamic Aspects of a New Hermetically Sealed Pipeline Compressor", Proceedings of 20th Turbomachinery Symposium, Texas A&M University, (1991).
3. Marriott, A., Ryrie J., Gilon, D., "Kompressor für Gaspipelinestation MOPICO", Technische Rundschau SULZER, 1/1991, pp 30-34.
4. Gilon D., "Design and Tests of a 6 MW, 10000 RPM Induction Motor" IEE Fifth International Conference on Electrical Machines and Drives, pp 6-10.
5. Casey, M.V., Dalbert, P., Schurter, E., "Radial Compressor Stages for Low Flow Coefficients", Fourth European Congress on Fluid Machinery for Oil, Petrochemical and Related Industries, The Hague, (1990).
6. Lorenzen, H., Niedermann, E.A., Wattinger, W., "Solid Couplings with Flexible Intermediate Shafts for High Speed Turbocompressor Trains", Proceedings of 18th Turbomachinery Symposium, Texas A&M University, (1989).

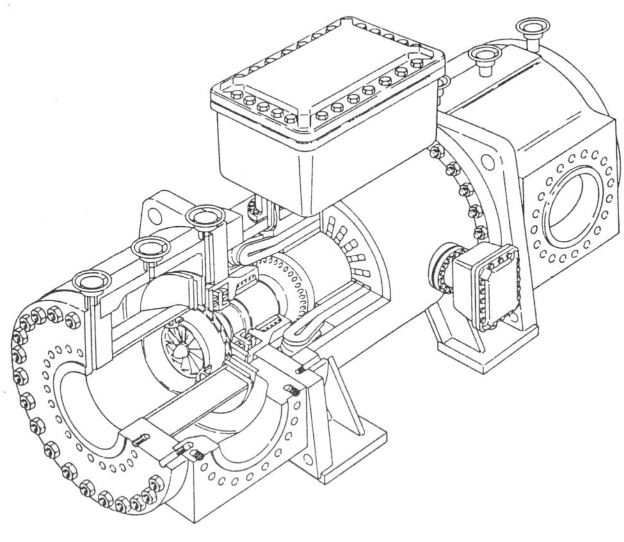

Figure 1 MOPICO pipeline compressor

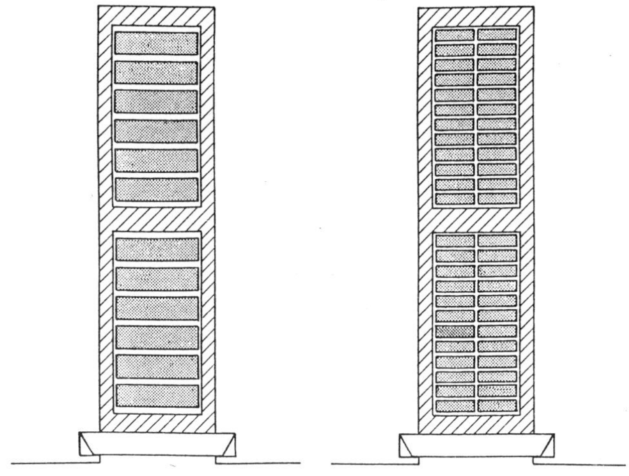

Figure 2 Arrangement of conductors in stator slots, left, conventional : right, HOFIM

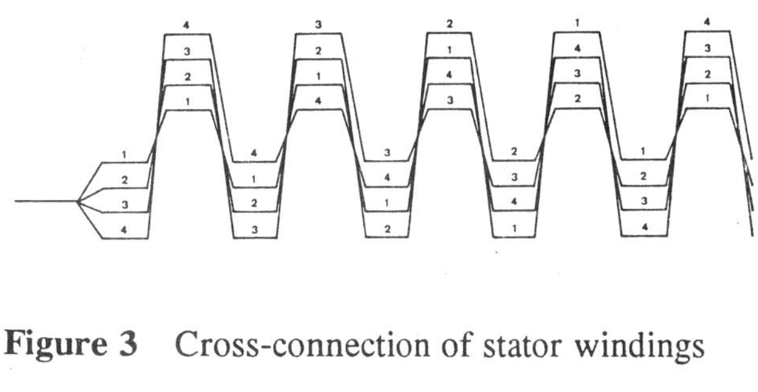

Figure 3 Cross-connection of stator windings

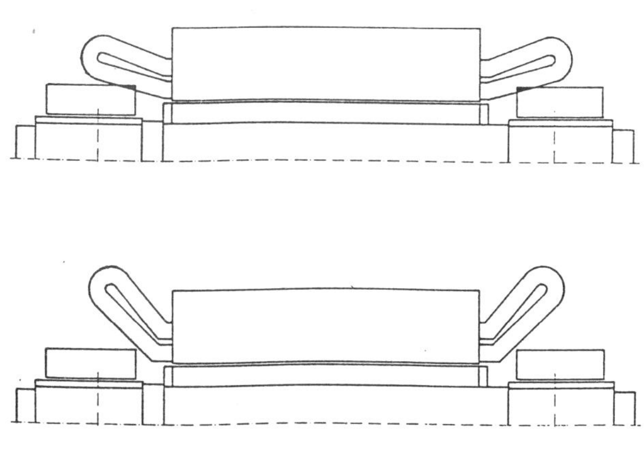

Figure 4 Stator end winding shape, top, MOPICO : bottom, HOFIM

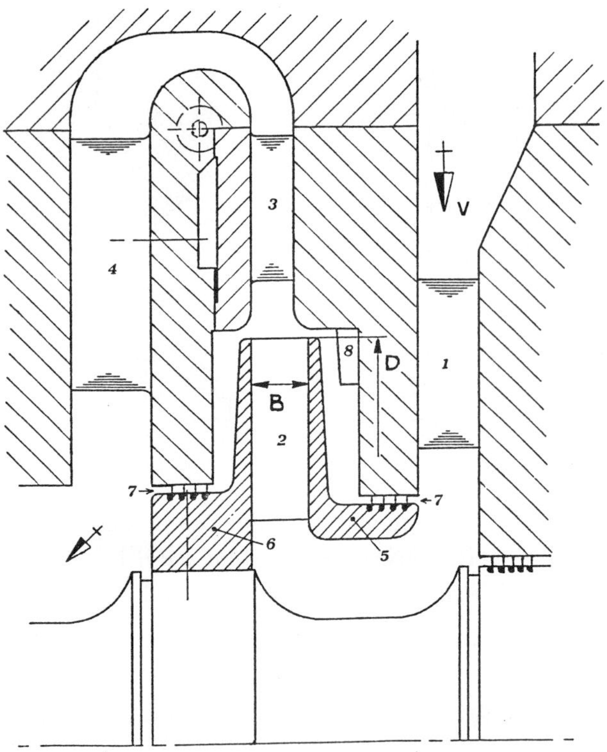

Figure 5 Geometry of low flow coefficient stage

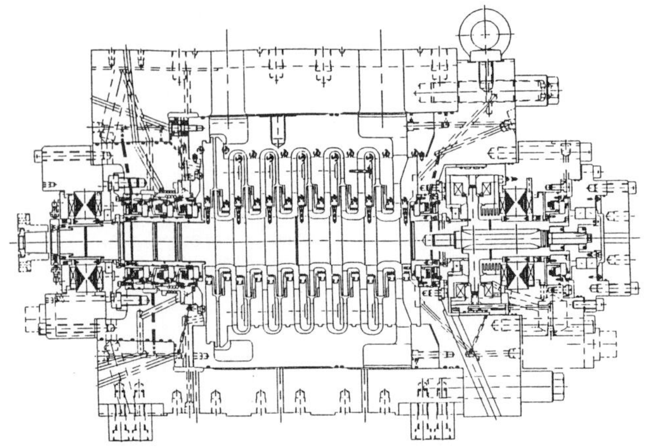

Figure 6 Cross section of compressor

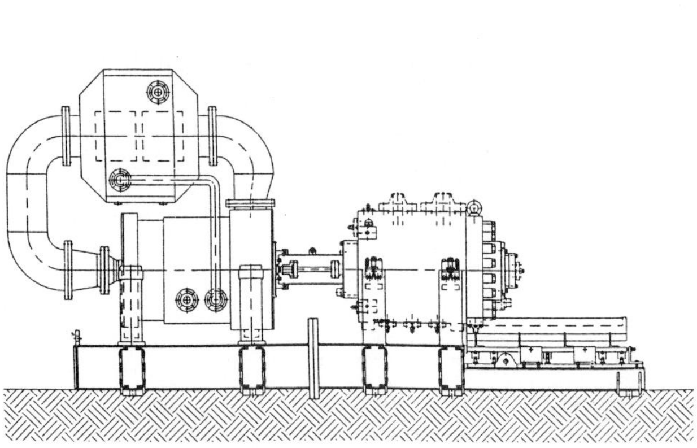

Figure 7 Arrangement of motor-compressor unit

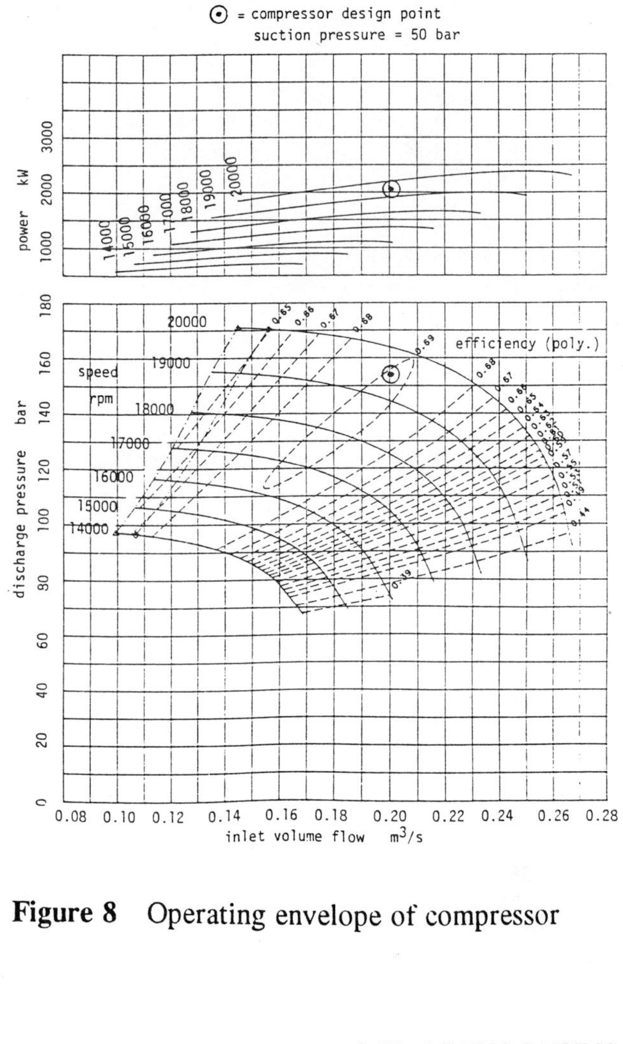

Figure 8 Operating envelope of compressor

ASME PTC-10 Class 1 performance test results correlated with Class III results

T C WILLEMSEN, H J AARNINK, Ir and J G H DERKINK
Delaval Stork, Hengelo, The Netherlands

ABSTRACT

This paper describes a series of Class I and Class III tests according to ASME Power test Code no. 10. The aerodynamic and mechanical test results are discussed, and conclusions and recommendations are given.

INTRODUCTION

The Offshore Industry, through its demands for full load testing of centrifugal compressors, has presented Delaval with the unique opportunity to compare the traditional no-load mechanical and Class III performance tests with full load and Class I performance tests. All of these tests are carried out in the Hengelo test facility.

ASME Power Test Code no. 10 Class I, II and III tests

In an effort to satisfy the needs of both users and manufacturers for an independent standard, the American Society of Mechanical Engineers established the ASME Power Test Code no. 10 (PTC-10) for compressor performance tests. Three classes of tests are defined.

Class I tests include all tests using the specified gas at the speed, inlet conditions and flow for which the compressor is designed to operate in service. The allowable departure from specified operating conditions for Class I testing is indicated in table 1.

Table 1 Allowable departure from specified conditions for class I test.

Variable	Departure (%)	(1)
(a) Inlet pressure	5	(2)
(b) Inlet temperature	8	(2)
(c) Specific gravity of gas	2	(2)
(d) Speed	2	
(e) Capacity	4	

NOTES
(1) Departures are based on the specified value where pressures and temperatures are absolute.
(2) The combined effect of items (a), (b), and (c) shall not produce more than 8 per cent departure in inlet gas density.

Class II and III tests are intended for use when a decision has been made to test the compressor on another gas than that specified for the actual service.
For these tests, methods are provided in PTC-10 to calculate aerodynamically equivalent test conditions. Normally this results in test conditions at lower pressures and powers.

The allowable departure from specified design parameters for Class II and III tests is shown in table 2.

Class II tests differ from Class III tests only in the methods provided for calculating results. Where the gas thermodynamic properties depart from perfect gas laws beyond limits specified by PTC-10, the calculation procedures for Class III testing are used. As the introduction of computers has facilitated the reduction of test data using real gas formulas, Class II tests have in general been replaced in the industry by Class III tests.

Table 2 Allowable departure from specified design parameters for class II and III tests.

Variable	Range of Test Values Limits- % of Design Value	
	Min.	Max.
Volume ratio	95	105
Capacity-speed ratio	96	104
Machine Mach Number		
0 to 0.8	50	105
Above 0.8	95	105
Machine Reynolds Number where the design value is		
Below 200,000 Centrifugal	90	105
Above 200,000 Centrifugal	10 (1)	200

NOTE
(1) Minimum allowable test Machine Reynolds number is 180,000

For all classes of tests the allowable fluctuations of the test readings during a test run are specified by PTC-10, as shown in table 3.

Table 3 Allowable fluctuation of test readings during a test run for all test classes

Measurement	Fluctuation (%) (1)
Inlet pressure	2
Inlet temperature	0.5
Discharge pressure	2
Nozzle differential pressure	2
Nozzle temperature	0.5
Speed	0.5
Torque	1.0
Electric motor input	1.0
Specific gravity test gas	.25

NOTE:
(1) Pressure and temperature fluctuations for the gas as expressed as percent of average absolute value.

A PTC-10 Class I test is essentially a full load test with the specified gas at the speed, inlet pressure and inlet temperature for which the compressor has been designed. However, in most cases it is not practically possible to obtain the actual gas and a mixture is therefore blended with thermodynamic properties that come very close to those of the actual or specified gas.

It is common practice to replace the higher molecular weight hydrocarbon components in a test gas with carbon dioxide. This is to avoid condensation in the test loop.

TEST PROCEDURES

Since conducting our first Class I test in 1986, we have completed seven compressor Class I tests. The compressor model and the test conditions can be found in table 4.

All compressors were class I tested in the HP test loop (as shown in figure 1) and class III tested in the LP test loop. In addition, some of the compressors were also class III tested in the HP test loop.
All Class I tests were executed with Hydro-carbon mixtures and therefore a Safety procedure has been developed to perform the Class I tests in a controlled manner.

figure 1 *Class I high pressure test stand with intercooled compressor package*

Loop filling procedure Class I test

The testgas mixture is blended within the required accuracy and is delivered in bottles by a supplier of industrial gases.

The purity of the gas mixture is 99.5 percent.
These bottles are stored, submerged, in water in containers close to the test stand.
To supply the required gas mixture to the test loop, it is important that the gas mixture in these bottles remains homogeneous and that the heavier components do not condense.
To prevent condensation, bottles in the container are kept at an appropriate temperature.
When filling the test loop with bottled gas, the bottle and feed line temperature is kept well above the temperature at which liquefaction could occur.

Data acquisition system

During the tests automatic data gathering systems are used for acquisition and reduction of both aerodynamic and mechanical data, as shown in figure 2.
The aerodynamic data acquisition system samples all parameters every 15 seconds. Each test point is based on 60 sample readings which are averaged linearly.
Calculations of suction volume flow, polytropic head and efficiency are executed on line.

Warnings are given by the computer system if a certain parameter reaches a pre-determined upper or lower limit. Graphics of each parameter versus time can be generated instantaneously to determine that test conditions have stabilised.

All vibration signals are recorded digitally and scanned with an acquisition system for data reduction. The system generates "Bode" plots, "Nyquist" plots and "Waterfall" diagrams automatically. Pressure pulsations are measured at the compressor stage drains and are processed and analysed.

Calibration

All measuring sensors are calibrated against Measurement Standards which are traceable to the Netherlands Measurement Institute (Nederlands Meetinstituut). For interpolation between the calibrated points best third order polynomial computerised curve fitting is used.

figure 2 *Data acquisition systems for aerodynamic and mechanical data*

TEST RESULTS

Aerodynamic performance

In comparing the Class III and Class I tests, the complete curve shape is considered in addition to the design point. The following parameters of the Class III and Class I tests are compared:

- Head at design flow
- Efficiency at design flow
- Head at surge
- Flow at surge
- Carry-out flow at 85% head.

Table 4 Summary of class I operating conditions

Nr.	Model	Speed [rpm]	Suction pressure [bara]	Discharge pressure [bara]	Mol. Wt.	Flow [kg/h]	Power [kW]	Driver
1	6B22	11953	42.5	121.7	23.98	38032	2080	EL.Motor
2	7B26	12016	8.86	48.84	23.88	77348	6321	EL.Motor
3	5B22	12228	45.42	116.7	21.50	77384	3246	EL.Motor
4	6B22	12779	54.63	151.60	19.15	77784	4774	EL.Motor
5	6BK22	11668	7.72	28.93	34.95	10784	927	EL.Motor
6	4B26	10922	26.0	75.84	26.4	210216	7923	EL.Motor
7	5B22	11965	71.2	180.0	25.2	163926	5098	EL.Motor

Table 5 Summary of class I and class III test data

TEST No	Compressor Type	Test Class (1)	Test Loop (2)	Speed RPM	Test Gas	P1 BARA	P2 BARA	Power KW	Re X 10^6	Equation Of State	Design (3)		Surge (3)		Carry-out At 85% Head (3)
											$\frac{\Delta H\%}{H}$	$\frac{\Delta\eta\%}{\eta}$	$\frac{\Delta H\%}{H}$	$\frac{\Delta Q\%}{Q}$	$\frac{\Delta Q\%}{Q}$
1.1	6B22	I	HP	11953	Nat. Gas	42.50	121.70	2080	8.50	LKP	Ref.	Ref.	Ref.	Ref.	Ref.
1.2		III	HP	5969	Freon 22	3.02	7.06	58	0.80	Dupont	-3.50	+3.20	-3.90	+0.40	-2.0
2.1	7B26	I	HP	12016	Nat. Gas	8.86	48.84	6321	4.90	LKP	Ref.	Ref.	Ref.	Ref.	Ref.
2.2		III	LP	6119	Freon 22	1.25	6.30	423	1.00	Dupont	-2.40	-0.30	-0.50	-0.10	-1.60
2.3		III	LP	6107	Freon 22	1.31	6.67	439	1.10	Dupont	-2.40	-0.70	-0.50	-0.30	-1.60
3.1	5B22	I	HP	12228	Nat. Gas	45.42	116.70	3246	6.20	LKP	Ref.	Ref.	Ref.	Ref.	Ref.
3.2		III	HP	12228	Nat. Gas	12.64	26.96	764	0.63	LKP	-1.90	-0.04	-	-	-
3.3		III	HP	10311	CO_2/H_e	8.16	21.83	499	0.66	LKP	-3.00	-2.10	-2.50	+0.20	-3.00
4.1	6B22	I	HP	12779	Nat. Gas	54.63	151.60	4774	6.90	LKP	Ref.	Ref.	Ref.	Ref.	Ref.
4.2		III	HP	12719	Nat. Gas	12.71	33.30	980	1.70	LKP	+0.90	+2.50	-	-	+2.00
4.3		III	HP	10756	CO_2/H_e	7.71	22.47	550	0.60	LKP	+0.70	+3.70	-1.50	+4.00	-0.20
4.4		III	LP	8059	CO_2	6.40	15.70	300	0.80	Reynolds	-0.10	+4.50	+0.10	+6.70	-3.30
5.1	6BK22	I	HP	11668	Nat. Gas	7.72	14.79	235	1.50	RKS	Ref.	Ref.	Ref.	Ref.	Ref.
5.2		III	HP	11607	CO_2	7.16	15.40	260	1.30	Reynolds	+2.10	+0.20	-	-	+2.70
5.3		I	HP	11668	Nat. Gas	13.67	28.93	692	2.00	RKS	Ref.	Ref.	Ref.	Ref.	Ref.
5.4		III	HP	11607	CO_2	14.10	33.85	895	2.50	Reynolds	-1.50	+1.40	+0.30	+0.30	-0.20
6.1	4B26	I	HP	10866	Nat. Gas	24.40	70.49	7609	12.00	RKS	Ref.	Ref.	Ref.	Ref.	Ref.
6.2		III	LP	9648	CO_2	2.15	6.83	670	1.00	Reynolds	-1.40	+0.00	-2.10	+0.20	-1.20
7.1	4B22	I	HP	11965	Nat. Gas	70.30	180.3	5255	12.4	RKS	Ref.	Ref.	Ref.	Ref.	Ref.
7.2		III	HP	11898	$CO_2/N_2/He$	73.2	200.3	5885	10.9	RKS	-2.9	-1.2	-0.9	+1.9	-2.8
7.3		III	LP	9015	CO_2	5.56	12.56	268	1.12	Reynolds	-2.9	+0.2	-1.4	+0.9	+0.1
										AVERAGE	-1.4	+0.90	-1.30	+1.40	-0.90

NOTES

(1) Testclass in accordance with ASME PTC -10-, 1965.

(2) LP is low pressure permanently installed test loop.
HP is high pressure test loop.

(3) The relative differences are calculated using the computation: $\frac{ResultClassIII - ResultClassI}{ResultClassI} \times 100\%$

Specifically we have investigated the relative differences between the results of Class III and Class I tests. The definition of these differences is shown in figure 3.
The results of all Class I and Class III tests performed on the compressors as listed in table 4 are summarised in table 5. The last 5 columns of this table show the relative differences in head, flow and efficiency between classes.
The average differences are shown at the bottom of each of these columns, and can be summarised as follows:

- The head at design flow in the Class III tests is in average 1.4% lower than in the Class I tests.

- The efficiency at design flow in the Class III tests is in average 0.9% higher than in the Class I tests.

- The head at surge in the Class III tests is in average 1.3% lower than in the Class I tests, which is almost the same as the decrease in head at design flow.

- The surge flow in the Class III test is in average 1.4% higher than in the Class I test.

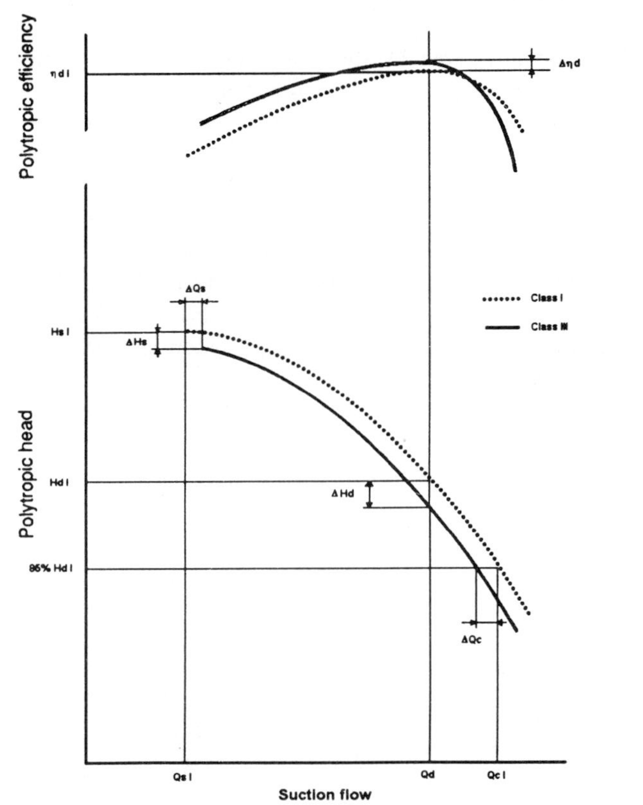

figure 3 *Definition of relative differences in results of Class I and Class III performance tests*

- The flow at 85% head in the Class III test is in average

- The flow at 85% head in the Class III test is in average 0.9% lower than in the Class I test.
- The test results of the carry-out flow at 85% head conform to the expectations related to the findings for head at design flow and head at surge: the correlation of head between Class I and Class III tests is very close over the complete characteristic curve.

A typical (test nr. 4) head- and efficiency curve as a function of suction flow is presented in figure 4. The match of the head-curves is very close. The efficiency curves show some spread in the test results.

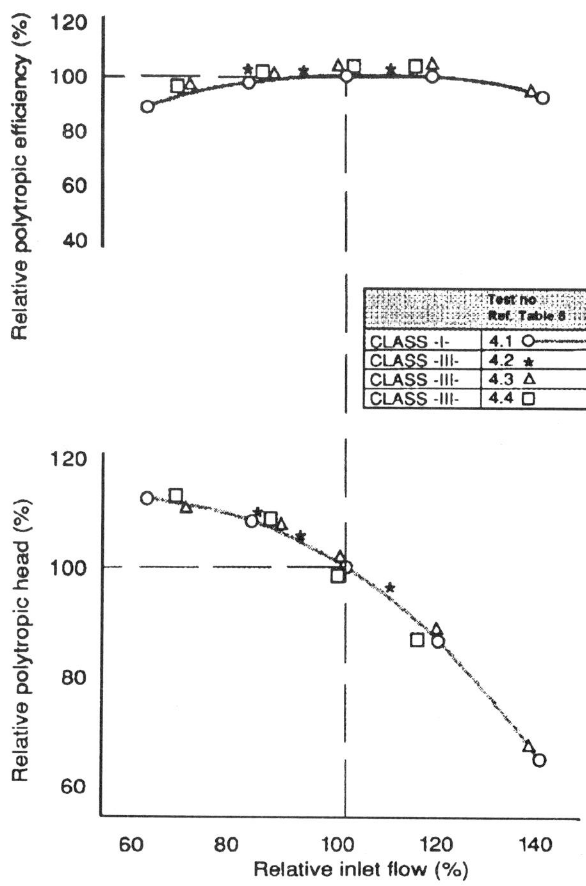

figure 4 *Typical head- and efficiency curve as a function of suction flow showing Class I and Class III performance test results.*

Stability

In addition to testing the aerodynamic performance the Class I compressor test provides the opportunity of investigating the performance of radial and thrust bearings, couplings and seals. Above all the Class I compressor test, at full load and full pressure, enables the verification of the vibrational behaviour and stability of the compressor at specified operating conditions. The no-load mechanical running test and the part load Class III test, although of great value, do not by definition allow investigation of operating characteristics at specified operating conditions.

Vibrations

Vibrations can be categorised as follows:
- synchronous vibrations
- subsynchronous vibrations

Synchronous vibrations
Synchronous vibration is a function of the amount and location of unbalance in the rotor, the bearing characteristics and the separation margin between operating speed and critical speed.

Modern rotor design philosophy, knowledge of bearing characteristics and high speed balancing facilities enable us to build rotors with very low synchronous vibration levels. As can be seen in table 6 the vibration levels are well within the vibration levels allowed by API 617. Synchronous vibrations are forced and in general do not become unstable.

Table 6 Synchronous vibration levels (compressor)

Test no.	Synchronous vibrations in micron peak-peak		
	No-load Mechanical	Class I	API 617 limit
2	DE 5.3 - 8.6 NDE 3.2 - 3.6	DE 8.1 - 8.3 NDE 2.8 - 2.9	30 30
3	DE 6.1 - 7.2 NDE 5.4 - 7.3	DE 2.4 - 3.5 NDE 4.5 - 5.0	30 30
4	DE 3.5 - 4.2 NDE 5.0 - 6.2	DE 2.0 - 2.2 NDE 9.0 - 11.0	29 29
5	DE 5.1 - 5.7 NDE 5.8 - 7.3	DE 2.9 - 3.2 NDE 7.8 - 12.8	31 31

NOTE : DE is Drive End
NDE is Non Drive End

Subsynchronous vibrations due to rotating stall
Flow instabilities inside the compressor, which can occur at low flow off-design conditions can cause vibrations. These instabilities, referred to as rotating stall, can exist in the impeller or the diffuser or both. Stall cells rotate at a frequency which is generally between 5% and 20% of synchronous speed as shown in figure 5. Rotor vibrations due to rotating stall may fluctuate in level, however they are forced vibrations and do not lead to rotor instability.
The mean vibration level is a function of the average density, or in other words the average pressure in the compressor. Tests in our single stage test compressor confirm that the vibration level is directly proportional to the pressure in the compressor, as shown in figure 6. Customers often find it preferable to require the vendor to demonstrate proper operation of the high pressure compressor by subjecting it to a Class I test.

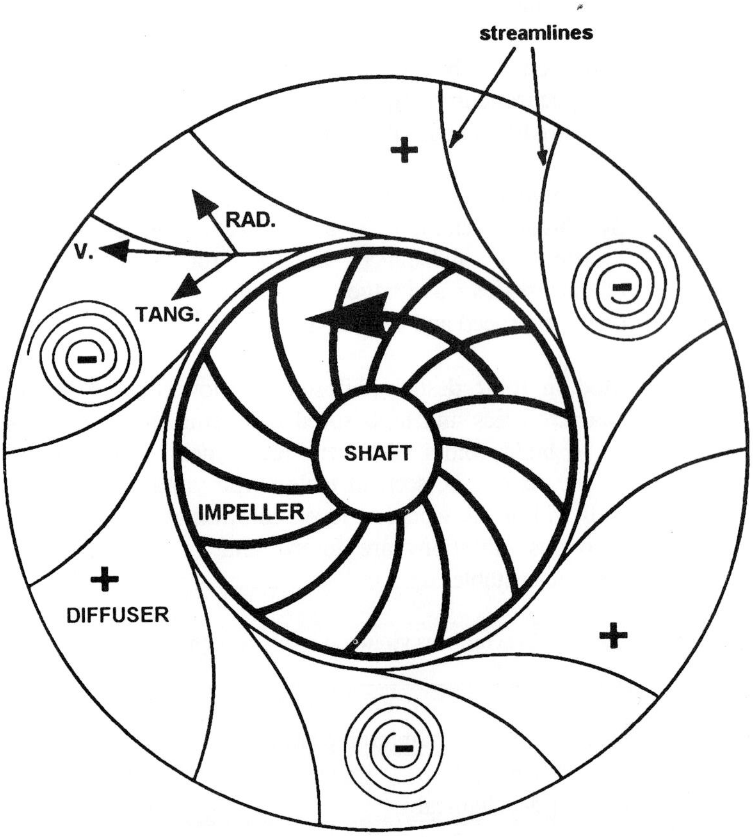

figure 5 *Rotating stall cells*

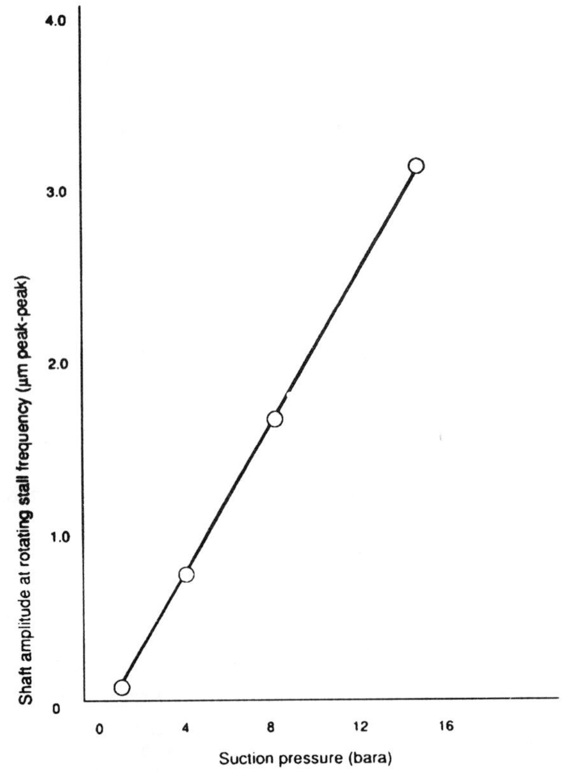

figure 6 *Shaft amplitude as function of compressor suction pressure at rotating stall frequency.*

Subsynchronous vibrations due to aero induced whirl

Aerodynamic forces exerted by swirling gas in the labyrinths of shafts, impellers and balance drums can cause the rotor to vibrate at a frequency typically 40-49% of running speed. If the frequency of whirl is near a rotor critical speed, this whirl will tend to lock on to the resonance frequency and thus cause rotor whip. This vibration, aero induced whirl, can become unstable.

The occurrence of this vibration increases with the average density of the gas in the compressor. Also other factors, such as pressure differential, the amount of gas swirl, labyrinth geometry and diameter, play a role in this process.
Subsynchronous vibrations due to rotating stall and aero-induced whirl are strongly influenced by the average density of the gas in the compressor.
The Class I test offers further opportunity to examine the occurrence of subsynchronous vibrations.

Vibration levels

To illustrate the three types of synchronous and subsynchronous vibrations discussed, figure 7 shows results from a full load mechanical test on a 30,000 kW compressor running at low flow off-design conditions, before and after modifications.

Vibrations are evident at the following frequencies:
- 7.5 % of synchronous (rotating stall)
- First critical (aero induced whirl)
- Synchronous / Harmonics

Figure 8 shows the Non Drive End vibration level of compressor no. 3 running close to surge during the full load test, at a discharge pressure of 126 bara.
The synchronous vibration is 6 micron peak-to-peak. There is no aero induced whirl and the vibration level at rotating stall frequency (8% of synchronous speed) is only 1 micron peak-to-peak.

DISCUSSION OF TEST RESULTS

As part of the discussions of test results, first the accuracy for the Class I and Class III performance tests need to be established.
This accuracy stems from:
- The applied equations of state
- The measuring accuracy
- The influence of Reynolds number

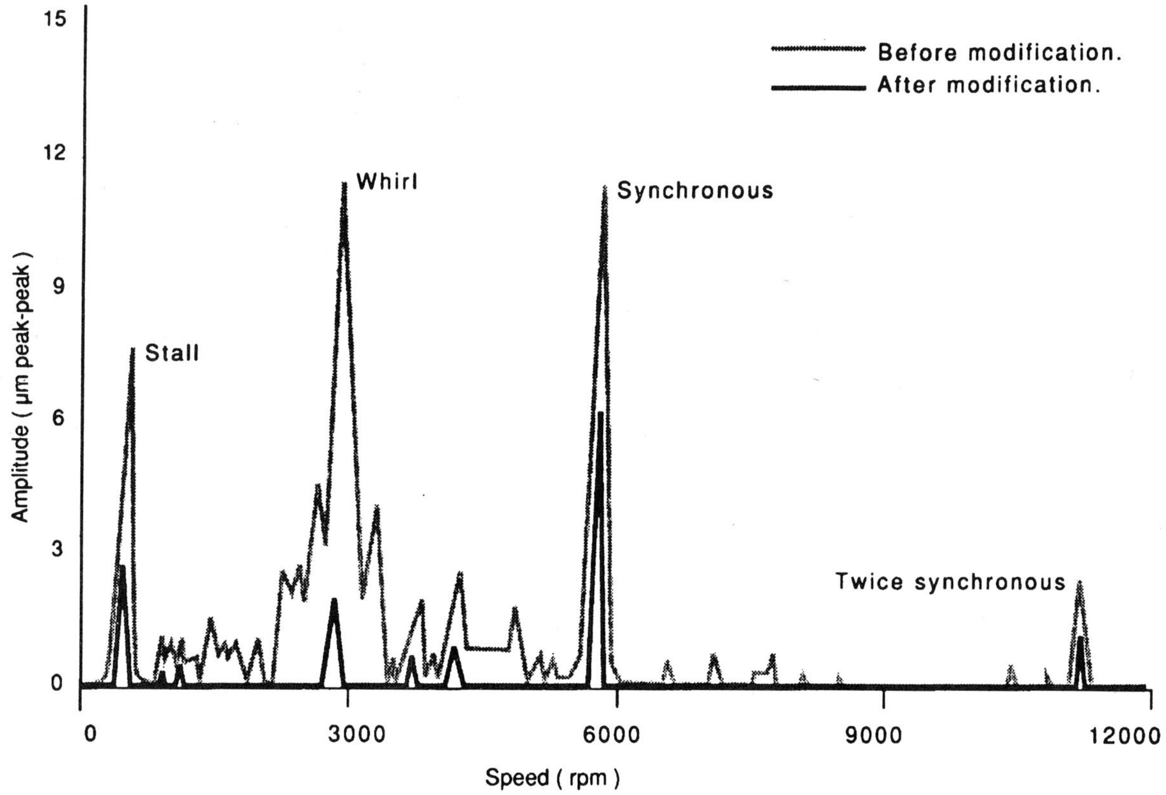

figure 7 *Frequency spectrum of a 30.000 KW compressor during full load mechanical running test operation at low flow off-design conditions, before and after modifications.*

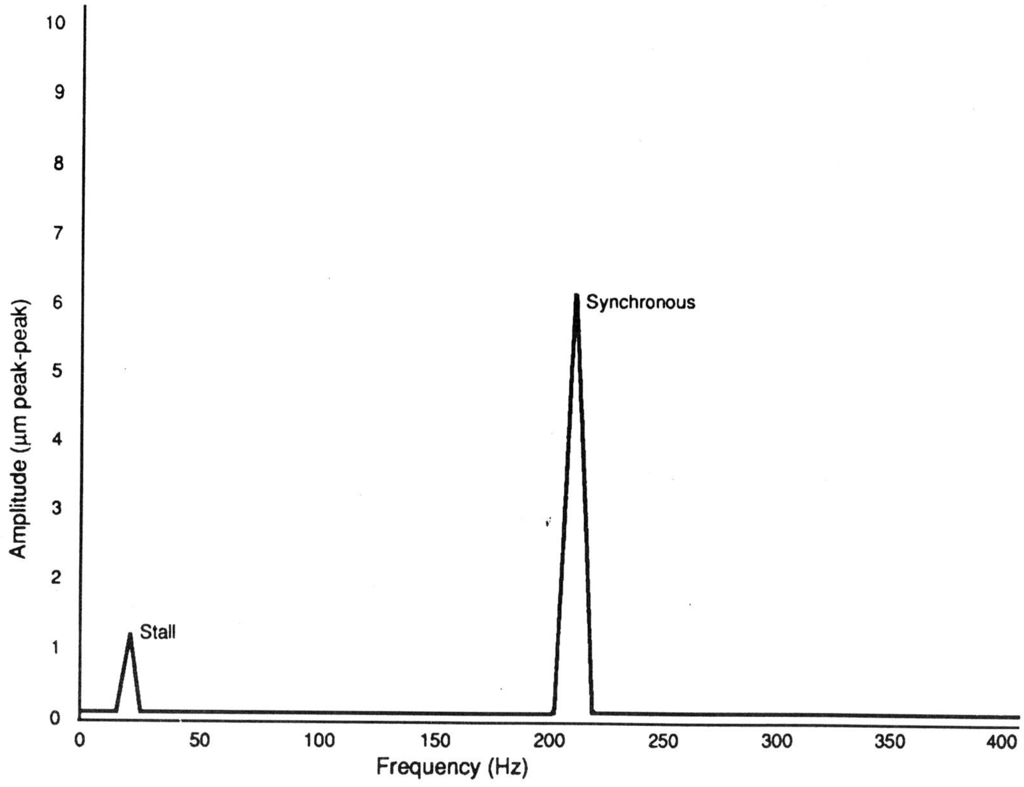

figure 8 *Frequency spectrum at the Non Drive End bearing of compressor no. 3 during full load Class I test at 126 bar discharge pressure and operating close to surge.*

Effect of the choice of the applied equations of state

The head and efficiency calculated starting with the same set of physical test data will vary with the selected type of equation of state.
The differences resulting from the applied equation of state for the test gases, freon 22, CO_2-Helium mixture, CO_2 and natural gas mixture can amount to 1.66% difference in head and 1.65% in efficiency. As an example the variations in computed head and computed efficiency for the class I test nr. 3 are shown in table 7, as a result of using various equations of state.
As customers sometimes require the use of a particular equitation of state, consideration of this effect should not be ignored.

Table 7 Influence of equations of state on head and efficiency for test no. 3.

Equation of state	Δ H %	Δ Eff. %
SRK Chemshare ®	-0.7	-0.2
LKP Chemshare ®	-1.1	-1.2
RSK API-Soave	+0.2	-0.3
LKP Delaval	-1.0	-0.1
RSK Delaval	Ref.	Ref.
PR Delaval	-3.5	-0.75
BWR-Joffe	+0.1	-3.8

Measuring accuracies

The measuring accuracies in the Class III LP test loop and the Class I HP test loop are referred to in table 8.

Influence of Reynolds number

The Reynolds number in the Class I test is higher than in the Class III test. Throughout the industry there is a common understanding that both head and efficiency will rise with increasing Reynolds number. Correction methods for head and efficiency have been established by the ICAAMC and PTC-10.
These methods are applicable within certain envelopes which are shown in figure 9. Test number 4 is plotted in these envelopes.
The corrections on head and efficiency for test no. 3.3 relative to test number 3.1, using the ICAAMC and PTC-10 methods are shown in table 9.

Table 8 Combined accuracies for test 3.3.

	Head (%)	Eff. (%)
Equation of state	1.66	1.65
Class III test loop	0.85	1.0
Class I test loop	1.1	1.2
Reynolds number	0.5	1.0
Combined accuracy	2.20	2.48

The correction arrived at applying PTC-10, 1965 is recognized in the industry as unrealistic. The correction factors applying ICAAMC 1990 and the PTC-10 draft 1985 for head are almost equal, but are not for efficiency.
The Class I test results also show an increase in head compared with Class III tests. The correction factor for efficiency is considered valid, but since the test results in this paper do not uniformly support this, further investigations are being conducted in this regard.

The combined accuracies of these major elements are presented in table 8.

Table 9 Reynolds number correction factors for head and efficiency for test 3.3.

	Correction factors (%)	
	Head	Eff
ICAAMC 1990	0.5	1
PTC-10 draft 1985	0.6	0.6
PTC-10 1965	8.0	8.0

The average spread in the test results presented in table 5 is well within the combined accuracy shown in table 8.

Operational security

Seven compressor units were Class I string tested utilising all contract train driver and driven elements, supports and auxiliary systems, including: gearboxes, electric motors, couplings,
3-point supported base plates on the resilient mounts, oil systems and unit control panels. It is evident that such an extensive test scope demonstrates all operational parameters, and offers a major security and operational reliability assurance for a complete compressor package.

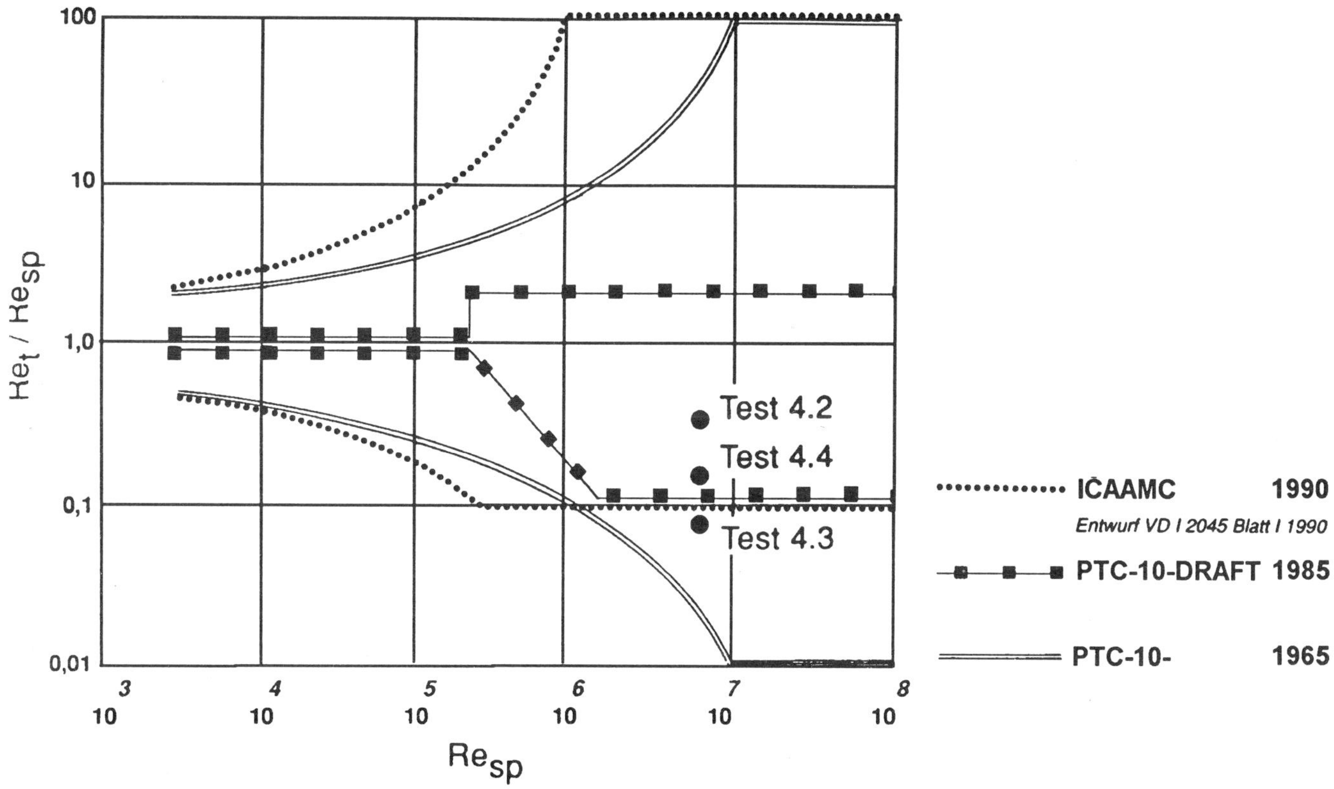

figure 9 *Applicable envelopes for Reynolds correction established by ICAAMC and PTC-10*

CONCLUSIONS AND RECOMMENDATIONS

Aerodynamic performance

Based on the test results the following conclusions are justified:

- The aerodynamic performance results of Class III and Class I tests do not deviate substantially. A Class III test is a true representation of the aerodynamic performance of a compressor under Class I test and specified operating conditions.
- Differences in the chosen equations of state play the most important role in measurement and computed result deviations.
- The influence of Reynolds number on test results is smaller than these contained in PTC-10, 1965.
  This finding supports recommendations included in the ICAAMC and ASME PTC-10 "Draft", 1985 guidelines in this respect.
  Correlation of measured and guideline efficiency corrections require further work to be done.

Mechanical performance

- The synchronous vibration levels are independent of load and are stable.
- Subsynchronous vibrations in compressors can lead to field problems and can sometimes limit operability. Class I testing is a valuable and reliable method of examining the occurrence of subsynchronous vibrations and yields results that are valuable in solving potential problems and deciding on modifications.

Reliability Assurance

For higher pressure compressor units in remote area's or in installations not easily accessible, such as offshore applications, some customers require that Class I shop string tests be carried out at conditions similar to those in actual service. In such tests rotor system integrity can also be checked out under changing temperature conditions. When carrying out Class I testing, it is also recommended that driver and auxiliary equipment should be used.
Sealing and seal control- and safeguarding systems should be demonstrated under service type conditions including start-up and transient modes.

References

1. "Power test code for compressors and exhausters" ASME PTC-10, 1965.
2. "Power test code for compressors and exhausters" ASME PTC-10 "Draft", 1985.
   At time of printing, this document had not been officially released. The data drawn from it may not represent that contained in its final official form.
3. Strub R.A. et al "Influence of the Reynolds number on the performance of centrifugal compressors", October 1987 in Vol. 109 of the Journal of Turbomachinery.

C449/001/93

Reliability improvements carried out on two recycle compressors operating in a polyethylene plant

B V DAKIN, MSc, CEng, MIMechE, FInstPet and B D R McCASKILL
Kemcor Australia, Altona, Victoria, Australia

SYNOPSIS Problems associated with fluidised bed reactor recycle compressors resulted in their very poor reliability, due to the operating conditions being incorrectly specified to the machinery manufacturer.
Investigations revealed that the main areas for concern were:-
Compressor build up; Ineffective labyrinth seals; Inability of the degas system to operate satisfactorily.
These issues resulted in dictating serious major maintenance overhauls every 2 months, which after being addressed and implementing the necessary modifications, brought about an increase in operational availability in excess of 24 months.

1. INTRODUCTION

The reaction section of every high density Polyethylene plant requires a recycle compressor, in order to maintain temperature, by keeping the reactor bed in a fluidised state through recycling unreacted ethylene in a low pass conversion process.

The recycle compressor circulates gasses from the top of the reactor, around a loop containing a heat exchanger, and back into the base of the reactor. (FIG 1)

Gas being circulated is a mixture of Ethylene, Nitrogen, and other co-monomers. Initially gas entered the compressor around 40 deg C. As a result of relocating the process gas cooler, this temperature rose to around 85 deg C. Modifications and improvements later carried out, resulted in the gas temperature rising to approximately 100 deg C.

The gas pressure is approximately 20 bar, and it contains up to 1% of polymer fines by weight. The fines range in size from 50 to 150 microns.

Compressors operating in this service are typically single stage centrifugal type, due to their relative low head and medium flow requirements.

Older designs (pre 1987) tended to be "between bearings" types, with horizontally split casings, employing radial gas inlet and discharge.

Modern high density polyethylene technology now employs recycle compressors of vertically split overhung wheel design, with axial gas inlet and radial discharge.

The two compressors which are the subject of this paper are type 1-H-50.3, and were manufactured by Delaval-Stork in Holland during 1971.

They first entered service in 1972, and were identical in design.

Each operates with it's own reactor, and they are independent of one another. They are located adjacently on the reactor structure for ease of piping and connection of utilities.

2. COMPRESSOR DETAILS (FIG 2)

Each compressor is horizontally split, with each half casing consisting of 3 main components. These are:-

1. A pressure casing, housing the impeller and all the working fluid.
2. A bearing casing, housing the drive end journal bearing and the shaft coupling.
3. A thrust bearing casing, locating the thrust bearing and the non drive end journal bearing.

The two bearing casings are located at each side of the main casing, and are bolted to it through a vertical joint.

Each bearing casing is also split horizontally, thereby providing quick and easy access to the internal components.

The two journal bearings are oil flooded pivot shoe type, and the thrust bearing is a Kingsbury double tilting pad.

Thermocouples are fitted in both the journal and thrust bearing pads. The thrust bearing is located furthest from the impeller.

Adjacent to the journal bearings, and towards the impeller, is a labyrinth seal. This seal is centre purged with Nitrogen, and serves to separate the seal oil from the lubricating oil and to minimise potential cross contamination.

Inboard from the labyrinth seals are the main process gas oil seals. These are the oil lubricated floating ring white metal type, with close running clearances.

The oil fed to these seals has to be of food grade quality, in case of contamination of the recycle gas. Due to this requirement, seal oil costs are high, and close control over seal oil losses is maintained.

Between the floating ring oil seals and the impeller, is another labyrinth seal. This is intended to minimise process gas leakage from the compressor, and as originally supplied, were centre purged radially.

The impeller, located centrally, is of welded steel construction.

Originally the compressors were supplied with impellers of 515 mm diameter. Towards the end of 1986, they were uprated and replaced by larger impellers of 528.5 mm diameter.

A balance piston is shrunk onto the discharge side of the impeller. The bore diameter of this is 154 mm, and the interference fit to the rotor is 0.125 mm. One side of this is connected back to the compressor inlet by means of a balance line pipe.

Initially the compressors were specified at:

Inlet Pressure	18.6 bar absolute.
Inlet Temperature	40 degrees C.
Flow Rate	Approx. 7000 actual cubic metres per hour.
Discharge Pressure	21.7 bar absolute.
Discharge Temperature	54 degrees C.

After uprating and relocation of the recycle cooler, these values changed to :-

Inlet Pressure	18.3 bar absolute.
*Inlet Temperature	85 degrees C.*
**Flow Rate	Increase of approximately 2 fold.
Discharge Pressure	20.5 bar absolute.
*Discharge Temperature	94 degrees C.*

* The rise in operating temperatures was primarily due to relocating the recycle gas cooler from upstream of the compressor to downstream.

** Some of the increased flowrate was due to a general debottlenecking and reduction in recycle loop pressure drop.

Each compressor is driven at 6600 rpm through an epicyclic speed increaser gearbox, from a 1305 kW, 1475 rpm, electric motor.

3. OPERATING PROBLEMS

There has been a history of problems since commissioning.

Initially, until their relocation, polymer fines separated out from the gas in the recycle gas cooler since there was no upstream separation facilities.

The design of the cooler left a lot to be desired, since each was a horizontally arranged air cooled exchanger, with header boxes at both the inlet and outlet ends. These header boxes were arranged in parallel with the tubes at right angles spacing them apart.

Gas would enter and discharge these header boxes vertically, and the flowstream had to turn through several right angles before leaving the cooler.

At each right angle bend, solids would be separated from the process gas stream, and would form on the respective cooler components.

The design also resulted in dead spots being formed at the ends of each header, and these dead spots inevitably packed with fines, which continued until the tubes blocked.

Average cooler life was 40 days between blockages, with 3 to 4 days downtime required to clean. Cleaning was performed by drilling out the headers and tubes to clear the marbleised fines.

In relocating the cooler downstream of the compressor, all fines passed through the machine, for which it was clearly not designed. Subsequently recycle gas at a higher temperature together with fines entered the compressor.

It is of no great surprise that the machine could not cope, since a reaction would eventuate due to the additional heat of compression and friction effects, between the unreacted ethylene and the fines.

Predominantly, the plant handles a variety of catalysts in order to manufacture the various product grades. It is usually the transition between grades whereby problems with equipment design were found wanting.

In 1982, one of the reactor recycle coolers was replaced by a water cooler. The second cooler was replaced in 1987.

The size of the water cooled units is such, that the tube bundles are 12.2 metres long, and each contains over 500 tubes.

Restrictions in the cooler, which still occasionally happened, presented a maintenance headache. The tubes had to be drilled from both ends due to their length, in order to remove the marble like contaminant. The whole exercise involved some 120 hours of round the clock working.

Occasionally, the drills would rupture a tube as they wandered, and depending on the percentage of capped tubes, a retube situation may have been required.

In later years with the advent of high pressure water lances, cleaning could be safely performed non destructively.

4. DETECTION

It should have been possible to detect early problems in the compressor by observing changes in the rotor position, as displayed by the axial proximity probe.

However, previous spurious alarms and shutdowns generated by this instrument, had resulted in a general loss of confidence in the displayed results, to such an extent that the shutdown facilities had been overridden. Consequently this was not utilised as a diagnostic tool, nor were the installed thrust bearing thermocouples.

When solid deposition did occur, it was unfortunately not recognised until the electric motor tripped on overload, and then refused to restart the seized rotor. Attempts to manually turn the rotor by all means also failed.

Later observations revealed that the motor current trace was a reliable indication of problems. This trace went from a relative straight line, to a saw tooth profile, with the peak to peak values increasing as deposition became more severe.

It should also have been possible to confirm rotor change of position, by observing an increase in thrust bearing temperature. However, this instrument was a display type only, without any trending of the results being done. Furthermore, as the readings were not logged by the operating department, only an instantaneous value was available.

5. INTERNAL INSPECTIONS

The system would be depressurised by the operating department, isolated, and made ready for maintenance activities to begin.

After the compressor casings had been removed, an attempt was made to remove the rotor. Often this was seized into the casing by fused or melted fines that were in the areas both behind the impeller back shroud, and the chamber immediately adjacent to, and downstream of, the balance piston.

The tendency was for the fines to plate out onto the compressor casing. These fines would slowly expand with time, accelerating as the increasing mass was able to absorb larger quantities due to the larger surface area.
(FIG 3)

As the fines developed, they would close the gap between the impeller shroud and the casing, and would then start to rub on the impeller shroud. The frictional heat from rubbing, would then fuse the fines and accelerate the rate of increase further.

This contact would result in an axial force being established, which acted in the same direction as the resultant dynamic forces, thereby adding to the load on the thrust bearing.

The fines also tended to propagate radially inwards towards the balance piston, encouraged by the gas flow from the impeller tip through that device back to suction.

Each compressor had occasions, whereby the balance piston was forced along the rotor shaft away from the impeller by these fines, causing extensive damage to the remaining components, and destroying the thrust bearings.

Cleaning of the casings was normally performed using abrasive and/or power tools. The rotor being changed out for the spare rotor in an attempt to minimise the downtime.

Even so, an outage required many days maintenance activities, which then resulted in additional days lost production due to re-establishing a reaction .

The frequency of problems was such, that at its worst, one compressor was undergoing maintenance every 12 weeks.

6. ASSOCIATED PROBLEMS

The labyrinth seal located between the floating ring white metal seal and the impeller, was initially radially purged with Ethylene, the flow dividing axially and sweeping the labyrinth teeth. It operated with a radial clearance of 0.2 mm to 0.25 mm.

The downstream side of this seal was directed to the degas system, whilst the impeller side was at balance piston discharge pressure. Differential pressure was calculated at 0.234 bar.

Flow distribution was calculated for the discharge side labyrinth to be 27 kg/hr towards the impeller, and 71 kg/hr into the degas system. For the suction side labyrinth, flows were 27 kg/hr away from the impeller, and 76 kg/hr towards it.

In practice, each seal flow rate was in excess of 120 kg/hr, with most of the purge flow to the discharge side labyrinth exiting towards the white metal seal, influenced by the pressure difference across the labyrinth.

Frequent complaints were generated by the operating department, of seal oil contamination of the product, excessive seal oil consumption, seal oil loss to the flare header, and fines contamination of the seal oil reservoir.

This latter event being the main reason for the excessive number of seal failures that took place.

7. SUMMARY OF PROBLEMS

Clearly the main problems associated with poor compressor availability can be summarised as follows:-

1) Deposition of solids from the process gas, both behind the impeller and in the balance piston chamber, thereby causing rapid and catastrophic machine failure.
2) High seal wear, excessive seal oil losses, and product contamination caused by ineffective labyrinths.
3) Failure of the degas system to manage the liquid/gas interface correctly.

8. IMPELLER FINES CONTAMINATION

Severe build up of fines in the chamber at the balance piston exit occurred, and it was initially thought that this build up was responsible for restricting the flow through the balance piston, and causing it to block.

Discussions with Delaval-Stork showed that the gas temperature flowing through the balance piston, was sufficient to cause softening of the fines, and form them into a sticky gooey deposit. This occurred between approximately 112 degree C & 130 degrees C, dependent on product type.

Further temperature increases, due to the restricted balance piston flow, then caused the fines to fuse.

It was the deposition of this substance at the contact point of the impeller shroud and the balance piston, that was responsible for generating the high axial forces that pushed off the balance piston.

The solution was to purge the area immediately behind the balance piston. This was completed in 1985, by modifying the existing labyrinth seal to direct a clean cool purge towards the balance piston chamber in an axial direction.

Results were discouraging, and it soon became clear, that an additional purge was required behind the impeller back shroud.

The total available quantity of clean purge gas was then limited to around 1,000 kg/hr for each machine, and the calculated balance piston flow was 1060 kg/hr with a radial clearance of 0.55 mm.

Clearly there was still a nett inflow of process gas and fines into the chamber and across the balance piston.

Initially this problem was addressed by reducing the balance piston clearance to 0.125mm (radial), at the expense of an associated temperature rise, and directing all of the purge gas into the area behind the impeller.

At the reduced clearances, the balance piston flow was calculated at 541 kg/hr.

In November, 1987, two 12.5 mm purge lines were fitted. Each was located in the top and bottom halves of the compressor casings, and arranged to discharge axially towards the impeller back shroud at approximately mid height. (FIG 2).

This method had some degree of success, but failed to completely eliminate the contaminated problem.

Notwithstanding, contamination incidents had now reduced to around 3 (total) per year.

Following several more failures over approximately 7 months, the proposal was then re-evaluated.

A stationary seal ring was then devised, that ran in close proximity to the impeller back shroud. This seal bolted onto the circumference of the modified balance piston labyrinth carrier.

The seal was manufactured from Bronze, such that it would be non sparking (flammable gas), and sacrificial due to it's reduced hardness. It was profiled to the appearance of an orifice, such that the inward or outward flow rates could be calculated. (FIG 4)

In order to maximise the effect of this seal, adjustments were made to the thrust bearing setting, such that with the compressor rotor barred hard towards the seal, a clearance of 0.025mm existed, and with the rotor barred hard the other way, a clearance of 0.279 mm existed.

Ethylene purge gas was boosted to 28 bar, and supplied at a rate in excess of 1,000 kg/hr. Balance piston demand had now increased to 587 kg/hr, due to slight changes in operating conditions.

The total thrust bearing load from the compressor is calculated at 7,550 Newtons, including the loading due to the impeller purge.

The thrust bearing is rated at 12,000 Newtons.

It was an instant success, with all further incidents being attributable to either loss or reduction of clean purge flow, dictated by external factors.

9. CONVERSION TO ABRADABLE SEALS

Having resolved the problems with impeller contamination, the next area of concern was the short life being obtained with the white metal oil seals.

As a result of the initial impeller purging, (no bronze ring), seal life was averaging around 6 months, i.e., two compressors overhauls per year.

Analysis of the problem seemed to indicate that fines contamination of the seal oil, was responsible for most of the failures.

Attention was then focussed on the labyrinth located between the white metal seals and the impeller.

Inspection of the two seals showed that the one on the discharge side was experiencing most of the failures, although both seals would be replaced on each occasion.

On stripping the compressor, a quantity of fines was always observed in the balance piston chamber. Closer inspection showed that these fines were also present in the labyrinth teeth, and a smaller quantity were evident in the area around the impeller side of the seal. Quantities were also found in the degas system.

The initial thoughts were that the purge demands of the impeller were now starving the two labyrinths, and various methods of reducing the total labyrinth purge requirements of nearly 250 kg/hr were sought.

Trials were done with closer clearance seals, and initially the results were encouraging.

However, it was only a matter of time before the clearances increased, due to wear, and then the problems returned.

Consultation was made with Delaval-Stork in July 1988, who designed close clearance abradable seals, utilising a steel carrier with Fluorosint 500 material inserts. (FIG 5)

The first seals were installed during the early part of 1989, with the second compressor following later that same year.

Abradable seal clearance was 0.075 mm radial, and they were designed for a reduced purge gas flow of 50 kg/hr each.

Concern was initially expressed over the durability of these seals to withstand a start up, whereby a great deal of axial rotor reversals were being observed. However, these fears were dismissed following two years satisfactory compressor operation.

Inspection of the seals on removal, failed to identify any measurable wear of the abradable lining.

It is thought that, due to temperature expansions of the respective components, a zero or interference running clearance occurs. This results in plastic flow of the material taking place, such that it has "self healing" properties.

Since installation of the abradable seals in 1989, there has been no recorded failures of the white metal seals, other than one incident whereby the seals were starved of oil during a start up.

10. DEGAS SYSTEM IMPROVEMENTS

10a. Original system (FIG 6)

The system consisted of :-

1. An isolation valve (1), by which means the trap could be isolated from the compressor.
2. A level switch (2), to detect a build up of seal oil in the line leading to the degas trap.
3. The degas trap (3), which separated the sour gas from the seal oil.
4. A sight glass (4), which gave visual indication that the trap was functioning correctly.
5. An isolation valve (5), located on the downstream side, to isolate one trap from the other.
   This was common to both sour gas outlets.
   The two isolating valves (5) were connected to the seal oil reservoir by means of a common line.
6. A cross tie valve (7), located upstream of the first isolating valve, (1), in order that all the sour gas could be routed to one trap, thereby allowing the other to be worked on.
   Gas vented from each trap was linked to a common flare line, through manual valves (8).

   The oil discharge from the white metal seal was directed through a sight glass (6), such that visual evidence of oil flow could be observed.
   A common line linked the two sight glasses back to the seal oil reservoir.

10b. Operating Problems

The degas traps were supplied with a small orifice in the oil discharge port. (FIG 7)

The orifice continued to block with fines, making the traps inoperable.

If this remained undetected during the 8 hourly inspections by the operating department, the drain line from the compressor to the trap would fill with a mixture of oil and fines.

The general unreliability of the level switch located above the trap, inhibited the detection of problems with the system. To assist in problem detection, a second sight glass was fitted above the switch.

Maintenance activities associated with the traps involved between one and two overhauls per week.

The design of the traps, in allowing continuous venting to flare, resulted in excessive losses of process gas. Large volumes of seal oil were also discovered in the flare header.

Initially, the flare connection was isolated, and the manual valve on the downstream side of the trap was shut. Once every shift, the operator would go and blow the traps down to the seal oil reservoir, thereby ensuring that the system remained partially operational. (FIG 6)

Failure to do this would result in the seals running on contaminated seal oil, and failure would result through fines ingression.

Unfortunately, this operating procedure resulted in the gradual build up of fines in the seal oil reservoir, with the obvious consequences, and also large volumes of process gas would be periodically vented to atmosphere through the seal oil reservoir vents.

A secondary problem was the process gas absorption by the seal oil, thereby affecting the lubricity and the flash point.

Commercial considerations in continuously venting approximately 100 kg/hr had to be considered, as did the government body dealing with hydrocarbon emissions.

11. MODIFIED SYSTEM (FIG 8)

The new system includes a degassing vessel with automatic level control. The control system automatically drains the vessel when it is 2/3 full, and maintains the oil/gas interface.

Piping was sized to take account of possible fines contamination.

The design was incorporated during 1988. It was arranged such that, sour gas leaving the compressor would flow through the following components in sequential order.

11a. Degas trap (1)

Each trap has a built in spare, and is arranged with appropriate valves, such that either may be worked on at any one time.

The traps incorporate several novel features, including, a sight glass to visually display the operating liquid level, and a handle that permits manual operation of the float to discharge the trap contents.

The units were designed with contaminants in mind, and were constructed with a 14 mm orifice in place of the original 3 mm design.

Oil flows from the underside of the trap to the seal oil reservoir, through a large degassing vessel (7), and a final trap (8).

The degassing vessel was designed to have 48 hours retention capacity of seal oil to allow full process gas release to take place. The top section of the vessel is connected to the flare header. A level switch and sight glass are connected to the vessel to warn of any liquid concerns.

The gas outlet from the trap (1) is connected to a demister (2).

11b. Demister (2)

This vessel is employed as a flash tank, to facilitate further separation of process gas from the seal oil, by pressure expansion and subsequent boiling off.

The gas outlet from this device is connected back to the compressor suction, through a control valve, which receives a reference signal from a ratio controller set at compressor discharge pressure.

The oil outlet from the demister, flows into a trap (9), and from there to the degassing vessel.

The gas side of this latter trap (9), connects with the gas outlet from the demister. This trap is also arranged with a level gauge.

12. EXPERIENCE TO DATE

Following modification of the two compressor degas systems during 1988/1989, there has been no recorded incidents of system fouling.

Seal oil losses have been minimised, and seal life is now in excess of 24 months.

Maintenance activities on the degas system have reduced to quarterly inspections of the traps. These inspections only reveal trace quantities of polymer fines. Components supplied and installed at the modification, are still in service.

Instrument reliability, (level gauges and switches), has considerably improved, such that confidence in these devices has now been restored.

There have been no further incidents involving seal oil contamination of the flare header.

Since modification during 1988/1989, the process gas emissions have fully complied with all government requirements, and commercial benefits have been obtained.

13. CONCLUSION

The experiences learned at this plant have shown, that it is possible to overcome major deficiencies in machinery wrongly specified, or operating outside of the original specifications as a result of plant upgrade, without incurring expensive modifications or replacement.

Furthermore, it is possible to incorporate such changes effectively in locations remotely situated from the original machinery manufacturer.

14. ACKNOWLEDGMENTS

The authors wish to acknowledge the help and assistance provided by:-

Delaval Stork
The Polymer Corporation

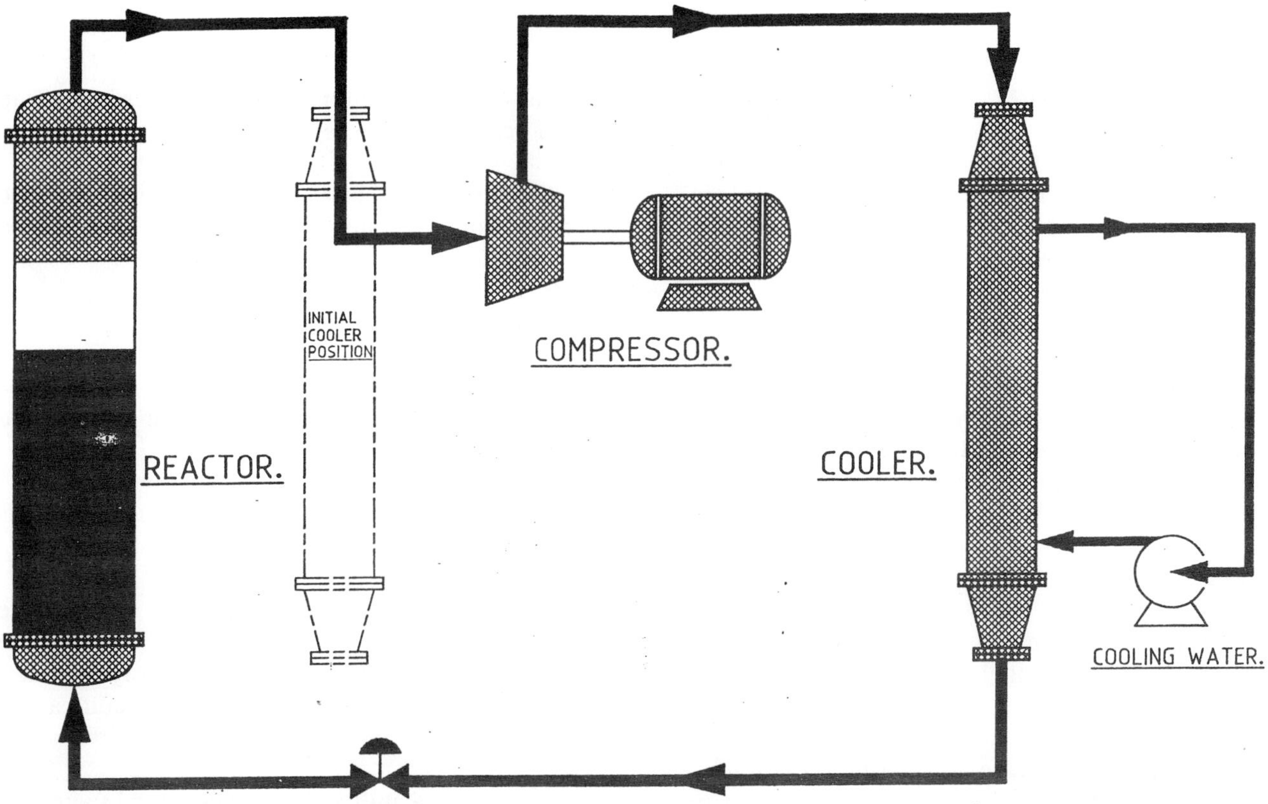

Fig 1 Recycle gas valve

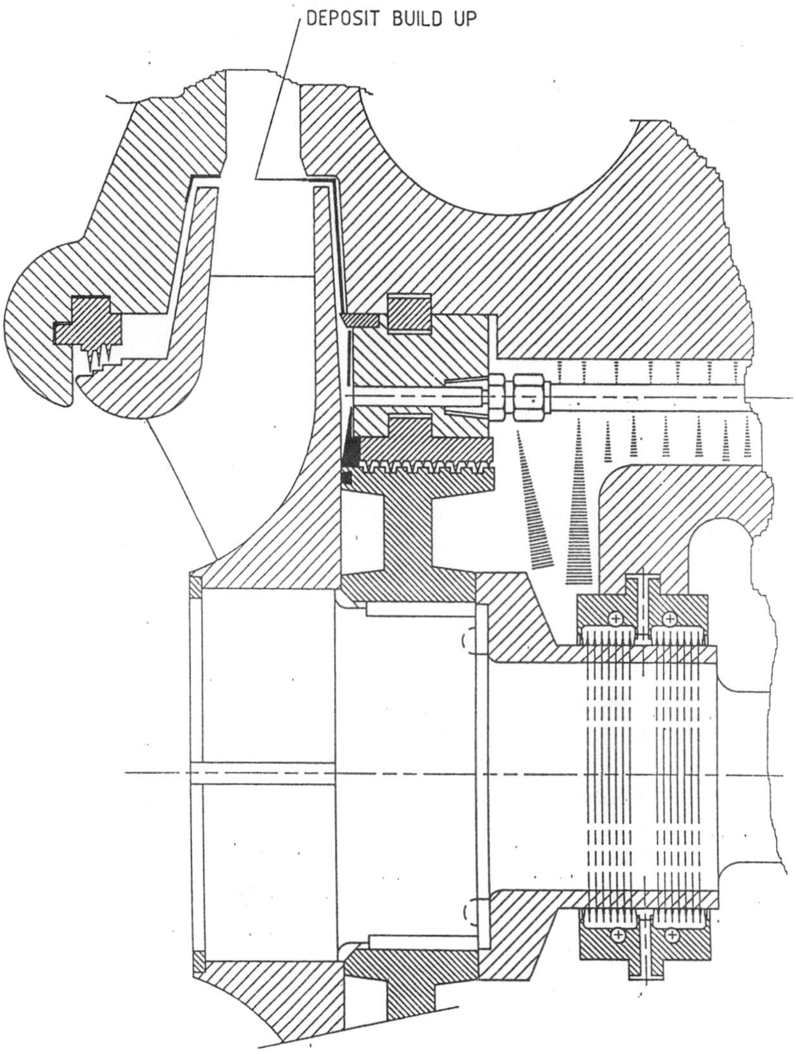

Fig 3

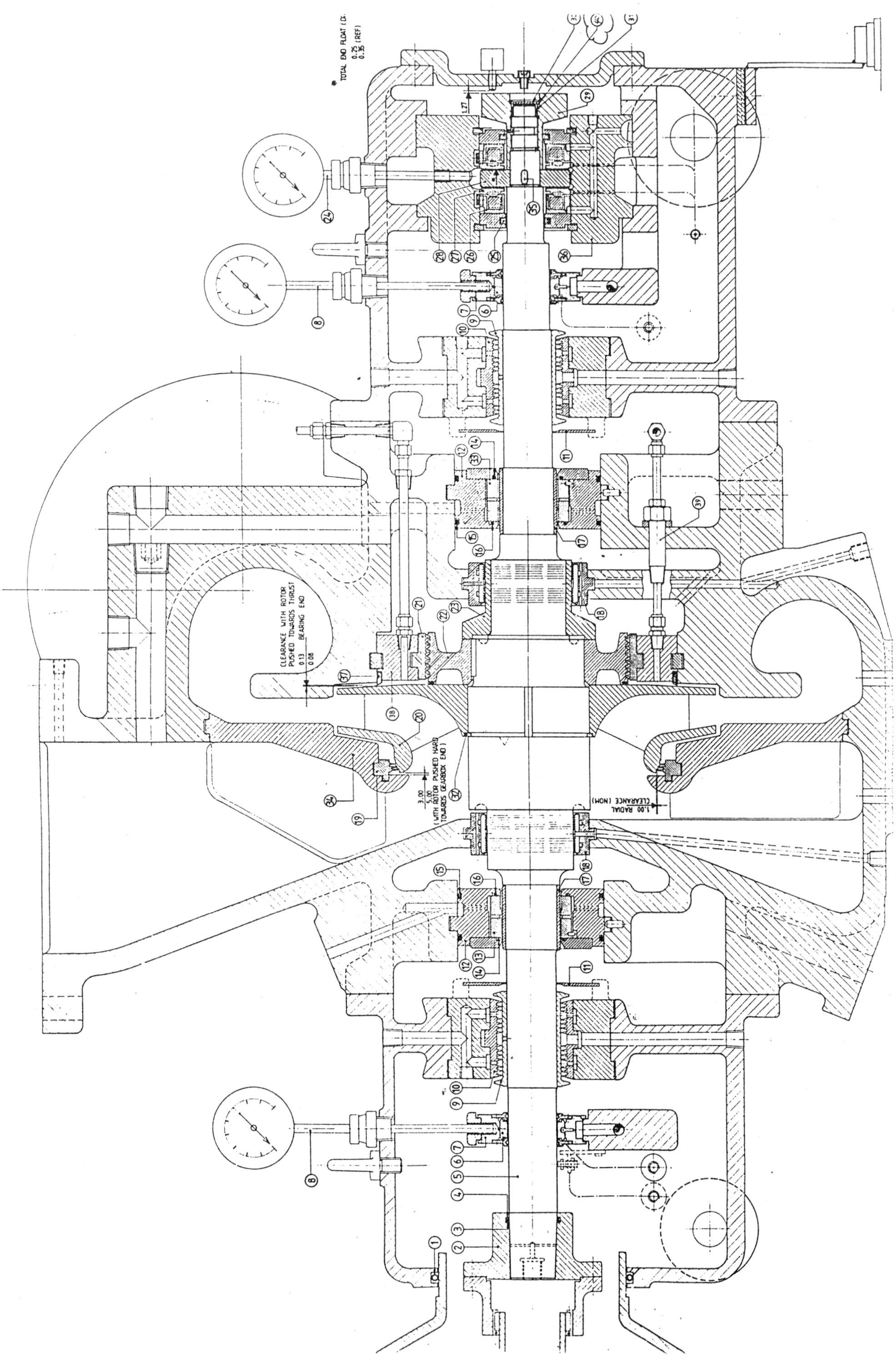
TOTAL END FLOAT
0.25
0.35 (REF)
CLEARANCE WITH ROTOR PUSHED TOWARDS THRUST BEARING END
0.13
0.08
3.00
5.00
(WITH ROTOR PUSHED HARD TOWARDS GEARBOX END)
1.00 RADIAL CLEARANCE (NOM)

Fig 2

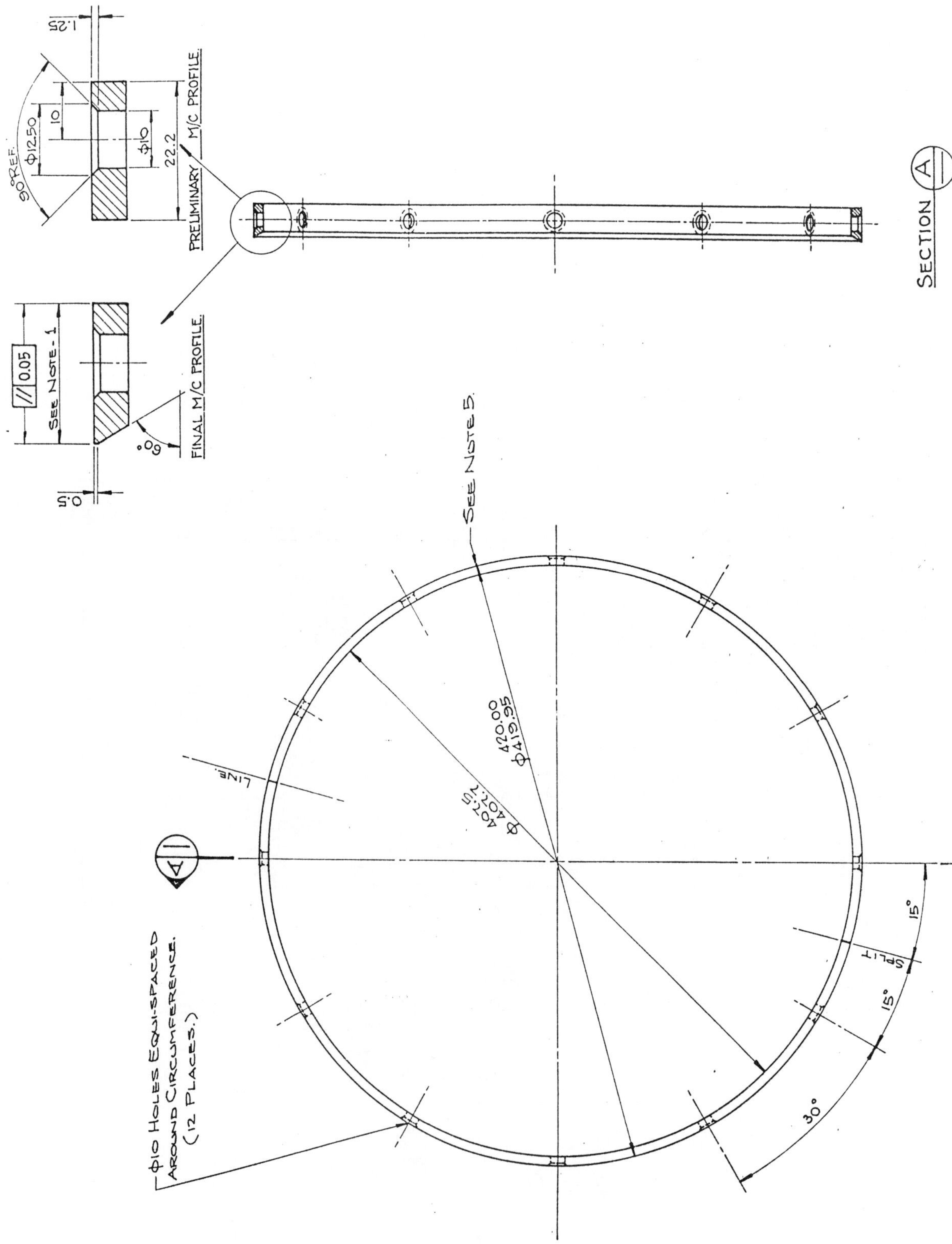
Φ10 HOLES EQUI-SPACED
AROUND CIRCUMFERENCE.
(12 PLACES.)
SEE NOTE 5.
420.00
Φ419.95
LINE.
SPLIT
15°
15°
30°
SECTION A
PRELIMINARY M/C PROFILE.
FINAL M/C PROFILE.
SEE NOTE-1
0.05
60°
0.5
1.25
22.2
Φ10
10
Φ12.50
90° REF.

Fig 4

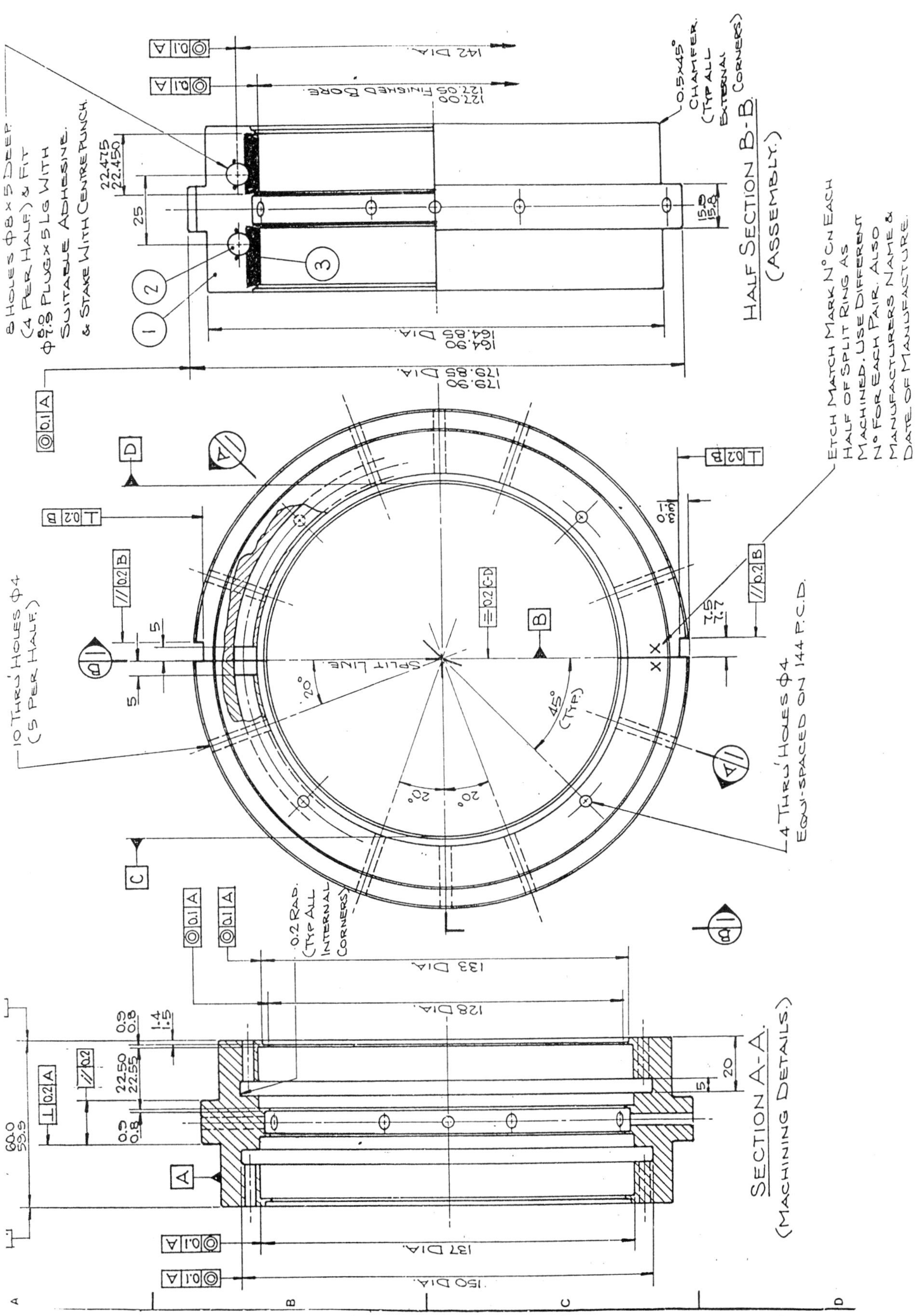
8 Holes φ8 x 5 Deep (4 Per Half) & Fit φ8.0/7.9 Plug x 5 Lg With Suitable Adhesive & Stake With Centre Punch.
142 DIA.
122.05/122.00 Finished Bore.
0.5x45° Chamfer (Typ All External Corners)
22.475/22.450
25
15.9/15.8
164.90/164.85 DIA.
179.90/179.85 DIA.
Half Section B-B. (Assembly.)
Etch Match Mark N° on Each Half of Split Ring As Machined. Use Different N° For Each Pair. Also Manufacturers Name & Date of Manufacture.
10 Thru' Holes φ4 (5 Per Half)
Split Line.
20°
45° (Typ.)
4 Thru' Holes φ4 Equi-spaced on 144 P.C.D.
0.2 Rad. (Typ All Internal Corners)
133 DIA.
128 DIA.
137 DIA.
150 DIA.
60.0/59.9
22.50/22.55
0.9/0.8
1.4/1.5
7.5/7.7
Section A-A. (Machining Details.)

Fig 5

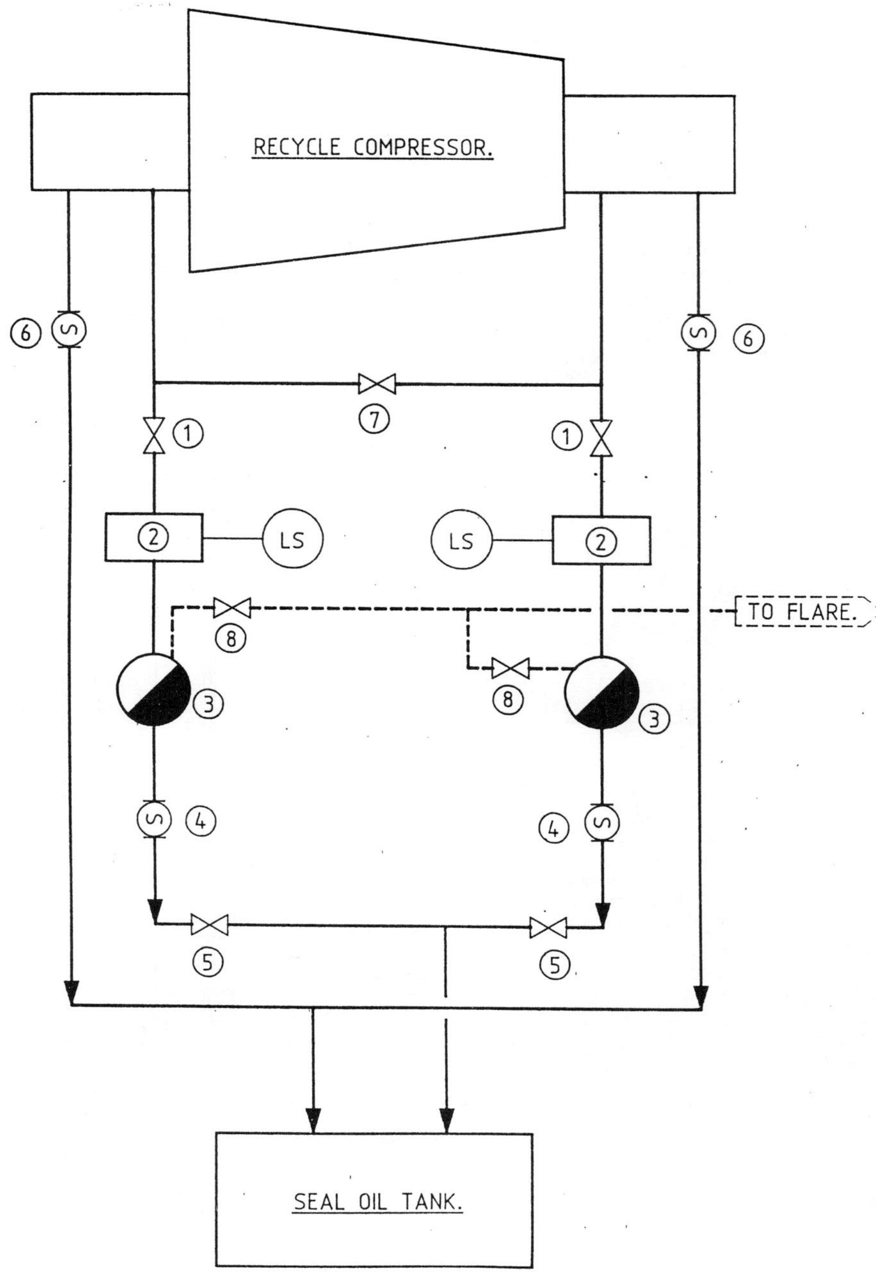

Fig 6 Degas system

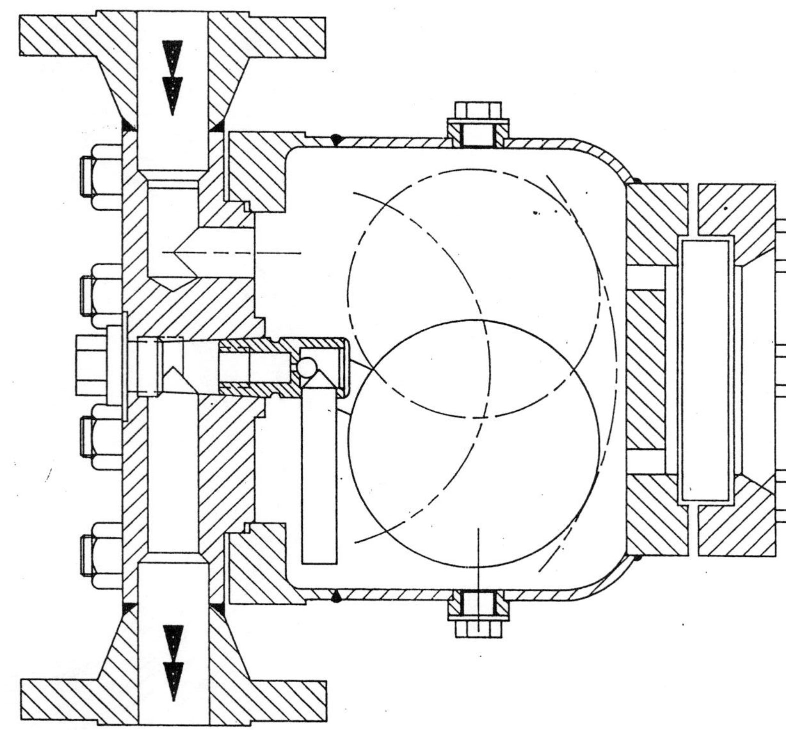
Fig 7 Original degas trap

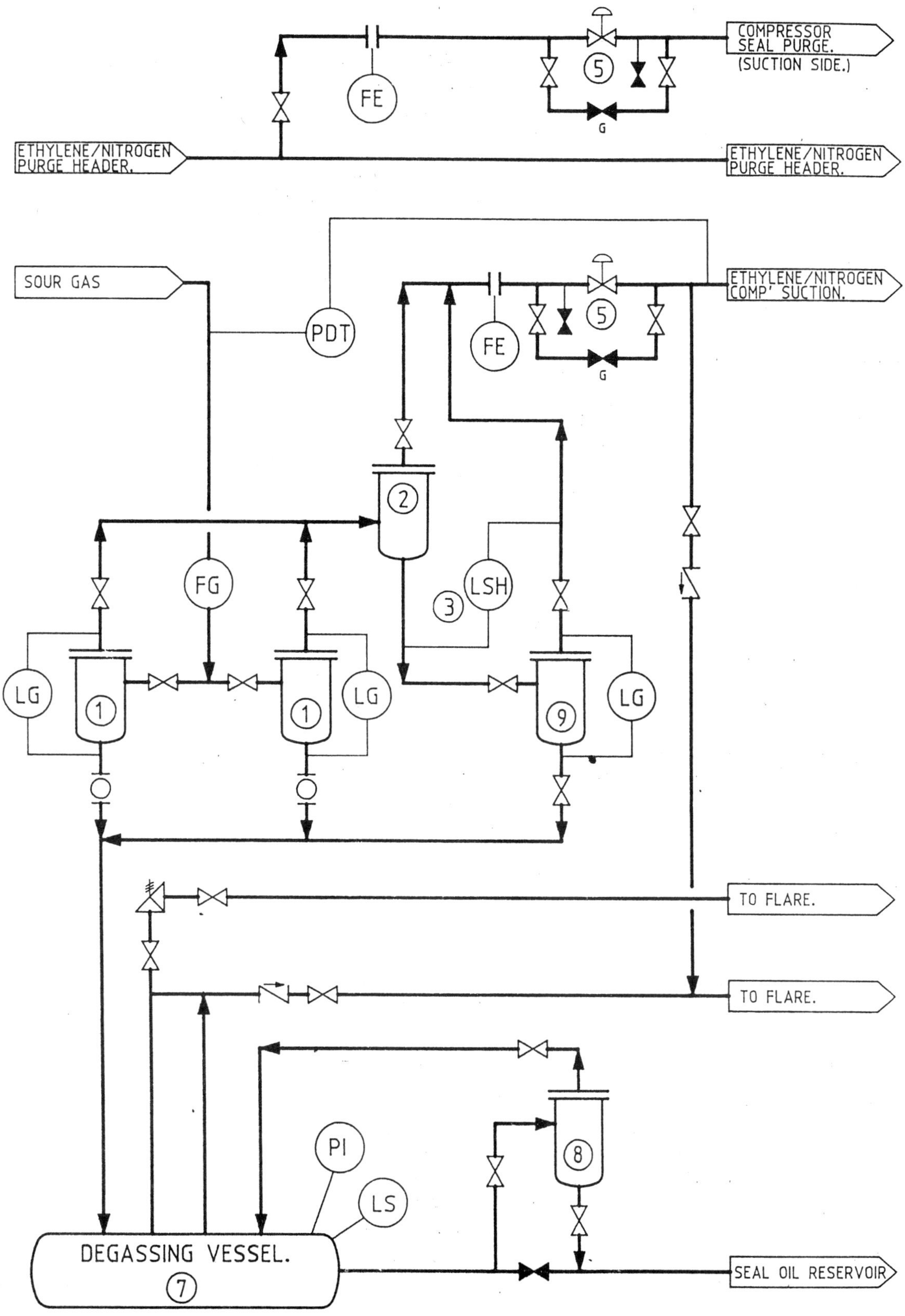

Fig 8 Details of modified system

C449/016/93

Predictive maintenance – a vision of the future

L J RUCKLEY, CEng, FIMechE, MINucE
BP Engineering Limited, Uxbridge, Middlesex

SYNOPSIS: Commercial, business, safety, and environmental pressures require Plant Management to be as efficient, competitive, safety conscious, and aware of the consequences as possible. Success for a Company Management can depend on how well it applies the various improvement tools now available. This means providing it with an early and correct diagnosis of a potential problem coupled with the life expectancy or prognosis of the equipment.

This paper concentrates on how a major Oil Company views the development of current predictive maintenance tools, and how they will inter relate with maintenance and production management systems in the future. Current and potential developments in reliability engineering, machinery and static plant monitoring are discussed, to show how in the future predictive maintenance will link with other information systems to provide a comprehensive Management overview.

INTRODUCTION

Decisions and judgements on any activity are only as good as the information available to make them. This applies equally to buying a new car, or deciding on maintenance activities, needs, and schedules.

Commercial, business, safety, and environmental pressures require Plant Management to be as efficient, competitive, safety conscious, and aware of the consequences as possible. How management approaches this makes the difference between being in the top quartile of the International Oil Industry or not. Success can depend on how well management applies the improvement tools such as maintenance surveys, computerised maintenance management systems, predictive maintenance tools, and productivity improvement programs in a co-ordinated strategy of asset management. The key to all this is providing management with an early and correct diagnosis of a potential problem coupled with the life expectancy or prognosis of the equipment.

This paper concentrates on the development of current predictive maintenance tools and how they will inter relate with maintenance and production management systems in the future. It begins with a brief review of predictive maintenance and its current status. Specific areas of reliability engineering, rotating equipment, and static plant are then examined, followed by progress on some current and potential developments in these fields. Finally there is a foresight of how predictive maintenance will link with other information systems to provide any Company Management with a comprehensive overview, to help it make decisions that will tend to retain it in the upper quartile of Industry.

As an example the developments in major machinery will demonstrate how future developments will enable links to be forged between data collection, prognosis, and maintenance management to provide readily assessable information for decisions to be made by the Senior Maintenance Management.

WHERE ARE WE NOW?

Predictive maintenance is at a very interesting stage of development. A few years ago it consisted of "Condition Monitoring". This incorrectly referred to vibration monitoring techniques as applied to machinery. Nowadays Condition Monitoring refers to any technique which enables the engineer to assess the health of the item of plant and its estimated life cycle.

The design engineer assesses the reliability of plant to enable better and more cost effective designs to be built. The rotating equipment engineer carries out Condition Monitoring in various forms such as vibration, performance, and lub oil analysis to decide on the health of the machinery and its life cycle.

In the design of offshore platforms in particular reliability engineering now assumes a major role. Statistical failure data collected from maintenance management systems and Condition Monitoring and other sources is analysed to determine the likely life of a component in a particular arrangement. From this estimates can be made as to whether the Operator's production targets can be met. Savings result when it is determined that for example a spare water injection pump can be eliminated from the original design. Savings are made not only from the pumpset but also the expensive platform on which it stands. The technique can also indicate the weakest point in the design to enable necessary changes to be made - where necessary.

Data collection on equipment has been limited by the inability of various computer data banks to inter communicate (figure 1). However the OREDA project in Europe has enabled a cross fertilisation of data to take place. Manual analysis of this data has meant that the techniques are limited to a few key items. Even so the potential benefits arising from this technique have assisted in the re-design of safety systems, and the redundancy of expensive capital equipment.

Small machinery which traditionally carries no permanent monitoring, until the last few years has had little attention to minimise its maintenance effort. Bearing and seal replacement was an accepted practice, and a spare pump was provided to maintain the production. This approach has changed dramatically with the advent of portable data collection and monitoring systems (figure 2). Machinery data is now collected on up to about 80% of the small rotating machinery in many refineries and offshore oil production installations on a weekly, 2 weekly, or monthly cycle. In general the other 20% of machines are not considered to be cost effective. The savings in machinery maintenance costs arising from the use of this technique range from about 15% to about 40% of the maintenance costs prior to its introduction - because of the prognosis available. This system enables potential faults to be detected early, cost effective maintenance planning to ensue. Skilled interpretation by engineers is required for this technique.

Major machinery has traditionally over the last 15 years had a large local monitoring panel with a suite of vibration detectors and permanent monitors with built in alarms and trips. The monitors and gauges are usually read by the operator every shift. The benefit to maintenance planning and prognosis has often been marginal since, they are rarely interpreted by the maintenance engineer unless an incident has occurred. Any event in general is therefore a corrective maintenance event - usually at great cost since a loss of production is involved.

Further there is a tendency for these local panels to become ever more complex, and with the addition of omputer based data collection systems it has meant that there is ever more data to analyse - making the work of the maintenance engineer ever more difficult.

Routine monitoring of static plant to provide prognostic information has long been undertaken - BUT using manual techniques. Consider the number of radiographic or ultrasonic thickness measurements Companies take on pipelines or vessels, and how they analyse them in a prognostic fashion manually. This is labour intensive and the amount of analysis possible is limited - by the skilled manpower available.

This then is the current status in many large Oil Companies, and some see the need to change to improve the strategy and prognostic tools in these areas with a view to becoming more efficient. How about You?

CURRENT DEVELOPMENTS

How then is the challenge being approached?

One aspect is a change in philosophy to consider the "Whole Life Cost" of plant equipment. Items are now not selected necessarily for the lowest capital cost, but the operating and maintenance costs over a lifetime are also considered in the selection. This makes the design engineer's and the project engineer's task even more difficult.

Pioneering use has been made of Artificial Intelligence (AI) to assist the Design Engineer in making his decisions regarding reliability. This approach has reduced the design time required in deciding the optimum plant layout for a desired plant availability. It also helps in machinery type selection, as well as process configuration for the specified availability.

Similarly operating cost models are being developed to give running cost data for various process configurations to assist the Project Engineer.

The introduction of a Condition Monitoring strategy for asset management is also a major change being introduced. All plant - and in particular machinery is assessed for its criticality, availability requirement, and maintainability. From this a maintenance strategy including Condition Monitoring is drawn up for each item of equipment. The site's Condition Monitoring strategy then results. Criticality rating techniques and cost benefit modelling techniques for rotating equipment have been developed under differing maintenance regimes. One company has produced a site user's guide to cost effective Condition Monitoring, which details to the site engineer - not necessarily experienced in Condition Monitoring - his probable principal significant faults and machines from a cost aspect, and the optimum monitoring techniques available.

In the small machinery monitoring field there is a move towards more reliable diagnosis, and for the diagnosis to be executed by lower skilled staff, as well as of improved prognosis. To this end AI systems are being developed. BP developed one of the first AI vibration diagnostic systems and the first lub oil AI diagnostic systems (figure 3). Both of these are designed to read the data from commonly used data collection systems directly, and are used by the technicians. The first has the capability to diagnose over 50 common machinery problems on a complete machinery train consisting of several items. More importantly with the open approach taken to data transfer it has the potential of directing the diagnosis directly to a Maintenance Management System to create the work request or to a reliability data base.

The AI system above has been in service at a Swedish refinery for over 2 years, and in a Far Eastern refinery for about 18 months.

Major machinery monitoring as was hinted at earlier is also undergoing a revolution. The day is foreseen in the not so distant future where the large local panel is replaced by the VDU with a small dedicated computer linked to the master process computer (figure 4). There is a trend though to monitor everything in sight using on line computerised data collection techniques. This is considered to be dangerous since there is often too much data to review - even for a computer, and the user is being swamped by excessive amounts of data to analyse.

Instead it is considered essential to forecast the faults - including loss of performance or performance optimisation - determine what parameters will best indicate the particular problems, and finally only monitor these. The processing of algorithms and their prediction is now a part of the equation. This approach minimises the data collection requirements to essential data and enables data processing to be concentrated. Current developments are aimed at optimising the operating point of a major machinery train such as a gas turbine driven compressor set on its combined performance map, and minimise the operating cost as well as its maintenance cost. In addition it enables the operator to better operate the machinery within its design envelope and thus reduce the maintenance costs.

This is only part of the story in the drive for better prognosis and better use of resources in the maintenance management picture. The next stage - already designed - is a modular AI system to carry out this prognosis and diagnosis ON LINE for the operator (figure 5). Initially it will be an advisory system but later in a control mode if required. This system will routinely scan the process data, check it for alarms, and if one is detected collect a data set for analysis. The AI system will then carry out a diagnosis using this data, and the historic data in its data bank, and the machinery history available from the Maintenance Management System prior to coming to as conclusion. This will then be presented to the operator in plain language such as "Increase the cooling water to the Lub oil cooler because the bearing temperatures are rising" or " The Inlet Air filter is getting dirty. A work order should be raised for it to be done in the next 2 weeks". It will also be possible for the maintenance engineer to interrogate the system for a more detailed analysis of a current or past event. The modular design will mean that the system can be applied to any machinery train merely by reconfiguration.

Static plant predictive maintenance techniques have moved at a slower rate, but it is forecast that they will follow the approach adopted for small machinery. Since commercial data collection and monitoring systems are not generally available for static plant BP has developed one suitable for various applications depending on its configuration. An intrinsically safe portable data collector is used. Applications such as operational log taking, lub oil sample trending, and the monitoring of service routines can be undertaken with it.

It has also been in the forefront of pioneering the use of portable data collection and computer based analysis techniques for ultrasonic corrosion thickness monitoring and for predicting the plant life remaining (figure 6). This system is now in service in some 37 locations world wide.

A recent review of the system has taken into account the knowledge gained in machinery monitoring, the special needs of the inspector, and the recent major changes to statutory requirements in Europe. The result is a system which concentrates the user only on the problems whilst retaining the data on the rest of the plant. This enables the Inspection Engineer to help minimise and target maintenance by providing a means to examine data in logical common groups, and provide a better prognosis of plant life cycle.

THE FUTURE?

To achieve efficient cost effective plant maintenance the engineer and management must receive the **RIGHT INFORMATION AT THE CORRECT TIME.**

Therefore there is a need to develop new systems or techniques where a problem is found to exist. The User Guide referred to earlier highlighted some areas where there were difficulties monitoring in a predictive non invasive mode certain faults and types of equipment. These have significant cost implications on plant operating costs. As a result the development of Condition Monitoring tools has been expedited in these areas.

For example the study has highlighted the predictive fault diagnosis of reciprocating compressors as a key area. It has joined forces with a third party and is currently engaged in the development of a new technique including fault diagnosis and prediction non invasively (fig 7) suitable for use by the technician.

It is also working with an international machinery shaft sealing Company to develop a new technique for non invasive predictive mechanical seal condition monitoring with the aim of ensuring that the seal operating environment is maintained in optimal condition.

In addition data capture and storage systems must be optimised. In the field of data capture already SMART intrinsically safe Radio Frequency data storage devices are becoming available which can be connected locally to a single transducer. Their use as passive identity TAGS, and developing them for interrogation by data collectors is being considered. It is conceivable that on command they can be interrogated from a central point by a device similar to the remote control on a video recorder and downloaded to the central system.

In new applications where cost effective the use of permanently installed multiplexing surveillance systems will be considered.

In addition MMS's, stores holdings systems, predictive maintenance information, and production profiles must be freely available and relevant data capable of being interchanged. The development of relational databases is helping in this area.

A key facet of any maintenance operation has yet to be discussed. This is the one of availability of the necessary spare parts prior to any maintenance activity. It is pointless scheduling the maintenance activity if the spare part is not available. Therefore the interlinking of the stores management system with the MMS is essential.

Predictive maintenance activities can be linked to the stores inventory system and the spares holdings of expensive capital parts can be optimised based on the frequency of use or predicted use.

Another key facet and in some way the most important from a commercial point of view, is the profit margin optimisation by the production engineer. It is obvious that to carry out essential maintenance when the margins are good is a bad management decision - unless it is unavoidable. This paper is concerned with the correct and optimum application of predictive maintenance in the future, and so it should be obvious that the greater lead time that can be given of an impending problem or activity gives the production manager more flexibility in his planning. It must therefore become possible in the future for the current and future margins on a product to carry weight on the timing of any maintenance activity. The only way this can be optimally done is for this information to be available when the scheduling operation of the activity is planned.

Take the case of the major machinery AI system above and referring to figure 8:

> Once the AI system detects a problem developing it would interrogate the MMS for the equipment history and the current and past operating history.
>
> If a maintenance activity becomes necessary the predictive system would predict the remaining life of the equipment, and generate a signal for a work request. The action would then move to the MMS. Communicate with the Stores System to ascertain the availability of the required spares, and the expected lead time for ordering would take place. The MMS would also ask for data regarding the margin profile on the product. Using this information, as well as the normal resource availability data the work would be provisionally scheduled and the key features provided with the work request. Upon authorisation the various systems would ensure that all the necessary resources are available for the work activity.
>
> At regular intervals the Stores systems should interrogate itself regarding usage and stock holding patterns, and present inconsistencies for the Plant management to resolve. This will minimise the capital tied up in spares, and maximise the benefits to plant operations with regard to future expected work.
>
> Similarly the MMS would be interrogated for maintenance event history as would the predictive AI system to transfer the relevant features to the reliability data base. This would enable the design engineers to maximise their knowledge in the design of the next generation of plant.

As you can see this presents a global overview of the maintenance management picture, but one which is important if one is to take part in today's competitive climate - and succeed!

It is not possible or sensible to make numerous changes or steps forward at the same time, therefore priorities and targets must be set. These can be established by knowing the start point in our respective organisations, and ensuring that the best use of existing systems is made, before stepping out into the unknown. Firm targets and goals should be established, with key milestones on the way identified.

To be in the lead with regard to minimising production costs and maximising the availability of plant means maximising the benefits available from such techniques as:

- Predictive maintenance
- AI to assist the operator and maintenance engineer
- Promoting the development where necessary of new techniques
- The inter action of relevant systems
- Ensuring the best information is available in simple terms to the Plant Management

Following this route plant management can make decisions based on as complete a picture as possible thus minimising risk, and ensuring the profits are maximised to its shareholders.

Where do **YOU** fit into this picture?

ACKNOWLEDGEMENTS

The author gratefully acknowledges the permission of BP Engineering Ltd and John Crane International Ltd for their assistance in the preparation of this paper, and for permission to publish it.

L J RUCKLEY

JOHN CRANE INTERNATIONAL LTD
CROSSBOW HOUSE
40 LIVERPOOL ROAD
SLOUGH
BERKS.
SL1 4QX

0753 531122

RELIABILITY BY DESIGN

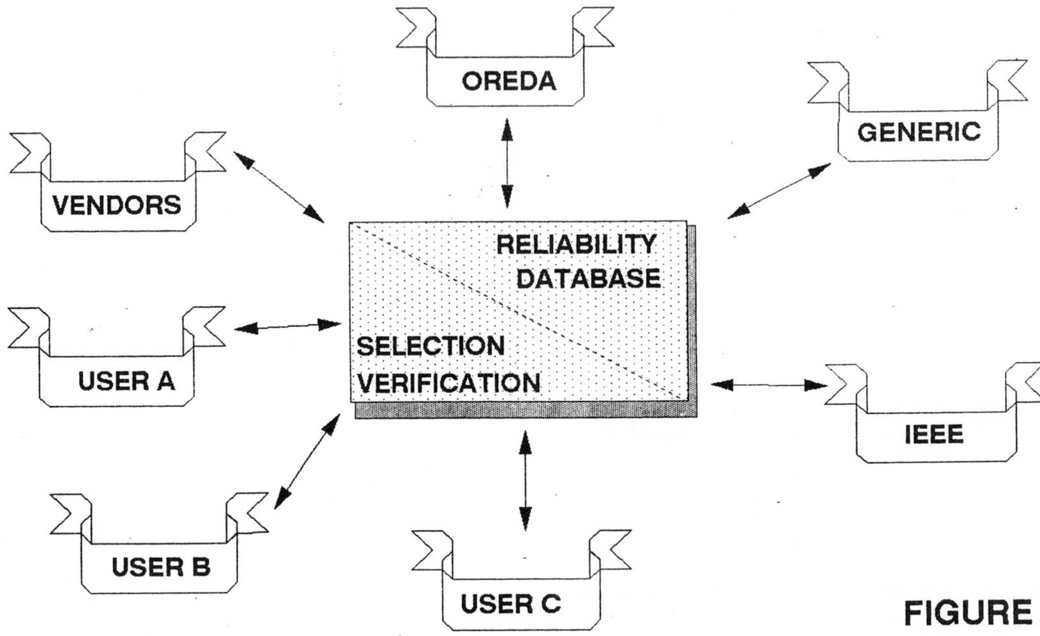

FIGURE 1

MODERN CONDITION MONITORING TECHNIQUES

SMALL MACHINERY

FIGURE 2

APPLICATION OF ARTIFICIAL INTELLIGENCE

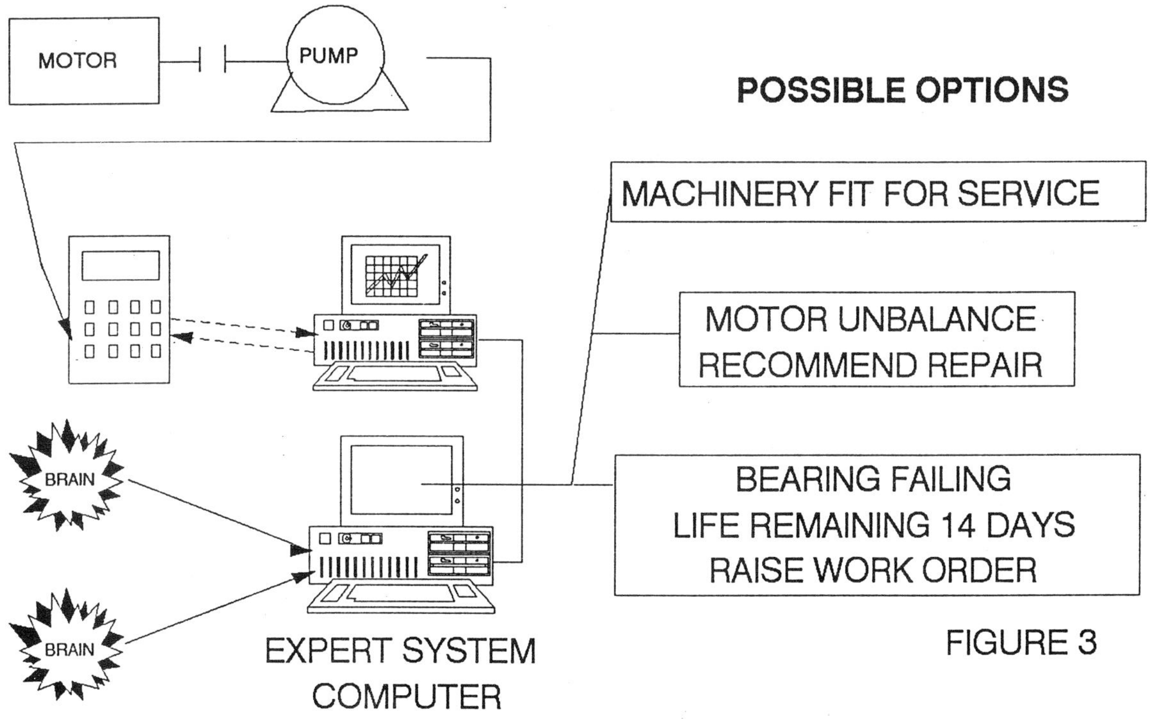

FIGURE 3

MAJOR MACHINERY

MODERN MONITORING TECHNIQUES

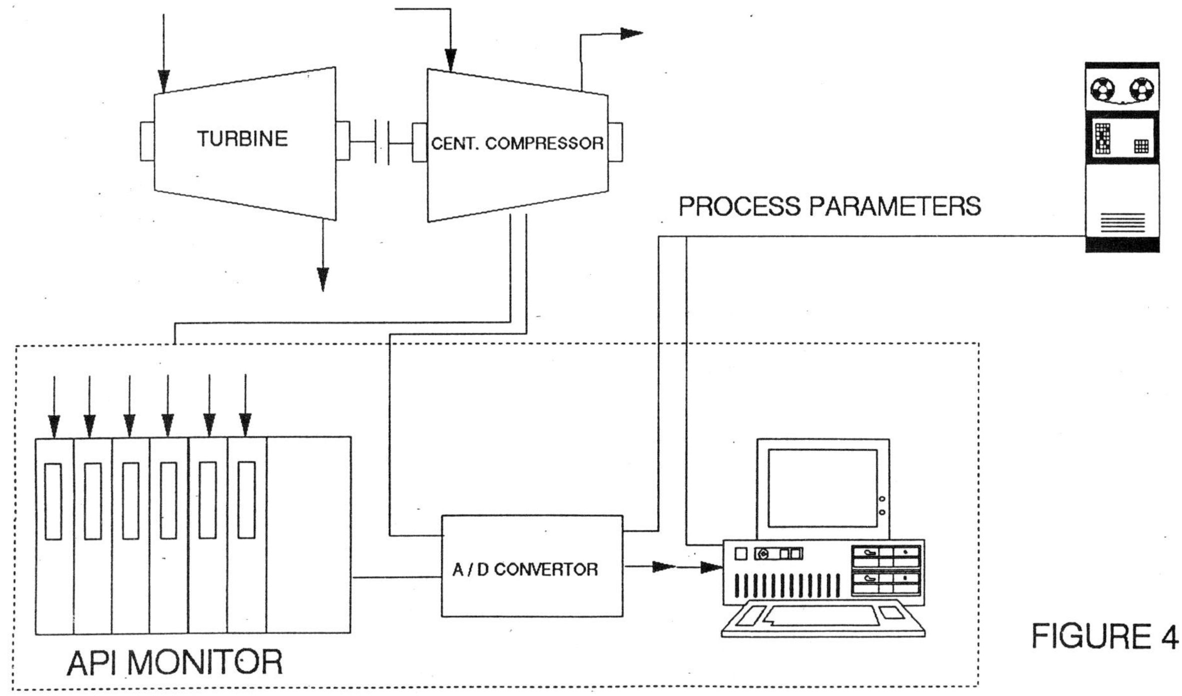

FIGURE 4

MAJOR MACHINERY
APPLICATION OF ARTIFICIAL INTELLIGENCE

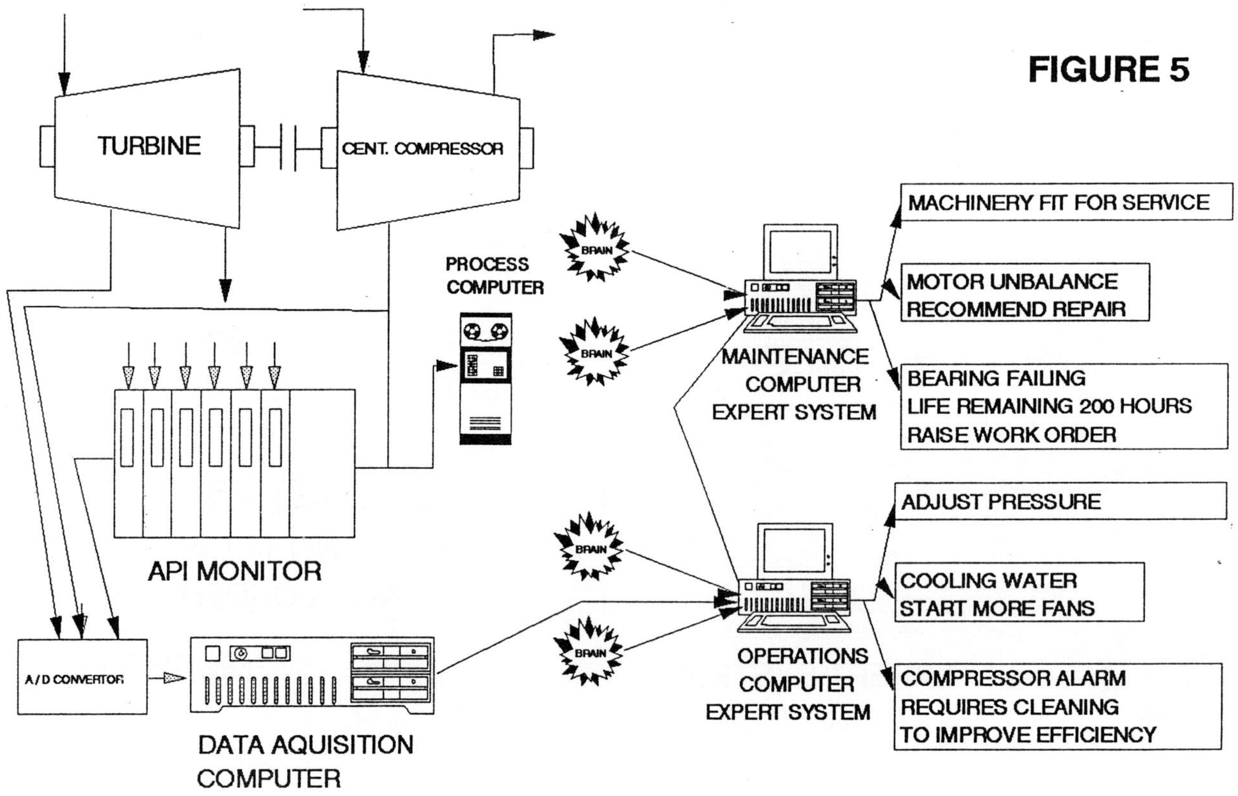

FIGURE 5

MODERN MONITORING APPROACH
PCMS FOR PLANT CORROSION MONITORING ACTIVITIES

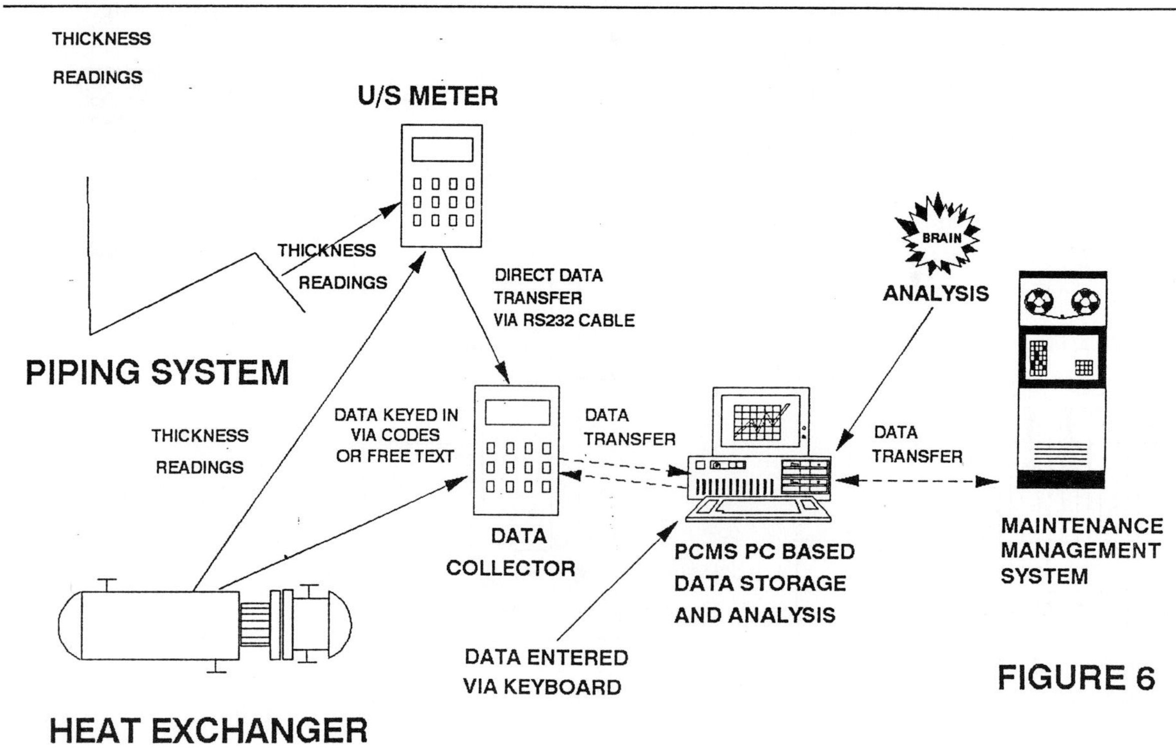

FIGURE 6

RECIPROCATING COMPRESSOR MONITORING

FIGURE 7

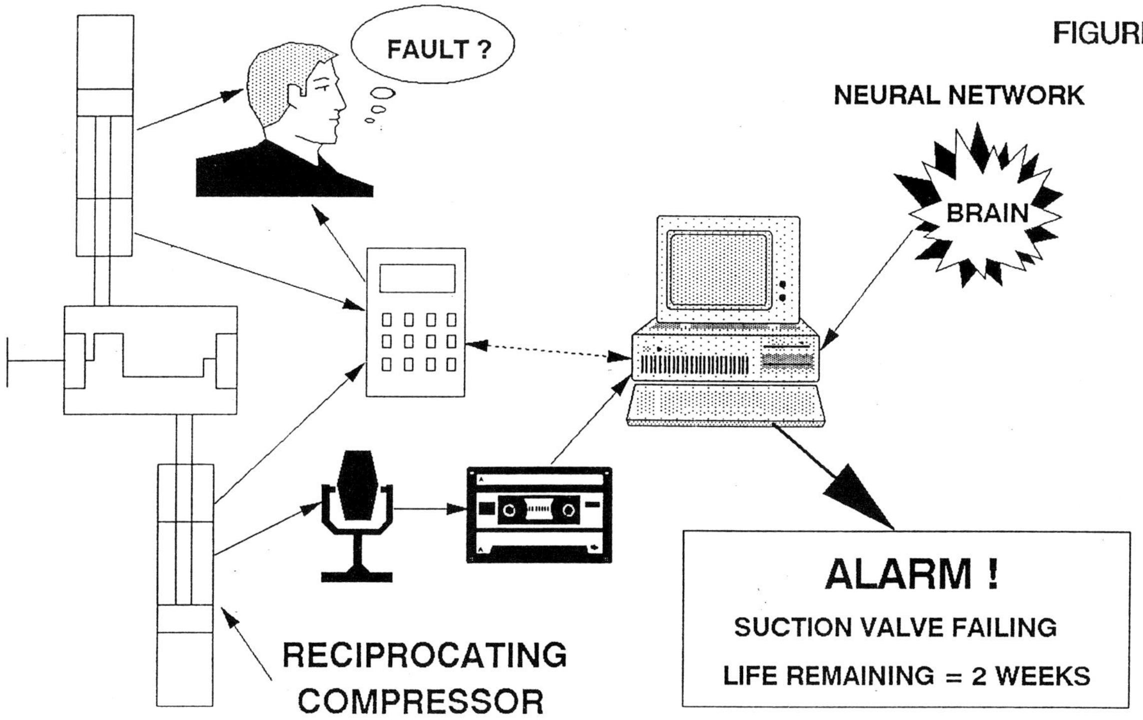

THE FUTURE

FIGURE 8

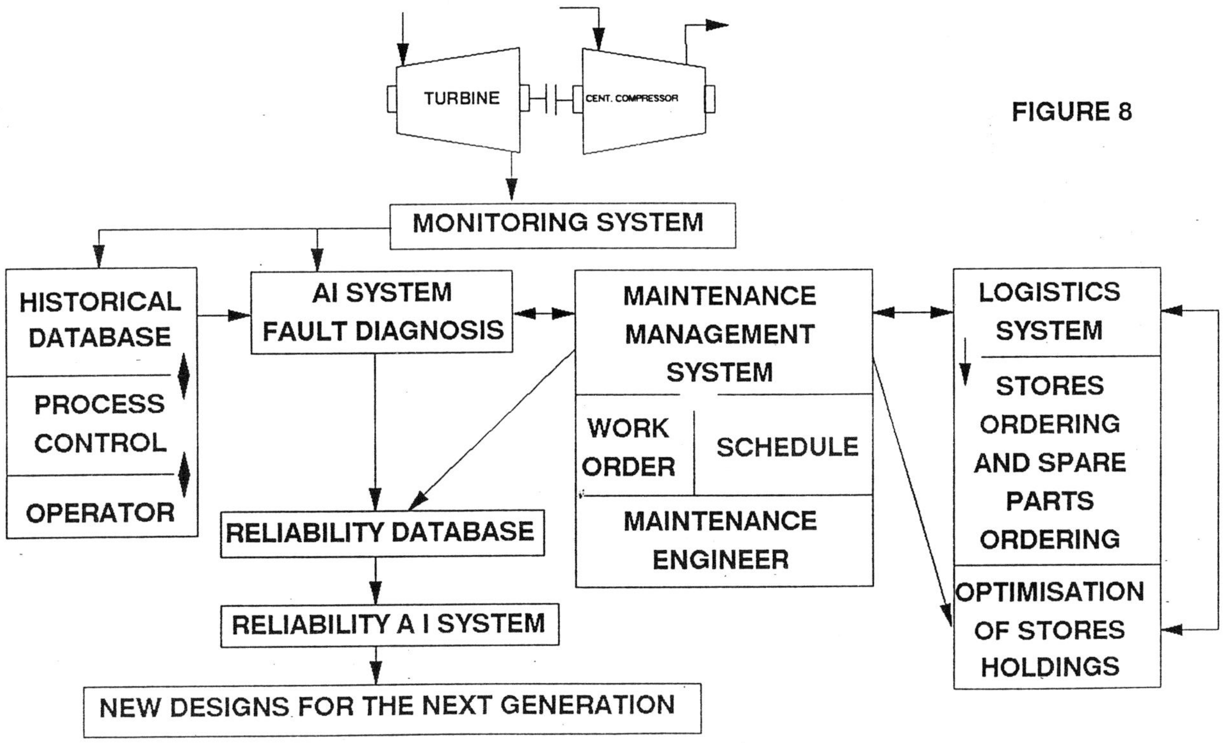

C449/013/93

Machinery safety – registration of critical machines

K G RAYNER, BSc, CEng, MIMechE
ICI Engineering, Northwich, Cheshire
H B CARRICK, MA, PhD, CEng, MIMechE
ICI Engineering, Billingham, Cleveland

Synopsis

This paper describes a user's procedure applied to high hazard machines for containment of process fluids and control of other machine hazards. The procedure is used to identify machines with greater hazard potential, and includes a verification, inspection and monitoring procedure for those machines.

The paper covers the reasons for developing the procedure, the basis for classifying and verifying hazardous machines, machine supplier requirements and the benefits obtained.

1. Introduction

This paper describes a procedure which has been established in the users company to control the hazards associated with machinery throughout the lifetime of the equipment. To do this, the design is subject to a special review, and requirements are established for the safe use of the equipment throughout its life. These requirements are formally recorded and maintained by an inspection agency, independent of production.

In the chemical industry, the hazards associated with loss of containment of chemicals are of particular concern, and their control is the main objective of the procedure. However, it is intended that other significant hazards associated with machines, including mechanical hazards such as overspeeding, shall also be identified and controlled in the process.

2. Critical Machine Registration

The first step in the procedure which we have adopted is called 'Classification'. This classifies machines into two groups: 'critical' and 'non-critical'. Those which are described as 'critical' are judged to offer greater potential for unacceptable events if a loss of containment were to take place.

In principle, hazard analysis could be used as a means of evaluating the significance of potential failures. However, comprehensive data is required for all possible failure modes, for the machine in question, as it will be used in that particular system. Such data is simply not available. Therefore at this stage the probability of a loss of containment is not evaluated, as we are trying to make a quick judgement about the relative need for further work. Thus this judgement will be based mainly on the nature of the process fluid within the machine. This is the same basis used by OSHA in the USA for their most recent regulations.

Additional factors which are taken into account in classification are the size of the system to which the machine is connected, the location of the machine and the likely exposure of people to any loss of containment. Thus if the system is sufficiently small, or if the location is such that people cannot be exposed then a 'non-critical' classification can be applied.

For the 'critical' machines, design information is assembled and the machine system is assessed for suitability for service, as far as the principal hazards, such as loss of containment , are concerned. We call this step 'Design Verification', by analogy with the requirements of our pressure systems procedures.

However, there are important differences between machines and pressure systems. Normally, for a machine, fitness for purpose cannot be obtained solely, or even primarily, by periodic examination. Because of the dynamic nature of the loading, control of the operating conditions is usually much more important. Maintenance of machines and pressure systems is frequently handled differently, and the need for maintenance QA & QC is often overlooked.

Because of these factors, we decided that requirements for machines need to be specified in operational control, in maintenance and in monitoring. These requirements are identified in the 'Design Verification' process, and the essential on-going arrangements are recorded in a 'List of Inspection requirements and Protective Systems', which is analogous to the pressure systems 'Scheme of Examination'. This 'List' can be modified throughout the life of the equipment, as experience is gained.

Where inspection requirements have been identified in the 'List' they are carried out by a nominated inspector who is required to be knowledgeable of the particular type of machine to be inspected. This means that the inspector will frequently be a maintenance supervisor working for the user, or perhaps a vendor's maintenance engineer. The inspector will then formally report on the condition of the machine prior to it being put back into service. The in-house pressure systems inspection service, which is independent of production, will be used to flag up when inspections are due, and to monitor conformance with the ' List of Inspection Requirements'.

3. Why Register Critical Machines?

Although the procedure for critical machine registration was brought into existence following one ICI incident in which there were two fatalities, there are several other events, inside and outside our company which indicate the need for a system to control the hazards associated with chemical plant machinery.

One example of this type occurred during the Piper Alpha investigation, where one line of enquiry related to the failure of cylinder head bolts on high pressure reciprocating compressors, essentially due to the lack of QA and QC in the maintenance process.

Another example was reported by the press in 1991, in which a chemical operator was sprayed with a toxic chemical (dichlorophenol) when the wrong type of seal was fitted to a pump. The company in question was criticised by the UK Health & Safety Executive for 'failing to keep records showing how and when the unsuitable seal was fitted'.

Fires in oil lubricated reciprocating air compressor systems have been another related problem. ICI has had a number of these over the years. In one incident at Wilton, UK, in 1987, the end cover of the compressor aftercooler was fractured and blown a short distance across the plant to impact with a wall. This followed a fire in the delivery system of the compressor, which melted tubes in the aftercooler, resulting in overpressurising the endcover. We know how to prevent such events, but we needed a management system to make it happen!

As well as the failures which point to the need for some system to control the hazards associated with machines, we have received additional clues from a routine engineering audit procedure, also begun in 1989. It was clear, that with the exception of guarding, which is a legal requirement, there were no corporate requirements for systems to control the hazards associated with machinery on our process plants. We were crucially dependent on people and their experience in operating and maintaining machines. The lack of any system of recording the design basis and for controlling locally determined policies and practices made us particularly vulnerable to movement of staff and to the reduction in specialist expertise which new ways of working required, Some statement of the requirements for safe operation of machines was urgently needed.

4. Ammonia injector failure

Into this background came the ICI event that we referred to earlier. It occurred on our Urea plant at Billingham, and has been reported separately in an Ammonia plant operators Symposium (ref 1). The key elements, relating to the machine, are referred to here. The event resulted in a major loss of containment of liquid ammonia, caused by the fatigue fracture of the crankshaft. When the crankshaft fractured it smashed the casing of the power end of the machine. This allowed the plungers to be pushed out of the cylinders, by suction pressure, leaving three large holes.

The primary conclusion reached by the investigation was that neither ICI nor the manufacturer had foreseen such a failure, despite the fact that crankshaft failures had occurred in several places before! A key issue therefore is the system of recording failures, the learning arising from them, and methods of disseminating them throughout the organisation.

Elsewhere, as a result of other crankshaft failures, the manufacturer had introduced a deflection probe and drive motor trip, which he now regarded as a key protective device. These were not fitted to this older machine. Another key issue is therefore the updating of best practice and learning from failures around the world.

Another protective system, a low suction pressure trip, had been removed early in plant commissioning, and this raises the issue of maintenance of design intent. Clearly such a change would be covered by a modification procedure in current practice, but if the intent of the protective system is not clear the risk of an uninformed change is much greater.

The crankshaft which failed had been subject to repair and maintenance in the previous year. Whilst it could be demonstrated that this did not cause the failure, the maintenance carried out raised many questions about the control of repairs on a machine which was pumping a hazardous chemical.

Other factors have reinforced the need for Machines Registration, including EC Machinery Safety Directive considerations and the requirements of Company Safety Guidelines. However, there is no doubt that this one event was a major reason for ICI embarking on this course of action at this time.

5. Critical Machine Registration - how is it achieved?

The critical machine classification and design verification procedure described earlier is incorporated into an ICI corporate management system. The system operates in the following way:

5.1 ICI Group SHE Standards

Corporate Safety Health and Environmental Standards apply to all ICI's operating facilities worldwide. Compliance with these standards is reported annually to the ICI Main Board by the business responsible for each facility. The corporate standards are statements of principle which require the review of hazards and routine inspection of plant for engineering integrity and containment of materials. The standards are supported by engineering guides which are implemented at each facility by local procedures. One such standard applies to critical machine classification, and the supporting engineering guide describes how critical machines should be classified, design verified, registered, and how on going inspection and records should be maintained.

5.2 Local Procedures

A local procedure to implement the corporate principles is needed to take account of the many and varied requirements of local legislation which apply to the 500 facilities operated worldwide by ICI. For example, in North America the recently enacted OSHA Regulations (29 CFR Part 1910) require a procedure to ensure mechanical integrity for process pumps handling specific highly hazardous chemicals. In the United Kingdom the Pressure Systems and Transportable Gas Containers Regulations (1989) include requirements for a scheme of inspection for certain machines. The relevant company local procedures for Critical Machines reflect these requirements.

6. The Machine Suppliers Role

The machine supplier has a key role in the critical machine registration procedure.

6.1 The Design Stage

At the design stage a detailed machine specification is developed by the user and the machine supplier. The user normally includes hazard studies as part of this procedure to ensure that the implications of all operating modes are assessed. The machine supplier will, in turn, identify the operating limits, and operating and maintenance requirements.

For a machine which has been 'classified' as having a potential for unacceptable events, if a loss of containment were to take place, a 'design verification' is required. Operating limits, together with process and mechanical data is used within ICI for the design verification of the machine. The containment envelope-the components which form the containment barrier- is identified and modes of failure are reviewed.

Appropriate protective systems, inspection or condition monitoring may be required as a result of this review. The conclusions of the study are recorded in a 'List of Inspection Requirements and Protection Systems' which forms part of the registered machine documentation. Figure 1 illustrates a sample page of such a list for a reciprocating compressor.

To meet the requirements of the EC Machinery Safety Directive the machine supplier will also be required to issue a 'Certificate of Conformance' to certify that the machine meets essential safety requirements.

6.2 Lifetime Experience

During the life of the machine the schedule will be reviewed and updated based on the user and supplier experience. Ongoing contact with the machine supplier is therefore encouraged for the life of the machine.

Any events which occur on registered machines are investigated, recorded and reported to the ICI corporate engineering group who will inform other ICI users of similar registered machines.

7. Benefits of Critical Machine Registration

The procedure described ensures that the user concentrates his efforts on the machines that matter. In a facility where there may be many hundreds of machines, some containing highly hazardous chemicals or with a potential for significant mechanical damage, the procedure enables the user to focus attention on the key components.

We have found that other advantages have emerged from the procedure. Some of the benefits are:

* Recognition of machinery hazards and application of appropriate protection for safe operation.

* Greater control in the machine selection process. Only those authorised for design verification in the user organisation may approve the machine selection.

* Clarification of the roles of supplier, operating and maintenance staff in ensuring fitness for purpose of machinery.

* Lifetime machine supplier contact. In the past contact between the user and machine supplier was often lost with the consequent loss of learning and experience transfer.

* Unnecessary maintenance work eliminated. The user is often not sure of which components are critical to machine safety, and the machine is totally dismantled, leading to unnecessary work.

* The registration process requires completion of all important documentation prior to operation. Without such formal registration this is too often forgotten and never completed.

8. Current Status

More than 10,000 machines have been reviewed to date under this procedure on ICI plants in the UK. Approximately 10% have been classified as critical. The cost of classification is trivial, and the cost of verification for new equipment is between 8 and 24 manhours.

The procedure is now being implemented on all of the company facilities worldwide. It is likely to have cost and reliability benefits, and have a significant effect on the Safety, Health and Environmental performance of the company's machines.

References

1. AIChemE Ammonia Symposium, San Diego, 1990

Figure 1 List of Inspection Requirements and Protection Systems for a reciprocating compressor (sample page)

	Containment Component	Failure Mode and Classification	Potential Causes	System for Prevention	System for Protection	Comments During Assessment
1	Cylinder & Liner	Fracture-MAJOR	Cylinder internal explosion	Operating procedures (Note 1)		
		Leak-Minor	Cylinder internal over temperature	Monitoring to detect valve failure	High temp. trip	
			Thermal shock (Note 2)			
2	Cylinder Head	Fracture-MAJOR	Jacket freeze-up	Operating procedure - (Note 3)		
	(including fastening	Total loss of	Piston rod wrongly fitted (Note 4)	Mtce QA & QC / good design	Vibration monitor	
	system)	Fasteners -	Loose piston securing nut	Mtce QA & QC / good design		
		MAJOR	Liquid ingestion	See ingestion panel report	HL.trip on KO Pot	
			Solid in cylinder eg. valve and piston part	Suction KO pot/filter; Mtce QA & QC		
			Cylinder internal over pressure	Discharge RV	High press. trip	
3	Valve Covers	Fracture-MAJOR	- liner wrongly fitted	Mtce QA & QC		
	(including unloaders	Total loss of	- suction valves in discharge ports	" " "		
	& fastening system)	Fasteners -	- inverter discharge valve	" " "		
		MAJOR	- pump valve cage obstructing flow or pump cavitation (Note 5)	Operating procedure	Suction P & T	
			Cylinder support failure (Note 6)	Mtce QA & QC	Alarm/trip	
			Big - end bolt failure (Note 7)			
			Bolting - head to cylinder and and cylinder to C/case			
			- wrong specification fasteners	Mtce QA		
			- wrong fitted fasteners	Mtce QC		
			- worn-corroded-loose fasteners (Note 8)	Inspection		
4	Screwed Plugs	Blow Out	As 1, 2 and 3 above plus			
	- cylinder head	MAJOR/minor	- wrongly fitted	Mtce QC		
	- cylinder	(Note 9)	- wrong/worn thread	Mtce QA & QC		
			- internal corrosion of plug	Inspection		